Biogas for Rural Areas

Biogas for Rural Areas

Editors

Ivet Ferrer
Stephanie Lansing
Jaime Martí-Herrero

MDPI • Basel • Beijing • Wuhan • Barcelona • Belgrade • Manchester • Tokyo • Cluj • Tianjin

Editors

Ivet Ferrer
Department of Civil and Environmental Engineering
Universitat Politècnica de Catalunya·BarcelonaTech
Barcelona
Spain

Stephanie Lansing
Department of Environmental Science & Technology
University of Maryland
College Park, Maryland
United States

Jaime Martí-Herrero
Biomass to Resources Group
Universidad Regional Amazónica
Ikiam
Tena
Ecuador

Editorial Office
MDPI
St. Alban-Anlage 66
4052 Basel, Switzerland

This is a reprint of articles from the Special Issue published online in the open access journal *Energies* (ISSN 1996-1073) (available at: www.mdpi.com/journal/energies/special_issues/Biogas_Rural_Areas).

For citation purposes, cite each article independently as indicated on the article page online and as indicated below:

LastName, A.A.; LastName, B.B.; LastName, C.C. Article Title. *Journal Name* **Year**, *Volume Number*, Page Range.

ISBN 978-3-0365-3237-0 (Hbk)
ISBN 978-3-0365-3236-3 (PDF)

Contents

About the Editors

Ivet Ferrer

Dr. Ivet Ferrer is a Full Professor at the Department of Civil and Environmental Engineering of the Universitat Politècnica de Catalunya·BarcelonaTech. She leads the Research Group of Environmental Engineering and Microbiology (GEMMA-UPC). Her main research topic is the optimization of biomass anaerobic digestion by applying pretreatments and co-digestion. She has also addressed the implementation and assessment of low-tech digesters for rural areas. Her research is now focused on the recovery of resources from waste streams, including high-value bioproducts and biogas.

Stephanie Lansing

Dr. Stephanie Lansing is a Full Professor and Associate Chair at the Department of Environmental Science & Technology at the University of Maryland, USA. She leads the Bioenergy and Bioprocessing Technology Laboratory. Her research focuses on anaerobic digestion, waste to energy, antimicrobial resistance, microbial electrolysis cells, nanotechnology, gasification, nutrient management and capture, and modeling. She has 20 years of experience in on-farm energy research, outreach education, and economic/sustainability analyses of waste to energy systems.

Jaime Martí-Herrero

Jaime Martí Herrero (PhD) is a Full Professor at the Universidad Regional Amazónica Ikiam (Ecuador) and belongs to the Biomass to Resources Research Group. In addition, he is an Associated Research Professor at the International Center of Numerical Methods in Engineering (CIMNE, Spain) at the Building Energy and Environment Research Group. His research focuses on the development, optimization and diffusion of biogas technology in rural areas of developing countries, with an emphasis on cold climates, psychrophilic anaerobic digestion and appropriate technologies. Currently, his research is focused on treatment of domestic wastewater through digesters for rural areas.

Preface to "Biogas for Rural Areas"

Bioenergy is renewable energy obtained from biomass—any organic material that has stored sunlight in the form of chemical energy. Biogas is among the biofuels that can be obtained from biomass resources, including biodegradable wastes like manure, sewage sludge, the organic fraction of municipal solid wastes, slaughterhouse waste, crop residues, and more recently lignocellulosic biomass and algae. Within the framework of the circular economy, biogas production from biodegradable waste is particularly interesting, as it helps to save resources while reducing environmental pollution. Besides, lignocellulosic biomass and algae do not compete for arable land with food crops (in contrast with energy crops). Hence, they constitute a novel source of biomass for bioenergy.

Biogas plants may involve both high-tech and low-tech digesters, ranging from industrial-scale plants to small-scale farms and even households. They pose an alternative for decentralized bioenergy production in rural areas. Indeed, the biogas produced can be used in heaters, engines, combined heat and power units, and even cookstoves at the household level. Notwithstanding, digesters are considered to be a sustainable technology that can improve the living conditions of farmers by covering energy needs and boosting nutrient recycling. Thanks to their technical, socio-economic, and environmental benefits, rural biogas plants have been spreading around the world since the 1970s, with a large focus on farm-based systems and households. There are several opportunities to introduce rural biogas plants in small and medium populations using wastewater, agriculture wastes, and organic municipal solid wastes. However, several challenges still need to be overcome in order to improve the technology, financial viability, and dissemination.

This Special Issue aims to gather research papers on recent developments for bioenergy supply in rural areas; highlight new insights on bioenergy production and utilization processes; detail the development of new efficient technologies for biogas production and utilization; present full-scale case studies; and feature environmental, energy, or economic assessments of decentralized biogas plants.

Ivet Ferrer, Stephanie Lansing, Jaime Martí-Herrero
Editors

Article

Simultaneous Synergy in CH_4 Yield and Kinetics: Criteria for Selecting the Best Mixtures during Co-Digestion of Wastewater and Manure from a Bovine Slaughterhouse

Zamir Sánchez [1,*], Davide Poggio [2], Liliana Castro [1] and Humberto Escalante [1]

1 Grupo de Investigación en Tecnologías de Valorización de Residuos y Fuentes Agrícolas e Industriales para la Sustentabilidad Energética (INTERFASE), Escuela de Ingeniería Química, Universidad Industrial de Santander—UIS, Carrera 27, Calle 9 Ciudad Universitaria, Bucaramanga 680002, Colombia; licasmol@saber.uis.edu.co (L.C.); escala@uis.edu.co (H.E.)

2 Energy 2050, Department of Mechanical Engineering, Faculty of Engineering, University of Sheffield, Sheffield S3 7RD, UK; d.poggio@sheffield.ac.uk

* Correspondence: zamir2178220@correo.uis.edu.co

Abstract: Usually, slaughterhouse wastewater has been considered as a single substrate whose anaerobic digestion can lead to inhibition problems and low biodegradability. However, the bovine slaughter process generates different wastewater streams with particular physicochemical characteristics: slaughter wastewater (SWW), offal wastewater (OWW) and paunch wastewater (PWW). Therefore, this research aims to assess the anaerobic co-digestion (AcoD) of SWW, OWW, PWW and bovine manure (BM) through biochemical methane potential tests in order to reduce inhibition risk and increase biodegradability. A model-based methodology was developed to assess the synergistic effects considering CH_4 yield and kinetics simultaneously. The AcoD of PWW and BM with OWW and SWW enhanced the extent of degradation (0.64–0.77) above both PWW (0.34) and BM (0.46) mono-digestion. SWW Mono-digestion showed inhibition risk by NH_3, which was reduced by AcoD with PWW and OWW. The combination of low CH_4 potential streams (PWW and BM) with high potential streams (OWW and SWW) presented stronger synergistic effects than BM-PWW and SWW-OWW mixtures. Likewise, the multicomponent mixtures performed overall better than binary mixtures. Furthermore, the methodology developed allowed to select the best mixtures, which also demonstrated energy and economic advantages compared to mono-digestions.

Keywords: anaerobic co-digestion; slaughterhouse wastewater; synergistic effects; kinetic modeling; biodegradability

Citation: Sánchez, Z.; Poggio, D.; Castro, L.; Escalante, H. Simultaneous Synergy in CH_4 Yield and Kinetics: Criteria for Selecting the Best Mixtures during Co-Digestion of Wastewater and Manure from a Bovine Slaughterhouse. *Energies* **2021**, *14*, 384. https://doi.org/10.3390/en14020384

Received: 19 October 2020
Accepted: 8 January 2021
Published: 13 January 2021

1. Introduction

The global meat industry consumes 24% of the total water used for food and beverage production. [1]. Beef production has one of the largest water footprints among all foods (15,400 m^3 t^{-1} of meat) [2]. Animal slaughter and meat processing are the main contributors to the footprint, in terms of water use and wastewater generation. Slaughterhouse wastewater volumes have been reported to be between 0.57 m^3 $bovine^{-1}$ [3] and 4.22 m^3 $bovine^{-1}$ [4]. These wastewaters are characterized by a chemical oxygen demand (COD) between 2000 mg L^{-1} [5] and 20,400 mg L^{-1} [6].

The slaughter bovine process varies depending on the available technologies; however, in general, it consists of four stages and generates similar wastewater streams: (i) cattle-yard wastewater (CWW), generated from the preliminary washing of livestock and yards, containing urine and feces; (ii) slaughter wastewater (SWW), which contains blood, rich in protein; (iii) paunch wastewater (PWW), generated in the removal of the digestive tract content, with structural carbohydrates in the form of lignocellulosic material; (iv) offal wastewater (OWW) from the cleaning of the white viscera, therefore containing particles of meat and fat. In middle- and high-income countries, slaughterhouse wastewater streams are generally treated before discharge into local watercourses or sewer systems. Primary treatments are the most common; however, they are costly and sometimes insufficient [7].

Anaerobic digestion is an efficient technology for waste treatment and valorization since compounds are degraded into a biogas (55–70% volume of CH_4) and a nutrient rich sludge [1]. In developing countries, tubular digesters are the most widely used in rural homes, farms and rural sector companies (agricultural and livestock) due to their simple construction and operation [8]. Furthermore, tubular digesters have demonstrated to be adequate for the anaerobic digestion of slaughterhouse wastewater [9]. However given the biochemical composition of animal slaughter waste (rich in lipids, proteins and lignocellulosic material), anaerobic digestion of these wastes can lead to several problems. During anaerobic digestion, proteins break down to NH_3 [10] while lipids hydrolysis produces long-chain fatty acids (LCFA) [11], which can inhibit the process and reduce the biogas production and waste treatment rates. The tolerance of the microbial consortia to inhibitors is characterized by an inhibition coefficient (K_{I50}), which indicates the concentration where the uptake rate is half the maximum [12]. Likewise, the lignocellulosic material from ruminal content presents a low hydrolysis rate coefficient (between 0.10 and 0.12 d^{-1}) [12,13] causing slow anaerobic degradation rates. Slow degradation kinetics require long hydraulic retention times (HRT) [14] and, for a given organic load rate, full scale results in a larger reactor volume [15]. This leads to a rise in investment, since more than 50% of the fixed costs correspond to the digester [16]. The above problems may limit the widespread tubular digester use for slaughterhouse wastewater treatment.

Anaerobic Co-digestion (AcoD) has been used as an approach to mitigate the afore-mentioned drawbacks, given the potential synergies between co-substrates towards the reduction of inhibition and increase of both the extent and rate of biodegradation. In this regard, most AD studies consider slaughterhouse wastewater as a single substrate (a mixture of CWW, SWW, OWW and PWW in the proportion of its generation). However, in a study by Jensen et al. (2014) [4], it was evidenced how each stream has particular characteristics and can be treated as an individual substrate. Moreover, bovine manure (BM) is an excellent base substrate (carrier) [17]. Therefore, an adequate mixture of these substrates can enhance the performance of the anaerobic digestion process, without requiring further external substrates. Nonetheless, to the best of authors' knowledge, the AcoD of different slaughter wastewater streams has not been explored in previous studies.

Usually, AcoD studies have evaluated the synergy between co-substrates focused on CH_4 yield [18,19] while the kinetics (rate of degradation) in most cases is evaluated with mathematical models, without determining whether there is synergy in the kinetic factors [17,20]. Thus far, three methodologies have been published to assess synergy in kinetic parameters. Pagés-Diaz et al. (2014) [21] implemented a mixture design to evaluate an AcoD process, then adjusted the results to statistical models and estimated the significance of the regression coefficients. This methodology is extensive and its precision depends on the correct selection of the statistical model to evaluate the synergy. Ebner et al. (2016) [22] proposed a co-digestion rate index (CRI) based on the ratio of the experimental apparent hydrolysis rate coefficient over its expected value. The authors demonstrated through numerical estimation, how the weighted geometric mean of the hydrolysis rates of the single substrates is the best estimate for the expected combined co-digestion rate. The numerical procedure added the curves of pairs of substrates, fitted the first-order model to the experimental co-digestion data and compared the resulting hydrolysis rate coefficient with different statistical means of the individual substrates. Thus, the application of the above methodology to other kinetic models (with more parameters compared to the first order model) could be too complex. Donoso-Bravo et al. (2019) [23] presented a simpler method that consists of the linear combination (weighted arithmetic mean) of the kinetic parameters, which could be applied to any model. However, this methodology does not consider the complexity of kinetic interaction and the error introduced by an approximation with arithmetic mean. Thus, the above approaches can be tedious or lead to uncertainties in the evaluation of the kinetic synergy.

Based on the above review of co-digestion studies and modeling, the main contributions of this study are: (i) the evaluation of the performance of AcoD of novel mixtures of bovine slaughterhouse wastewater streams and manure, with a focus on reducing potential inhibition and biodegradability problems; (ii) the development of a methodology to assess the synergy between co-substrates, which considers both CH_4 yield and kinetics in a practical and accurate way, (iii) the application of the methodology to select the best mixtures between slaughterhouse wastewater streams and BM. The methodology proposed in the current study differs from those reported in the literature since the synergy was evaluated directly from the expected biochemical methane potential (BMP) curves without approximations (statistical models, arithmetic mean or geometric mean), which reduces the errors in parameters estimation. In addition, the energy and economic feasibility of the AcoD of synergistic mixtures was evaluated from the results of the BMP assays and the modeling.

2. Materials and Methods

The current study employed a four-part methodology: (1) experimental evaluation of AcoD of slaughterhouse wastewater streams and BM, (2) implementation and evaluation of kinetic models, (3) evaluation of synergistic effects and (4) energy and economic analysis of the implementation of AcoD in slaughterhouses. For the first part, the substrates and inoculum were collected. Then, a statistical mixture design was applied to prepare different combinations of wastewater streams and BM, which were tested by BMP assays to obtain the ultimate experimental specific CH_4 yield (B_o). The theoretical specific CH_4 yield (B_{oth}) was calculated from the composition of the mixtures; thereafter, the extent of degradation (f_d) was calculated from the value of B_o and B_{oth}. In the second part, both the first-order and the modified Gompertz models were calibrated against the BMP experimental data, and the most suitable kinetic model was selected based on fit. In the third part, the synergistic effects of AcoD on CH_4 yield and kinetics were evaluated by a comparison between the experimental and the model-based expected values. Finally, taking as a case study a Colombian slaughterhouse, the energy and economic feasibility of AcoD of synergistic mixtures was evaluated by the electrical and thermal potentials, the payback period (PBP), Net Present Value (NPV) and Internal Rate of Return (IRR).

2.1. *Evaluation of Anaerobic Co-Digestion (AcoD)*

2.1.1. Substrates and Inoculum Origin

Fresh bovine manure (BM) and samples of slaughter wastewater (SWW), offal wastewater (OWW) and paunch wastewater (PWW) were obtained from a Colombian slaughterhouse (Floridablanca-Santander: Latitude 7°3′14.82″ N and longitude 73°7′55.82″ W). The OWW stream comes from the cleaning of white viscera (intestines and stomachs). The wastewater from the cleaning of red viscera (liver, heart, tongue, lungs, kidney and spleen) makes up the SWW stream. The main operational characteristics of the case study slaughterhouse are presented in Table 1.

The substrates were characterized by measuring pH, total solids content (TS), volatile solids content (VS), chemical oxygen demand (COD), total alkalinity (TA), total volatile fatty acids (TVFAs) and biochemical composition (carbohydrates, lipids and proteins) (Table 2).

The reactors were inoculated with mesophilic sludge from a small biogas plant located in an organic farm (Floridablanca-Santander, Colombia: Latitude 7°01′0.07″ N and longitude 73°08′13.3″ W). The main characteristics of the inoculum used were: 33.70 ± 0.11 kg TS m^{-3}, 19.95 ± 0.14 kg VS m^{-3}, 8.09 ± 0.03 pH, TA of 2.57 ± 0.10 kg $CaCO_3$ m^{-3}, TVFAs of 1.42 ± 0.12 kg CH_3COOH m^{-3}, specific methanogenic activity (SMA) of 0.035 ± 0.005 kg COD kg^{-1} VS d^{-1} and a coefficient of inhibition by NH_3 ($K_{I50\text{-}NH3}$) of 18.53 ± 0.34 mg L^{-1}. The same inoculum source has been utilized in previous studies [24].

Table 1. Operational characteristics of the case study slaughterhouse.

Parameter [a]	Unit	Value
Average slaughter capacity	Bovines d^{-1}	327
Flow of SWW	m^3 d^{-1}	45.34
Flow of OWW	m^3 d^{-1}	111.60
Flow of PWW	m^3 d^{-1}	139.50
Flow of BM	t d^{-1}	7.70
Thermal energy consumption	kWh d^{-1}	8594.90
Electrical energy consumption	kWh d^{-1}	4743.28

[a] SWW: slaughter wastewater; OWW: offal wastewater; PWW: paunch wastewater; BM: bovine manure.

Table 2. Characteristics of slaughterhouse wastewater streams and BM. Results are reported as an average of three measurements (±95% confidence interval).

Parameter [a]	Unit	SWW [b]	OWW [b]	PWW [b]	BM [b]
pH	-	6.72 ± 0.08	6.90 ± 0.08	7.80 ± 0.08	7.38 ± 0.06
TS	kg m^{-3}	8.28 ± 0.12	12.53 ± 0.22	18.23 ± 0.93	242.14 ± 1.04
VS	kg m^{-3}	7.63 ± 0.21	10.96 ± 0.23	15.99 ± 0.98	154.22 ± 1.50
COD	kg m^{-3}	9.39 ± 0.04	9.75 ± 0.08	8.35 ± 0.14	37.06 ± 1.66
TVFAs	kg CH_3COOH m^{-3}	0.72 ± 0.00	0.88 ± 0.07	1.25 ± 0.07	2.40 ± 0.00
TA	Kg $CaCO_3$ m^{-3}	0.80 ± 0.00	1.38 ± 0.18	1.75 ± 0.05	3.25 ± 0.35
Lipids	%VS	26.5	38.6	4.1	3.3
Proteins	%VS	69.3	36.1	11.6	12.1
Carb	%VS	4.2	12.0	8.8	21.4
Cell	%VS	-	2.3	21.9	24.8
Hem	%VS	-	6.4	32.0	22.1
Lig	%VS	-	4.6	21.6	16.3

[a] Carb: non-structural carbohydrates; Cell: cellulose; Hem: hemicellulose; Lig: lignin. [b] SWW: slaughter wastewater; OWW: offal wastewater; PWW: paunch wastewater; BM: bovine manure.

2.1.2. Experimental Mixture Design

In order to eliminate the randomness of blending, the assay was based on a simplex lattice design *{4,3}* augmented with the overall centroid. Mixtures were based on the organic load expressed in VS. The mixture design was created using STATGRAPHICS Centurion XVI (StatPoint Technologies, Inc. Warrenton, VA, USA) and represented graphically as a tetrahedron made up of a triangular base and three triangular faces called simplex (Figure 1). Each simplex consisted of 10 points (mixture ratios) where vertices corresponded to ratios with 100% single substrate. The upper vertex of the tetrahedron was the pure BM ratio. Vertices on the base of tetrahedron comprised pure ratios of 100% SWW, 100% OWW, and 100% PWW. Points on the axis corresponded to binary mixtures. Interiors points on each simplex corresponded to ternary mixtures. Additionally, there is a central point in the tetrahedron, for a total of 21 mixtures (Table 3).

2.1.3. Ultimate Experimental Specific CH_4 Yield (B_o)

In order to determine the ultimate experimental specific yield B_o, BMPs assays were run according to the protocol presented by Holliger et al. (2016) [25] for organic material in 100 mL digesters (60 mL working volume). Assays were prepared with an inoculum to substrate ratio (ISR) of 2 (based on the amount of VS). For all the assays, the initial pH was between 7.0 and 8.0 and the buffer capacity, expressed as the ratio of total volatile fatty acid and total alkalinity (TVFAs/TA) [26], ranged from 0.2 to 0.4; these values are within the recommended range by the BMP protocol, and therefore, no buffers were added to adjust them. The digesters were flushed with pure N_2 and sealed using butyl rubber and an aluminum cap. Blanks, containing inoculum and deionized water to replace the substrate, were used to estimate the endogenous CH_4 production of the inoculum. All digesters were incubated at 37 ± 2 °C and mixed by manual inversion once per day. The CH_4 production was quantified by the volumetric displacement of an alkaline solution. The accumulated volume of CH_4 displaced was adjusted to standard temperature and pressure conditions

(STP: 273 K and 1 atm) and the specific CH_4 yield was expressed on the basis of VS added (m^3 CH_4 kg^{-1} VS) [27]. A separate positive control was conducted using cellulose resulting in a CH_4 yield of 0.364 ± 0.013 m^3 kg^{-1} VS (88% of the theoretical specific CH_4 yield of cellulose). All tests, blanks and control were performed in triplicate. The BMP assays were terminated once the daily CH_4 production for all mixtures decreased below 1% of the accumulated volume during three consecutive days, which resulted in a duration of the assays of 30 days.

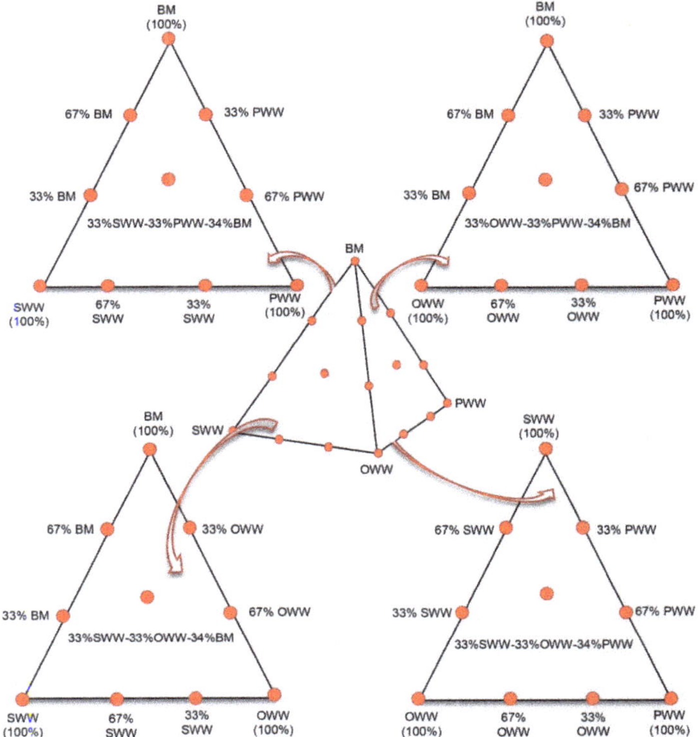

Figure 1. Simplex-lattice mixture design tested for anaerobic co-digestion (AcoD) of slaughterhouse wastewater streams (SWW: slaughter wastewater; OWW: offal wastewater; PWW: paunch wastewater) and bovine manure (BM).

Table 3. Mixture design applied in the evaluation of AcoD.

Mixture	SWW [a] (% VS)	OWW [a] (%VS)	PWW [a] (%VS)	BM [a] (%VS)	Mixture Type
S100	100	0	0	0	Single Substrates
O100	0	100	0	0	
P100	0	0	100	0	
B100	0	0	0	100	

Table 3. *Cont.*

Mixture	SWW [a] (% VS)	OWW [a] (%VS)	PWW [a] (%VS)	BM [a] (%VS)	Mixture Type
S33:P67	33	0	67	0	
S33:B67	33	0	0	67	
O67:P33	0	67	33	0	
O67:B33	0	67	0	33	
O33:P67	0	33	67	0	
O33:B67	0	33	0	67	
P67:B33	0	0	67	33	
P33:B67	0	0	33	67	
S33:O33:P34	33	33	34	0	Ternary
S33:P33:B34	33	0	33	34	
S33:O33:B34	33	33	0	34	
O33:P33:B34	0	33	33	34	
S25:O25:P25:B25	25	25	25	25	Quaternary

[a] SWW: slaughter wastewater; OWW: offal wastewater; PWW: paunch wastewater; BM: bovine manure.

2.1.4. Theoretical Specific CH_4 Yield (B_{oth})

The theoretical specific yield B_{oth} allows the prediction of the maximum CH_4 production from a specific waste. This can be calculated from the knowledge of the composition of substrates and mixtures in terms of their biochemical fractions (i.e., carbohydrates, proteins lipids) [28], as shown in Equation (1):

$$B_{oth} = 0.415\ x\_Carbohydrates + 0.496\ x\_Proteins + 1.014\ x\_Lipids \quad (1)$$

The biochemical fractions (x) are given in VS and B_{oth} in STP m^3 CH_4 kg^{-1} VS; the carbohydrate fraction includes both non-structural and structural carbohydrates.

2.1.5. The Extent of Degradation (f_d)

The level of anaerobic biodegradability of a waste can be determined by comparing the ultimate experimental specific CH_4 yield B_o with the theoretical value B_{oth}, as shown in Equation (2) [29]:

$$f_d = \frac{B_o}{B_{oth}} \quad (2)$$

where f_d is a key parameter used to indicate the fraction of the waste that may be transformed into CH_4.

2.1.6. Analytical Procedures

TS, VS, COD, pH, total Kjeldahl nitrogen and lipids (Soxhlet) were determined conforming to standard methods [30]. TA and TVFAS were measured according to the method of Lahav and Morgan (2004) [31]. TA was quantified by titration of the sample with a 0.1 N HCl solution to a pH endpoint of 3. Then, the sample was boiled lightly for 3 min to completely remove the dissolved CO_2. Thereafter, the amount of NaOH solution 0.1 N required to elevate the pH from 3 to 6.5 was recorded to calculate TVFAs. Cellulose hemicellulose and lignin were determined from fiber fractions: neutral detergent fiber (NDF), acid detergent fiber (ADF) and lignin. The hemicellulose and cellulose contents were calculated as the differences between NDF and ADF and between ADF and ADL respectively [32]. Protein composition was calculated from the ratio of 6.25 g protein per g of organic nitrogen. Organic nitrogen was determined by the subtraction between Kjeldahl nitrogen and ammoniacal nitrogen [33]. Non-lignocellulosic carbohydrates (e.g., sugars starch and pectin) were obtained by difference. SMA and $KI_{50\text{-}NH3}$ of the inoculum were determined following the procedure by Astals et al. (2015) [34]. NH_4^+ concentration was measured by a test (Spectroquant ammonium test Merck) analogous to APHA 4500-NH_3 F [30]. NH_3 concentration [mg NH_3-N L^{-1}] was determined by Equation (3), where TAN [mg N L^{-1}] is the total ammonia nitrogen in the forms of NH_3 and NH_4^+, K_a is the acid-base equilibrium constant and γ_1 is the activity coefficient [35]:

$$NH_3 - N = \frac{K_a.TAN.\gamma_1}{K_a.\gamma_1 + 10^{-pH}} \quad (3)$$

$$TAN = NH_3 - N + NH_4^+ - N \quad (4)$$

At the BMP assays temperature (37 °C) K_a is 1.27 × 10⁻. The values of γ_1 were obtained from Equations (5) and (6) [35]:

$$\log \gamma_1 = -0.5.z_i^2.\left(\frac{\sqrt{I}}{1+\sqrt{I}} - 0.20.I\right) \quad (5)$$

$$I = \frac{1}{2}\sum z_i^2.C_i \quad (6)$$

where z_i is the valence of the ion i, I is the ionic strength [mol L^{-1}] and C_i is the concentration of the ion i [mol L^{-1}]. For the calculations, the only ion considered was NH_4^+.

2.2. Kinetic Modeling

The first-order model (Equation (7)) and the modified Gompertz model (Equation (8)) were compared based on their fitting to the BMP curves from AcoD of slaughterhouse wastewater streams and BM. The first-order model has been used in previous studies to describe the cumulative CH_4 production of various organic wastes [20,36] when the hydrolysis step is rate-limiting:

$$B_s = P\,(1 - \exp(-k_h.t)) \quad (7)$$

where B_s [m^3 CH_4 kg^{-1} VS] is the simulated specific CH_4 yield at time t [d], P [m^3 CH_4 kg^{-1} VS] is the simulated ultimate specific CH_4 yield and k_h is the apparent hydrolysis rate coefficient [d^{-1}]. In cases where biogas production is proportional to the microbial activity, the modified Gompertz model is more suitable than the first-order model [37]:

$$B_s = P\exp\left(-\exp\left(\frac{R_{max}.e}{P}(\lambda - t) + 1\right)\right) \quad (8)$$

where λ is the lag-phase [d], R_{max} is the maximum specific CH_4 production rate [m^3 CH_4 kg^{-1} VS d^{-1}] and e is exp (1) = 2.7183.

The models were fitted to curves from BMP assays in Aquasim 2.1d (Swiss Federal Institute of Aquatic Science and Technology—Eawag). Parameters were estimated by a weighted least square method, minimizing the cost function shown in Equation (9) [38]:

$$\chi^2 = \sum_{i=1}^{n}\left(\frac{B_{m,i} - B_{s,i}(r)}{\sigma_{m,i}}\right)^2 \quad (9)$$

where $B_{m,i}$ is the ith measured value of the accumulated CH_4 volume, assumed to be a normally distributed random variable, $B_{s,i}(r)$ is the model prediction, a function of the set of parameters r to be estimated, at the time corresponding to ith data point and $\sigma_{m,i}$ is the standard error of the measurement $B_{m,i}$, calculated from the values of the replicates, which weights each term of the sum. The standard errors of the measurements were calculated according to Holliger et al. (2016) [25] (Equation (10)):

$$\sigma_m = \sqrt[2]{(\sigma_{blank})^2 + (\sigma_{substrate})^2} \quad (10)$$

As a minimization technique, the Secant Algorithm implemented in Aquasim was used. The tolerance for convergence in the objective function was 4 × 10^{-3}. In order to check the convergence of the algorithm to the same optimum parameter values, different initial guesses of target parameters were used. The confidence interval of the estimated parameters was expressed as standard error, as calculated by the Secant Algorithm in Aquasim.

The accuracy of model predictions with respect to the experimental results was analyzed by the regression coefficient (R^2), and the normalized root mean square error (NRMSE):

$$NRMSE = \frac{\sqrt{\frac{\sum_{i=1}^{n}\left(B_{s,i}-B_{m,i}\right)^2}{n}}}{\overline{B_m}} \tag{11}$$

where $B_{s,i}$, $B_{m,i}$ and $\overline{B_m}$ are the simulated, measured and the mean specific CH_4 yields, respectively, and n is the number of experimental data points.

2.3. Evaluation of Synergistic Effects

The synergistic effects were evaluated for both the yield and the kinetic of the CH_4 production (Table 4). The ϕ factors were calculated following the approach of Castro-Molano et al. (2018) [39].

Table 4. Equations applied to evaluate the synergistic factors ϕ in AcoD of slaughterhouse wastewater streams and BM.

Synergistic Factor [a]	Equation	Evaluation
ϕy	$\left(\frac{B_o - B_{oexpected}}{B_{oexpected}}\right)100$	$\phi y, k_h, R, \lambda > 0$: the mixture has a synergistic effect. $\phi y, kh, R, \lambda < 0$: the mixture has an antagonistic effect. $\phi y, kh, R, \lambda = 0$: the mixture does not affect the performance of the substrates.
ϕk_h	$\left(\frac{k_h - k_{hexpected}}{k_{hexpected}}\right)100$	
ϕR	$\left(\frac{R_{max} - R_{max\,expected}}{R_{max\,expected}}\right)100$	
$\phi\lambda$	$\left(\frac{\lambda_{expected} - \lambda}{\lambda_{expected}}\right)100$	

[a] ϕy: synergy for CH_4 yield; ϕk_h: synergy for the apparent hydrolysis rate coefficient; ϕR: synergy for the maximum specific CH_4 production rate; $\phi\lambda$: synergy for the lag-phase.

The expected values of the parameters used were determined from predictive BMP curves (B_P) of co-digestion, calculated from the BMP curves (B_m) of the single substrates and assuming that the CH_4 production in co-digestion would be the weighted production of the single substrates. For all mixtures, B_P was, therefore, calculated as the summation of the products of the experimental B_m of single substrates j by their respective VS fraction in the mixture (α_j), as shown in Equation (12):

$$B_P = \sum_{j=1}^{n} B_{m,j} \cdot \alpha_j \tag{12}$$

The expected B_o was taken as the ultimate CH_4 yield of the predictive curve, whereas the expected kinetic parameters λ, R_{max} and k_h were obtained from the calibration of the modified Gompertz and first-order models against the values of the predictive BMP curves B_P.

2.4. Energy and Economic Considerations

In order to evaluate the feasibility of implementing the AcoD of slaughterhouse wastewater streams and bovine manure, an energetic and economic study was performed based on the results of the BMP assays and modeling. Moreover, the technical and economic advantages of synergistic mixtures were compared to a monodigestion-only scenario. The electrical (P_{EE}) and thermal (P_{TE}) energy potentials [kWh m^{-3}] were calculated by Equations (13) and (14) [40]:

$$P_{EE} = VS.B_o.P_c.\eta_E \tag{13}$$

$$P_{TE} = VS.B_o.P_c.\eta_T \tag{14}$$

where VS is the mixtures volatile solids content [kg m^{-3}], B_o is the ultimate specific CH_4 yield [STP m^3 CH_4 kg^{-1} VS] obtained from the previous analyses, P_c is the lower heating value of CH_4 (10 kWh m^{-3}) and η_E and η_T are the electric and thermal efficiencies, which were assumed to be 25% (electric generator) and 80% (boiler), respectively [41]. Based on

P_{EE} and P_{TE}, an economic evaluation was performed considering the design assumptions, CAPEX (capital expenditures), OPEX (operational expenditures) and Benefits, shown in Table 5.

Table 5. Parameters and assumptions for the economic study.

	Unit	Value
Design Assumptions [a]		
Flow of SWW to be treated	$m^3\ d^{-1}$	4.5
Flow of OWW to be treated	$m^3\ d^{-1}$	11.2
Flow of PWW to be treated	$m^3\ d^{-1}$	14.0
Flow of BM to be treated	$t\ d^{-1}$	0.8
Operational volume of digester (liquid fraction)	%	75 [b]
CAPEX		
Anaerobic digester	US$ m^{-3}	96
Electricity generator	US$	5640 [c]
OPEX		
Labour	US$ $year^{-1}$	4380
Electricity generator maintenance	US$ MWh^{-1}	14.82 [d]
Benefits		
Electricity saving	US$ kWh^{-1}	0.114
Natural gas saving	US$ m^{-3}	0.323
Wastewater treatment saving	US$ m^{-3}	1.30

[a] SWW: slaughter wastewater; OWW: offal wastewater; PWW: paunch wastewater; BM: bovine manure. [b] Data from Escalante et al. (2017) [40]. [c] Corresponding to a 20-kW biomass electric generator [42]. [d] Data from González-González et al. (2014) [43].

The waste flow values correspond to 10% of the total generated streams in the slaughterhouse considered as a case study (Table 1). The volume of the digester (VD) [m^3] for CAPEX was calculated from waste flows (Q) [$m^3\ d^{-1}$] and the *HRT* [d], considering an operational volume of 75% of the total digester volume (Equation (15)) [8]:

$$V_D = Q.HRT.0.75^{-1} \quad (15)$$

HRT was estimated as the difference between the duration time of the BMP assays and the λ obtained from the modified Gompertz Model [14]. The cost of the digester was calculated based on the volumes and prices available on the Colombian market for plastic tubular digesters. Slaughterhouses need steam and hot water for cleaning, so usually, they have boilers for this purpose. Therefore, the economic analysis did not consider further CAPEX costs for the conversion of CH_4 to thermal energy. In the OPEX, the labor costs correspond to the payment of a legal Colombian minimum wage, corresponding to the one worker that is needed to operate the anaerobic digestion system (8 h a day, 6 days a week, 1.52 US $ h^{-1} including social benefits). Regarding the benefits, the electricity and natural gas prices and cost of wastes treatment were supplied by the case study slaughterhouse.

The aforementioned data allowed to calculate the payback period (PBP), net present value (NPV) and internal rate of return (IRR). An equipment lifetime of 10 years was considered, with a discount rate of 10% and inflation of 3.85%.

2.5. Statistical Analysis

A one-way ANOVA (Analysis of Variance) facilitated the data analysis and detection of significant differences between mixtures with respect to variables B_o and f_d (p-values < 0.05), and allowed to estimate the standard deviation.

3. Results and Discussion

3.1. Ultimate Experimental Specific CH_4 Yield (B_o) of Single Substrates

The results from the BMP assays of the single substrates are shown in Figure 2. Depending on the prevalent biochemical composition of the substrates, it is possible to divide the results into two groups. The first group includes the substrates with lignocellulosic nature, namely the Paunch Wastewater (PWW) and Bovine Manure (BM), which had low CH_4 production due to their high content in scarcely degradable lignocellulose (Table 2) from the start of the BMP assay until day 12, the cumulative CH_4 yields of both substrates were almost similar (Figure 2). However, from day 12, the increase of the PWW yield slowed down and approached its plateau, whereas the BM yield continued to rise until reaching its stable value from approximately day 25. The above behaviors are similar to those found in previous studies on digestion of bovine manure [44] and PWW [13] showing a relatively higher rate of degradation of PWW compared to manure.

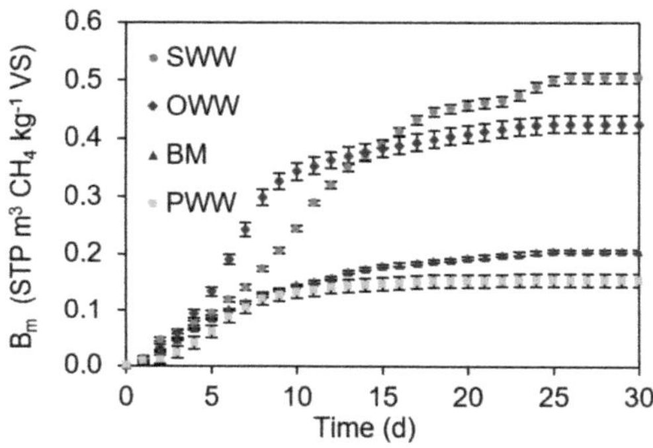

Figure 2. Accumulated CH_4 production of wastewater streams (SWW: slaughter wastewater; OWW: offal wastewater; PWW: paunch wastewater) and manure (BM) from a bovine slaughterhouse.

BM resulted in a B_o, at 30 days, of 0.206 ± 0.003 m^3 CH_4 kg^{-1} VS and an f_d of 0.46 ± 0.00, which are in the range of B_o values reported for dairy manure (0.089–0.303 m^3 CH_4 kg^{-1} VS) [44,45] and close to the biodegradability published in previous studies (0.54) [22]. PWW resulted in a B_o and an f_d of 0.154 ± 0.011 m^3 CH_4 kg^{-1} VS and 0.34 ± 0.01 respectively. These values are lower than those found for PWW in Australian slaughterhouses (0.309 m^3 CH_4 kg^{-1} VS and 0.84) [13]. Since the composition of ruminal content depends on how long the grass remains in the stomachs of animals [46], the above differences can be attributed to variations in the animals handling before slaughter. According to Australian regulation, animals must stay 24 h in yards before slaughter to be checked and to ensure that they are healthy [47]. However, in Colombian slaughterhouses, animals can be slaughtered 6 h after arrival [48].

The second group is formed by Offal Wastewater (OWW) and Slaughter Wastewater (SWW), which, contrary to the first group, are richer in lipids and proteins (Table 2) resulting in a relatively higher CH_4 production (Figure 2). During the first 3 days, the CH_4 yield of OWW and SWW did not present significant differences ($p > 0.05$). However, from day 4 to 10, the CH_4 yield of OWW increased at a higher rate than SWW and then slowed down from day 11 until it reached a steady-state at about day 25. On the other hand, in the case of SWW, the CH_4 yield presented an almost constant increase until about day 17, where it declined and achieved a plateau on day 25. Previous studies have shown how anaerobic digestion of wastes with high lipid concentrations result in a long lag period, due to LCFA accumulation and inhibition. For instance, Jensen et al. (2014) [4] reported a lag period of 18 days during anaerobic digestion of lipid-rich wastewater (10 g/L). In turn, Harris et al. (2018) [49] evidenced 7 days of lag period for anaerobic digestion of DAF (dissolved air flotation) sludge (10.5 g lipid/L). Likewise, Andriamanohiarisoamanana et al. (2017) [17 found that the BMP curve of crude glycerol presented an atypical shape (constant increase in the first 5 days followed by a slow CH_4 production until day 15 and then an exponential behavior) due to LCFA inhibition. On the contrary, in the current study, the BMP assays of SWW and OWW started CH_4 production from the first day, their curves had a typical behavior and their lipids concentration was lower than 10 g/L. This indicates how LCFA is unlikely to be a source of inhibition during anaerobic digestion of the tested slaughterhouse wastewater streams.

Ammonia is another potential cause of inhibition, which results from substrates with high protein content. In this regard, the BMP assay with SWW presented a final NH_3 concentration of 21.12 ± 0.25 mg L^{-1}, which is higher than the measured inhibition coefficient K_{I50}-$_{NH3}$ of the inoculum (18.53 ± 0.34 mg L^{-1}). Various studies investigated ammonia inhibition effects on BMP assays and reported experimental curves that were qualitatively similar to the present study. For instance, Nielsen and Angelidaki (2008) [50] evaluated the anaerobic digestion in BMP assays of cattle manure, with different initial total-N concentrations. The ammonia inhibition was evidenced in the slope of the cumulative CH_4 curves, which decreased with increasing initial nitrogen. In particular, samples with a total-N concentration of 3.0 g L^{-1} and 3.5 g L^{-1} achieved the same ultimate CH_4 yield. However, the former sample reached 80% of its ultimate CH_4 yield at 13 days while the latter reached 80% at 21 days; this result also highlights how ammonia inhibition follows a threshold behavior [35]. Similarly, Cuetos et al. (2017) [51] investigated the effect of active carbon addition in the anaerobic digestion of poultry blood (which is similar to the slaughter wastewater of this study). The experiments with lower activated carbon contents resulted in NH_3 inhibition and a significantly lower rate at the beginning of the BMP curve (specifically, during the first 13 days). The aforementioned analysis and studies confirm the likelihood of NH_3 accumulation and inhibition during the mono-digestion of SWW.

SWW and OWW BMP assays resulted in a B_o of 0.505 ± 0.008 and 0.425 ± 0.015 m^3 CH_4 kg^{-1} VS, respectively. Although OWW has the highest lipids content, it presented lower B_o than SWW due to the concomitant presence of lignocellulosic material (Table 2). The B_o of SWW was close to the values of 0.500 and 0.570 m^3 CH_4 kg^{-1} VS reported in the studies of Jensen et al. (2014; 2015) [4,52], while the f_d resulted in a value of 0.80 ± 0.01, which is close to the results of a similar BMP study investigating blood biodegradability (f_d of 0.77) [12]. On the other hand, the B_o of OWW is lower when compared to studies investigating similar substrates. For instance, Jensen et al. (2014) [4] found a B_o between 0.721 and 0.931 m^3 CH_4 kg^{-1} VS for an offal wastewater stream. Nevertheless, this wastewater also contained the waste stream from the cleaning of red viscera, resulting in a higher lipid concentration (up to 11.64 kg m^{-3}) compared to the OWW stream in the current study, thus explaining the relatively higher B_o. Regarding the f_d from OWW (0.63 ± 0.02), to the best of the author's knowledge, there is no available comparison in the literature.

3.2. Experimental Ultimate Specific CH_4 Yield of AcoD

Figure 3 shows the composition (lipids, proteins and carbohydrates) and the ultimate experimental yield B_o of the different AcoD mixtures evaluated (the BMP curves are depicted in Supplementary Data Figure S1). On the whole, for both binary and multicomponent mixtures, the B_o increased directly with the proportion of lipids and decreased with the proportion of carbohydrates. Therefore, the highest B_o corresponds to the binary mixtures of SWW and OWW (S33:O67 and S67:O33) and the ternary mixtures where SWW and OWW were present simultaneously (S33:O33:B34 and S33:O33:P34).

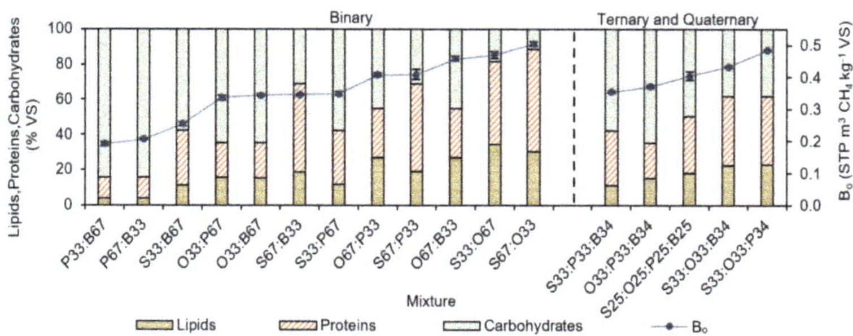

Figure 3. Biochemical composition of the different AcoD mixtures (left axis) and the resulting ultimate specific CH_4 yield (right axis). On the X-axis, the letter represents the waste stream (S: slaughter wastewater; O: offal wastewater; P: paunch wastewater; B: bovine manure) and the number its %VS in the mixture.

The ternary and quaternary mixtures had significantly higher B_o ($p < 0.05$) than binary mixtures with a similar biochemical composition. For instance, the combinations with the mixing ratio of S33:B67 and S33:P33:B34 have almost the same composition (~11%VS lipids, ~31%VS protein and ~58%VS carbohydrates); however, the latter mixture showed a B_o 40% higher than the former. Likewise, the ternary mixture O33:P33:B34 exhibited a B_o 10% higher than binary mixtures O33:P67 and O33:B67, despite having similar compositions (~15%VS lipids, ~20%VS protein and ~65%VS carbohydrates). When comparing the ternary mixtures with the highest B_o (mixtures S33:O33:B34 and S33:O33:P34) to the binary mixture with the highest B_o (S67:O33), the ternary mixtures have similar B_o (4–14% difference), while having 33% fewer proteins and 25% fewer lipids than the binary mixture. The above evidence a higher synergy between macromolecules on CH_4 production in multicomponent mixtures than in binary mixtures. This result is in agreement with the study by Astals et al. (2014) [12], who suggested that in addition to macro-composition, the structure of the substrates also affects their interaction. In this sense, there are differences in carbohydrates structure between PWW and BM and the kind of proteins between SWW and OWW.

The effects of AcoD on the reduction of initial lignocellulosic material composition and final NH_3 concentration (see Supplementary Data Table S1 for NH_3 calculation details) are shown in Table 6, taking biodegradability (f_d) as an indicator. In the case of BM and PWW, the co-digestion with OWW and SWW in binary or multicomponent mixtures allowed to achieve mixtures with relatively lower lignocellulosic content; this reduced the recalcitrant character of the mixture and as a consequence increased the biodegradability f_d above the values of both BM and PWW mono-digestion (0.46 and 0.34, respectively). On the contrary, the binaries AcoD between BM and PWW presented a high lignocellulosic composition, which resulted in an f_d around 0.44. Previous studies have demonstrated that the AcoD with lignocellulosic residues is an alternative to enhance the C/N ratio of animal manure, however, this requires pretreatment [53].

In the case of OWW, all its mixtures presented higher f_d than its mono-digestion (0.63), since fatty wastes are suitable co-substrates to lignocellulosic and protein wastes [12]. In turn, SWW showed the highest degradability of individual substrates (0.80) due to its content of soluble proteins in the blood (e.g., albumin and globulin), which are hydrolyzed fast and then converted to CH_4 while producing NH_3. In the case of SWW, AcoD offers the opportunity to reduce the risk of ammonia inhibition, through mixtures with substrates with lower protein content. For instance, the addition of PWW to SWW in binary mixtures allowed to reduce the inhibition risk by NH_3 and achieved an f_d around 0.7. The ternary mixture with a mixing ratio S33:O33:P34 exhibited an f_d (0.83) higher than SWW mono-digestion, which is consistent with its balanced composition of carbohydrates, lipids and proteins (Figure 3).

Table 6. Evaluation of AcoD of slaughterhouse wastewater streams and BM. Results are reported as an average of three measurements (±95% confidence interval). Mono-digestions are presented as a reference.

Mixture [a]	Initial Lignocellulosic Material (%VS)	Final NH_3 (mg/L)	Reduction of Lignocellulosic Material Composition [b]	Reduction of Inhibition Risk by NH_3 [a]	f_d [c]
S100	0.0	21.82 ± 0.25	n/a	n/a	0.80 ± 0.01
O100	13.3	10.62 ± 0.27	n/a	n/a	0.63 ± 0.02
P100	75.5	7.48 ± 0.25	n/a	n/a	0.34 ± 0.01
B100	63.2	7.24 ± 0.21	n/a	n/a	0.46 ± 0.00
S67:O33	4.4	15.89 ± 0.37	+	+	0.78 ± 0.01
S67:P33	25.2	16.73 ± 0.35	+	+	0.72 ± 0.03
S67:B33	***21.1***	***23.73 ± 0.33***	+	-	***0.61 ± 0.01***
S33:O67	8.9	15.71 ± 0.37	+	+	0.71 ± 0.02
S33:P67	50.4	9.91 ± 0.25	+	+	0.68 ± 0.01
S33:B67	***42.1***	***22.01 ± 0.33***	+	-	***0.50 ± 0.01***
O67:P33	34.1	8.64 ± 0.37	+	+	0.68 ± 0.01
O67:B33	29.9	6.32 ± 0.34	+	+	0.77 ± 0.01
O33:P67	54.8	6.62 ± 0.37	+	+	0.64 ± 0.01
O33:B67	46.6	10.22 ± 0.34	+	+	0.66 ± 0.01
P67:B33	***71.4***	***10.48 ± 0.33***	-	+	***0.45 ± 0.01***
P33:B67	***67.3***	***8.37 ± 0.33***	-	+	***0.43 ± 0.01***
S33:O33:P34	29.6	2.30 ± 0.44	+	+	0.83 ± 0.00
S33:P33:B34	46.2	5.43 ± 0.41	+	+	0.70 ± 0.01
S33:O33:B34	25.5	2.16 ± 0.42	+	+	0.74 ± 0.00
O33:P33:B34	50.7	1.79 ± 0.42	+	+	0.71 ± 0.01
S25:O25:P25:B25	38.0	3.56 ± 0.49	+	+	0.73 ± 0.01

[a] The letter represents the waste stream (S: slaughter wastewater; O: offal wastewater; P: paunch wastewater; B: bovine manure) and the number its %VS in the mixture. [b] Positive and negative effects are indicated by + and − signs, respectively. Mixtures in bold and italic resulted in either high lignocellulosic content or ammonia inhibition. [c] f_d: the extent of degradation.

On the other hand, important inhibition risk occurred during binary AcoD mixtures between BM and SWW, as indicated by the final NH_3 concentration being higher than $K_{I50\text{-}NH3}$, which led to a significantly lower f_d ($p < 0.05$) than the other AcoD mixtures of SWW. A similar result was presented by Andriamanohiarisoamanana et al. (2017) [17], who investigated the AcoD of meat and bone meal and manure in BMP assays. This study showed how the increase of meat and bone meal content from 10% to 66%VS caused inhibition by NH_3 and, as a consequence, the conversion rate of meat and bone meal to CH_4 was reduced. In the current study, the inhibitory effects between SWW and BM were mitigated in ternary and quaternary mixtures by dilution with OWW and PWW. Similarly, previous studies have highlighted lignocellulosic as a suitable co-substrate for anaerobic digestion of blood. For instance, López et al. (2006) [54] evaluated the AcoD of ruminal content and blood in batch digesters. The results showed an organic matter degradation from 55 to 70% when ruminal content/blood ratio (on a TS basis) varied between 2 and 8; the authors highlighted how during AcoD blood generates extra buffer capacity and brings micronutrients to the system. Cuetos et al. (2013) [55] conducted batch experiments on AcoD of poultry blood with maize residues. When maize concentration increased from 15% to 70% (VS basis), the CH_4 production raised from 0.130 to 0.188 $m^3\ kg^{-1}$ VS. Similarly, also in CSRT digesters, the AcoD of blood and organic fraction of municipal solid waste has been implemented in order to achieve stable operations, with a CH_4 yield between 0.200 and 0.289 $m^3\ kg^{-1}$ VS [56].

Because of the aforementioned drawbacks, the mixtures between BM and PWW and between BM and SWW can lead to low values of biodegradability and instabilities, respectively, in the digestion process (see bold/italic values in Table 6). Hence, these mixtures were excluded from the following sections to focus on the seemingly synergistic mixtures.

3.3. Kinetic Model Selection

The goodness of fit of the Gompertz and first-order models, and the respective estimated kinetic parameters, are summarized in Table 7. The best model was selected based on two statistical criteria: the normalized root mean square error (NRMSE) and the regression coefficient (R^2). NRMSE is the standard deviation of the prediction errors (residuals). Thus, NRMSE is a measure of how far the experimental points are from the simulated curves. R^2 provides a further measure of how well the model can reproduce the experimental data. For all mixtures, the Gompertz model resulted in a better fit of the experimental data compared to the first-order model. In particular, the ranges of NRMSE and R^2 were 0.011–0.044 and 0.992–0.999, respectively, in the modified Gompertz model and 0.037–0.134 and 0.918–0.988, respectively, in the first-order model. The confidence interval of the estimated parameters for Gompertz (reported as standard error, and shown in Supplementary Data Table S2), is in all cases below 3% for the simulated ultimate yield P and below 4% for the maximum specific CH_4 production rate R_{max}. For the lag-phase λ, the average error is 17%, with the highest value of 70% in the case S33:P33:B34, due to the smallest estimated value of the lag-phase (0.152 days). Given the better goodness of fit and the acceptable parameter identifiability, the Gompertz kinetics was selected for the following model-based analysis of the AcoD synergy (Section 3.4).

Figure 4 shows a selection of six AcoD BMP experimental data, together with the fitted Gompertz and first-order model; the complete set of curves is shown in Supplementary Data Figure S2. Figure 4a–c show three experiments which resulted in the smallest differences in the goodness of fit between the two models, with all cases achieving high values of the regression coefficient ($R^2 > 0.98$). These experiments correspond to the AcoD mixtures S33:P67; S33:P33:B34 and O33:P33:B34; it can be noted how they all have relevant content of the lignocellulosic substrates manure (BM) and paunch (PWW). In these cases, hydrolysis is significantly the rate-limiting step in the CH_4 production [53]. For first-order models, the hydrolysis rate coefficient of these mixtures resulted in the range 0.06–0.12 d^{-1}, which is similar to the value of 0.1 d^{-1} reported for paunch content by Jensen et al. (2016) [13].

On the other hand, Figure 4d–f shows the three experiments that presented the greatest deviation from the first-order model, namely, S33:O67, O67:P33 and O67:B33. It can be noted how these cases have a relevant content of lipid-rich offal wastewater (OWW). The lipid content from these mixtures caused an initial low CH_4 production, which is reflected in a significant value of the lag-phase (λ) between 2 and 3 days. After the lag-phase the CH_4 production occurred at a relatively high rate (R_{max} between 0.036 and 0.044 m^3 CH_4 kg^{-1} VS d^{-1}), which is comparable to the other mixtures. Similar behavior is reported by Astals et al. (2014) [12] in the anaerobic digestion of olive oil; the authors attributed the behavior to an initial LCFA absorption onto the surface of the microorganisms, which is followed rapidly by conversion to CH_4.

In general, ternary and quaternary AcoD mixtures had lower λ values (range 0.152–1.466 days; average 0.95 days) compared to binary mixtures (range: 0.281–2.982 days; average: 1.61 days) (Table 7). The λ range obtained in the current research is lower than values reported in previous research on slaughterhouse wastewater anaerobic digestions, with the work of Jensen et al. (2014) [4] reporting values of up to 18 days for lipid-rich streams. There is limited information on R_{max} in the anaerobic digestion of slaughterhouse wastewater. Hernández-Fydrych et al. (2019) [57] analyzed the CH_4 production kinetics of pretreated combined slaughterhouse wastewater by BMP assays. The authors fitted a Gompertz model and calculated a R_{max} of 0.0125 and 0.0140 m^3 CH_4 kg^{-1} VS d^{-1} for autoclaving and mechanical pretreatment, respectively. These values are lower than those found in this study (0.022–0.044 m^3 CH_4 kg^{-1} VS d^{-1}). Therefore, the possibility of controlling the mixture ratios of slaughterhouse wastewater streams in anaerobic co-digestion can have kinetics advantages, when compared to the digestion of the wastewaters' individual streams or combined as a whole.

Table 7. Kinetic parameters of the models fitted to the curves of the biochemical methane potential (BMP) assays.

Model	Mixture												
	S67:O33	S67:P33	S33:O67	S33:P67	O67:P33	O67:B33	O33:P67	O33:B67	S33:O33:P34	S33:P33:B34	S33:O33:B34	O33:P33:B34	S25:O25:P25:B25
First Order													
P	0.637	0.560	0.904	0.424	0.580	0.558	0.453	0.401	0.657	0.395	0.715	0.473	0.555
k_h	0.057	0.047	0.030	0.068	0.051	0.069	0.060	0.095	0.055	0.126	0.040	0.062	0.055
NRMSE	0.075	0.072	0.134	0.047	0.113	0.102	0.115	0.068	0.082	0.037	0.097	0.058	0.086
R^2	0.969	0.974	0.918	0.987	0.930	0.937	0.919	0.968	0.965	0.988	0.958	0.981	0.961
Modified Gompertz													
P	0.494	0.435	0.486	0.349	0.417	0.451	0.342	0.344	0.504	0.363	0.450	0.377	0.411
λ	1.734	0.697	2.142	0.281	2.152	2.982	1.690	1.187	1.145	0.152	1.466	1.045	0.934
R_{max}	0.036	0.022	0.037	0.023	0.036	0.044	0.034	0.036	0.033	0.036	0.030	0.026	0.029
NRMSE	0.022	0.044	0.011	0.024	0.026	0.027	0.029	0.025	0.033	0.018	0.031	0.016	0.026
R^2	0.998	0.992	0.999	0.998	0.997	0.997	0.995	0.996	0.995	0.997	0.996	0.999	0.997

P: Maximum specific CH_4 yield [STP m^3 CH_4 kg^{-1} VS]; λ: Lag-phase [d]; R_{max}: Maximum specific CH_4 production rate [STP m^3 CH_4 kg^{-1} VS d^{-1}]; k_h: Apparent hydrolysis rate coefficient [d^{-1}]; NRMSE: Normalized root mean square error; R^2: Correlation coefficient.

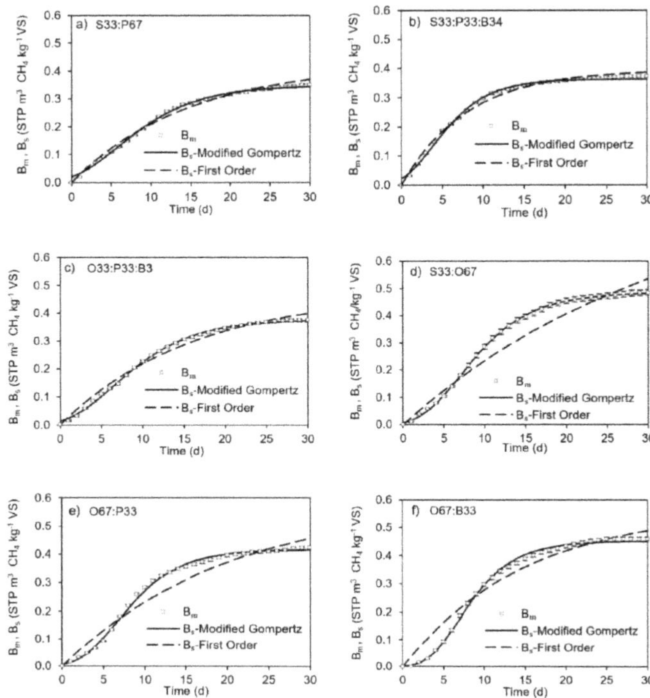

Figure 4. Accumulative CH_4 production from experimental data (B_m) and calibrated model (B_s) of the mixtures with the smallest deviations (**a**–**c**) and the greatest deviations (**d**–**f**) from the first-order model. The letter represents the waste stream (S: slaughter wastewater; O: offal wastewater; P: paunch wastewater; B: bovine manure) and the number its %VS in the mixture.

3.4. Evaluation of Synergy Effects

Figure 5 represents the synergistic effects of AcoD based on CH_4 yield (ϕy), lag-phase (ϕλ) and CH_4 production rate (ϕR). The predictive BMP curves along with the modified Gompertz plots are depicted in Supplementary Data Figure S3. All mixtures resulted in an experimental CH_4 yield higher than the expected (ϕy > 0). This result agrees with the evaluation presented in Table 6 and reaffirms the AcoD ability to reduce the inhibition risk by NH_3 and to improve the biodegradability of slaughterhouse wastewater and manure Regarding the kinetic synergy, antagonistic effects were observed in some mixtures (left side of Figure 5). Four AcoD mixtures resulted in a negative synergy with respect to the lag-phase (ϕλ < 0); these mixtures were characterized by a relatively high lipid proportion (23–34%VS), which slowed down the production of CH_4 during the first 2 or 3 days (Table 7). This observation is in agreement with the study on AcoD of dairy manure, meat, bone meal and crude glycerol carried out by Andriamanohiarisoamanana et al. (2017) [17], where an increase of glycerol proportion from 13%VS to 37%VS doubled λ. Additionally, antagonistic effects for R_{max} (ϕR < 0) were presented in four AcoD experiments.

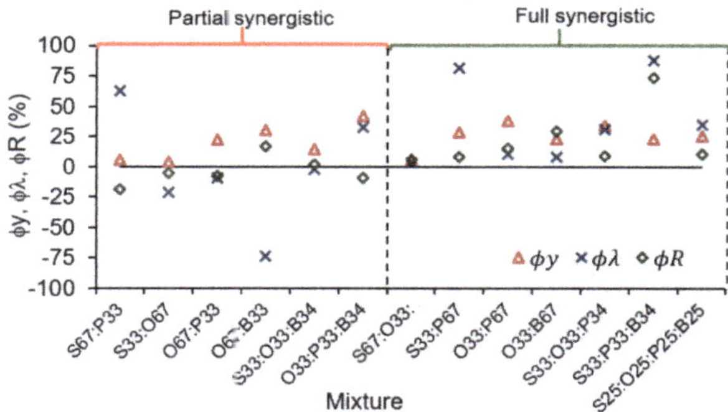

Figure 5. Synergistic effects of AcoD of slaughterhouse wastewater streams and bovine manure. The left side represents the mixtures that presented an antagonistic effect, while the right side indicates the mixtures with synergy in all the parameters (ϕy > 0; ϕλ and ϕR > 0). The letter represents the waste stream (S: slaughter wastewater; O: offal wastewater; P: paunch wastewater; B: bovine manure) and the number its %VS in the mixture.

Comparing the binary and multicomponent AcoD, greater synergy was observed in the latter. The binary mixtures exhibited synergistic factors between 4.2% and 38.0% for ϕy, between 3.4% and 81.5% for ϕλ and 5.6% and 29.5% for ϕR. Meanwhile, the ternary and quaternary mixtures showed synergistic factors between 14.5% and 41.9% for ϕy, between 31.1% and 87.9% for ϕλ and 2.1% and 73.9% for ϕR. This highlights the advantage of multi-component AcoD over binary ones, both in the final CH_4 yield and in the kinetics of production. Similar findings were found by Ara et al. (2015) [18] during AcoD of organic fraction of municipal solid waste, primary sludge and thickened waste activated sludge; the ternary mixtures exhibited CH_4 yields between 12 and 27% higher than binary mixtures. Additionally, Castro-Molano et al. (2018) [39] observed higher ϕy factors in ternary mixtures (25–167%) than binary mixtures (5–68%) when chicken manure was co-digested with industrial wastes.

The results showed seven mixtures in which all three synergistic factors were positive (ϕy > 0, ϕλ > 0 and ϕR > 0); these mixtures were considered fully synergistic and depicted on the right side of Figure 5. However, the synergistic effects in the AcoD with the mixing ratio of S67:O33 were relatively small, with values below 10%; these small values of synergy are generally considered not significant in AcoD studies [23]. Furthermore, the binary mixtures with significant synergy presented the BM or PWW as main substrates. This analysis suggests that when wastes with potential high CH_4 yield (e.g., SWW and OWW) are combined with the wastes with lower potential (e.g., BM and PWW), strong positive interactions are generated; on the other hand, weaker interactions occur when mixing wastes with similar characteristic (e.g., SWW with OWW and BM with PWW). Similar evidence can be found in the literature, such as in a study by Astals et al. (2014) [12], where the AcoD of DAF sludge and blood did not present significant synergy in CH_4 production; however, when DAF sludge was blended with paunch waste, the resulting CH_4 yield was 15% higher than expected. Likewise, Pagés-Diaz et al. (2014) [21] found antagonist effects in CH_4 production rate and no significant interaction in CH_4 yield when manure was co-digested with various crops (green fruit, vegetable residues and straw). Nevertheless, the AcoD of manure with slaughterhouse wastes presented significant synergy in both the production rate and yield of CH_4.

The six mixtures with significant full synergy correspond to the combinations: S33:P67; O33:P67; O33:B67; S33:O33:P34; S33:P33:B34 and S25:O25:P25:O25. These AcoD presented a lipids composition relatively lower (11–23%VS) than the rest of the mixtures (19–34%VS), while the carbohydrates and proteins did not show noticeable differences. Thus, it seems that the lipid concentration is the one that most influences the AcoD of slaughterhouses wastewater streams and BM, since a high concentration can improve CH_4 yield; however it negatively affects the kinetics. The aforementioned fully synergistic mixtures could improve the anaerobic digestion performance of slaughterhouse wastewater streams and manure in tubular digesters. In this sense, the current results are a starting point for a second stage of investigation where the synergistic mixtures will be tested in semi-continuous laboratory trials. This will allow to determine the effect of operational variables HRT and OLR and compare the synergistic effects achieved in the batch test with the synergy in semi-continuous processes, using the same model-based analysis described in this paper. The semi-continuous operation may result in the adaptation of the microbial community to inhibitors, therefore changing the absolute value of the synergistic effects while maintaining a similar qualitative evaluation of the synergy as achieved through batch tests [58].

3.5. Energy and Economic Feasibility

Table 8 shows a summary of the energy and economic study for the implementation of anaerobic digestion of the slaughterhouse wastewater streams and BM in mono-digestion and AcoD scenarios (see Supplementary Data from Tables S3–S8 for complete data). Mixtures present 27% more energy potential than single substrates as a consequence of the synergistic effect on methane yield (ϕy). Likewise, the anaerobic digestion of the mixtures would need almost 30 m^3 less digester volume compared to anaerobic digestion of the single substrates. This is due to the synergistic effects on kinetics, which reduce the estimated HRT on average by 3 days.

Table 8. Results of the economic study for the implementation of anaerobic digestion of the slaughterhouse wastewater streams and BM.

Scenario [a]	Unit	CH_4 for Thermal Energy Production	CH_4 for Electrical Energy Production
Mono-digestion			
Potential	kWh m^{-3}	33.74	10.54
Total volume of digesters	m^3	888	888
PBP	years	5	5
NPV	US$	50,894.00	56,962.88
IRR	%	22.77	23.28
AcoD			
Potential	kWh m^{-3}	42.69	13.34
Total volume of digesters	m^3	858	858
PBP	years	4	4
NPV	US$	70,636.35	79,675.98
IRR	%	27.71	28.48

[a] PBP: payback period; NPV: Net Present Value; IRR: Internal Rate of Return.

According to the energy potentials, the treatment of slaughterhouse wastewater streams and BM through anaerobic digestion would allow an energy saving between 0.91 and 1.21 US$ m^{-3} of waste in the mono-digestion scenario and between 1.16 and 1.53 US$ m^{-3} of waste in the AcoD scenario. These values added with the saving related to the avoided costs of current waste treatment (1.30 US$ m^{-3} of waste) result in an economic benefit from 2.21 to 2.51 US$ m^{-3} of waste and from 2.46 to 2.83 US$ m^{-3} of waste for mono-digestion and AcoD scenarios, respectively. The economic assessment shows that the CH_4 transformation into electric energy leads to higher NPV and IRR compared to the transformation into thermal energy. This is due to the low price of natural gas (0.026 US$ kWh^{-1}) compared to electricity (0.114 US$ kWh^{-1}). However, in both cases (electrical and thermal generation), the PBP is lower than the equipment lifetime (10 years), NPV is positive and IRR is higher than the discount rate (10%). These results confirm the energetic and economic feasibility of anaerobic digestion of slaughterhouse wastewater streams and manure. Moreover, the economic parameters (PBP, NPV and IRR) are better in the AcoD scenario than the mono-digestion scenario. This demonstrates that the synergistic effects of the mixtures also translate into economic advantages.

In developing countries, most slaughterhouses are located in small towns and supply only the local demand for meat (rural population mainly) [7]. Therefore, these slaughterhouses have low income, which limits their investment capacity in technology. In this sense, the tubular digester is a suitable alternative for waste treatment, given its low capital cost (compared to other kind of reactors), its simplicity of operation and lack of energy requirements for its operation [8]. Additionally, this type of waste management and renewable energy projects can access green financing. For instance, the Latin American banking sector has been developing a series of green products to finance projects that mitigate global warming [59]. Regarding Colombia, the country will issue green bonds in 2021 directed to finance sustainable and environmentally friendly projects [60].

4. Conclusions

The current results show that, except for binary mixtures between slaughter wastewater (SWW) and bovine manure (BM) and between BM and paunch wastewater (PWW), the AcoD enhanced the biodegradability and reduced the inhibition risk by NH_3 compared to the mono-digestion of slaughterhouse wastewater streams and BM. The synergy evaluation evidenced stronger positive effects when combining substrates with low methane potential (BM and PWW) with substrates with high potential (SWW and offal wastewater (OWW)) compared to binary mixtures BM-PWW and SWW-OWW. Likewise, the multicomponent mixtures performed better overall than the binary mixtures. The applied methodology allowed to select the mixtures with the best anaerobic digestion performance based on the CH_4 yield and kinetics criteria, which also present energetic and economic advantages over the single substrates. Therefore, the treatment of slaughterhouse wastewater streams and manure by AcoD in tubular digesters would be feasible. For small slaughterhouses, the implementation of the anaerobic digestion technology could be financed through green products offered by the banking sector.

Supplementary Materials: The following are available online at https://www.mdpi.com/1996-1073/14/2/384/s1, Figure S1: Experimental (B_m) and predictive (B_p) accumulative CH_4 production of AcoD of slaughterhouse wastewater streams and bovine manure, Figure S2: Experimental (B_m) and simulated (B_s) accumulative CH_4 production of AcoD of slaughterhouse wastewater streams and bovine manure, Figure S3: Predictive (B_p) accumulative CH_4 production of AcoD of slaughterhouse wastewater streams and bovine manure with the Modified Gompertz model fit, Table S1: Summary of NH_3 calculation data, Table S2: Standard error of the estimated parameters for the first-order model and the modified Gompertz model, Table S3: Energetic evaluation for the mono-digestion scenario, Table S4: Economic evaluation for electrical energy generation in the mono-digestion scenario, Table S5: Economic evaluation for thermal energy generation in the mono-digestion scenario, Table S6: Energetic evaluation for the AcoD scenario, Table S7: Economic evaluation for electrical energy generation in the AcoD scenario, Table S8: Economic evaluation for thermal energy generation in the AcoD scenario.

Author Contributions: Conceptualization, Z.S., L.C. and H.E.; methodology, Z.S.; software, D.P.; validation, D.P., L.C. and H.E.; formal analysis, Z.S. and D.P.; investigation, Z.S.; resources, L.C. and H.E.; data curation, L.C.; writing—original draft preparation, Z.S.; writing—review and editing, D.P., L.C. and H.E.; visualization, Z.S.; supervision, L.C. and H.E.; project administration, Z.S.; funding acquisition, H.E. All authors have read and agreed to the published version of the manuscript.

Funding: This research received no external funding.

Institutional Review Board Statement: Not applicable.

Informed Consent Statement: Not applicable.

Data Availability Statement: The data presented in this study are available in supplementary material available online at https://www.mdpi.com/1996-1073/14/2/384/s1. Additional data may be requested from the corresponding author.

Acknowledgments: This work was financially supported by the Colombian Departamento Administrativo de Ciencia, Tecnología e Innovación (Colciencias)-convocatoria 785 de 2017 and Universidad Industrial de Santander (UIS). The authors are grateful to ColBeef slaughterhouse for facilitating the technical and economic information and samples for the development of this study. Davide Poggio is grateful for the financial support from British Council Newton Fund (Institutional Links grant 527677146).

Conflicts of Interest: The authors declare no conflict of interest.

References

1. Bustillo-Lecompte, C.; Mehrvar, M. Slaughterhouse wastewater characteristics, treatment, and management in the meat processing industry: A review on trends and advances. *J. Environ. Manag.* **2015**, *161*, 287–302. [CrossRef] [PubMed]
2. Mekonnen, M.M.; Hoekstra, A.Y. A Global Assessment of the Water Footprint of Farm Animal Products. *Ecosystems* **2012**, *15*, 401–415. [CrossRef]
3. Geraghty, R. *Sustainable Practices in Irish Beef Processing*; Enterprise Ireland: Dublin, Ireland, 2009.
4. Jensen, P.; Sullivan, T.W.; Carney, C.; Batstone, D.J. Analysis of the potential to recover energy and nutrient resources from cattle slaughterhouses in Australia by employing anaerobic digestion. *Appl. Energy* **2014**, *136*, 23–31. [CrossRef]
5. Caixeta, C.E.; Cammarota, M.C.; Xavier, A.M. Slaughterhouse wastewater treatment: Evaluation of a new three-phase separation system in a UASB reactor. *Bioresour. Technol.* **2002**, *81*, 61–69. [CrossRef]
6. Saddoud, A.; Sayadi, S. Application of acidogenic fixed-bed reactor prior to anaerobic membrane bioreactor for sustainable slaughterhouse wastewater treatment. *J. Hazard. Mater.* **2007**, *149*, 700–706. [CrossRef]
7. World Bank. *Global Study of Livestock Markets, Slaughterhouses and Related Waste Management Systems*; Final Report; World Bank: Tokyo, Japan, 2009.
8. Kinyua, M.N.; Rowse, L.E.; Ergas, S. Review of small-scale tubular anaerobic digesters treating livestock waste in the developing world. *Renew. Sustain. Energy Rev.* **2016**, *58*, 896–910. [CrossRef]
9. Martí-Herrero, J.; Alvarez, R.; Flores, T. Evaluation of the low technology tubular digesters in the production of biogas from slaughterhouse wastewater treatment. *J. Clean. Prod.* **2018**, *199*, 633–642. [CrossRef]
10. Wang, H.; Zhang, Y.; Angelidaki, I. Ammonia inhibition on hydrogen enriched anaerobic digestion of manure under mesophilic and thermophilic conditions. *Water Res.* **2016**, *105*, 314–319. [CrossRef]
11. Zonta, Z.J.; Alves, M.M.; Flotats, X.; Palatsi, J. Modelling inhibitory effects of long chain fatty acids in the anaerobic digestion process. *Water Res.* **2013**, *47*, 1369–1380. [CrossRef]
12. Astals, S.; Batstone, D.; Mata-Alvarez, J.; Jensen, P. Identification of synergistic impacts during anaerobic co-digestion of organic wastes. *Bioresour. Technol.* **2014**, *169*, 421–427. [CrossRef]
13. Jensen, P.; Mehta, C.M.; Carney, C.; Batstone, D.J. Recovery of energy and nutrient resources from cattle paunch waste using temperature phased anaerobic digestion. *Waste Manag.* **2016**, *51*, 72–80. [CrossRef] [PubMed]
14. Kafle, G.K.; Kim, S.-H. Kinetic Study of the Anaerobic Digestion of Swine Manure at Mesophilic Temperature: A Lab Scale Batch Operation. *J. Biosyst. Eng.* **2012**, *37*, 233–244. [CrossRef]
15. Fogler, S. Conversion and reactor sizing. In *Elements of Chemical Reaction Engineering*, 5th ed.; Prentice Hall: Kendallville, IN, USA, 2016; pp. 31–68.
16. Devuyst, E.; Pryor, S.W.; Lardy, G.; Eide, W.; Wiederholt, R. Cattle, ethanol, and biogas: Does closing the loop make economic sense? *Agric. Syst.* **2011**, *104*, 609–614. [CrossRef]
17. Andriamanohiarisoamanana, F.J.; Saikawa, A.; Tarukawa, K.; Qi, G.; Pan, Z.; Yamashiro, T.; Iwasaki, M.; Ihara, I.; Nishida, T.; Umetsu, K. Anaerobic co-digestion of dairy manure, meat and bone meal, and crude glycerol under mesophilic conditions: Synergistic effect and kinetic studies. *Energy Sustain. Dev.* **2017**, *40*, 11–18. [CrossRef]
18. Ara, E.; Sartaj, M.; Kennedy, K. Enhanced biogas production by anaerobic co-digestion from a trinary mix substrate over a binary mix substrate. *Waste Manag. Res.* **2015**, *33*, 578–587. [CrossRef]
19. Wang, X.; Yang, G.; Li, F.; Feng, Y.; Ren, G.; Han, X. Evaluation of two statistical methods for optimizing the feeding composition in anaerobic co-digestion: Mixture design and central composite design. *Bioresour. Technol.* **2013**, *131*, 172–178. [CrossRef]
20. Dennehy, C.; Lawlor, P.G.; Croize, T.; Jiang, Y.; Morrison, L.; Gardiner, G.E.; Zhan, X. Synergism and effect of high initial volatile fatty acid concentrations during food waste and pig manure anaerobic co-digestion. *Waste Manag.* **2016**, *56*, 173–180. [CrossRef]
21. Pagés-Díaz, J.; Pereda-Reyes, I.; Taherzadeh, M.J.; Sárvári-Horváth, I.; Lundin, M. Anaerobic co-digestion of solid slaughterhouse wastes with agro-residues: Synergistic and antagonistic interactions determined in batch digestion assays. *Chem. Eng. J.* **2014**, *245*, 89–98. [CrossRef]
22. Ebner, J.H.; Labatut, R.A.; Lodge, J.S.; Williamson, A.A.; Trabold, T.A. Anaerobic co-digestion of commercial food waste and dairy manure: Characterizing biochemical parameters and synergistic effects. *Waste Manag.* **2016**, *52*, 286–294. [CrossRef]
23. Donoso-Bravo, A.; Ortega, V.; Lesty, Y.; Bossche, H.V.; Olivares, D. Addressing the synergy determination in anaerobic co-digestion and the inoculum activity impact on BMP test. *Water Sci. Technol.* **2019**, *80*, 387–396. [CrossRef]
24. Mendieta, O.; Castro, L.; Rodríguez, J.; Escalante, H. Synergistic effect of sugarcane scum as an accelerant co-substrate on anaerobic co-digestion with agricultural crop residues from non-centrifugal cane sugar agribusiness sector. *Bioresour. Technol.* **2020**, *303*, 122957. [CrossRef] [PubMed]

25. Holliger, C.; Alves, M.; Andrade, D.; Angelidaki, I.; Astals, S.; Baier, U.; Bougrier, C.; Buffière, P.; Carballa, M.; De Wilde, V.; et al. Towards a standardization of biomethane potential tests. *Water Sci. Technol.* **2016**, *74*, 2515–2522. [CrossRef] [PubMed]
26. Callaghan, F.; Wase, D.; Thayanithy, K.; Forster, C. Continuous co-digestion of cattle slurry with fruit and vegetable wastes and chicken manure. *Biomass Bioenergy* **2002**, *22*, 71–77. [CrossRef]
27. Angelidaki, I.; Alves, M.M.; Bolzonella, D.; Borzacconi, L.; Campos, J.L.; Guwy, A.J.; Kalyuzhnyi, S.; Jenicek, P.; Van Lier, J.B. Defining the biomethane potential (BMP) of solid organic wastes and energy crops: A proposed protocol for batch assays. *Water Sci. Technol.* **2009**, *59*, 927–934. [CrossRef] [PubMed]
28. Maya-Altamira, L.; Baun, A.; Angelidaki, I.; Schmidt, J.E. Influence of wastewater characteristics on methane potential in food-processing industry wastewaters. *Water Res.* **2008**, *42*, 2195–2203. [CrossRef]
29. Raposo, F.; Fernández-Cegrí, V.; De La Rubia, M.; Ángeles Borja, R.; Béline, F.; Cavinato, C.; Demirer, G.N.; Fernández, B.; Fernández-Polanco, M.; Frigon, J.C.; et al. Biochemical methane potential (BMP) of solid organic substrates: Evaluation of anaerobic biodegradability using data from an international interlaboratory study. *J. Chem. Technol. Biotechnol.* **2011**, *86*, 1088–1098. [CrossRef]
30. APHA-AWWA-WEF. *Standard Methods for the Examination of Water and Wastewater*, 23rd ed.; American Public Health Association: Washington, D.C., USA, 2017.
31. Lahav, O.; Morgan, B. Titration methodologies for monitoring of anaerobic digestion in developing countries? A review. *J. Chem. Technol. Biotechnol.* **2004**, *79*, 1331–1341. [CrossRef]
32. Van Soest, P.; Robertson, J.; Lewis, B. Methods for Dietary Fiber, Neutral Detergent Fiber, and Nonstarch Polysaccharides in Relation to Animal Nutrition. *J. Dairy Sci.* **1991**, *74*, 3583–3597. [CrossRef]
33. Passos, F.; Ortega, V.; Donoso-Bravo, A. Thermochemical pretreatment and anaerobic digestion of dairy cow manure: Experimental and economic evaluation. *Bioresour. Technol.* **2017**, *227*, 239–246. [CrossRef]
34. Astals, S.; Batstone, D.; Tait, S.; Jensen, P. Development and validation of a rapid test for anaerobic inhibition and toxicity. *Water Res.* **2015**, *81*, 208–215. [CrossRef]
35. Astals, S.; Peces, M.; Batstone, D.; Jensen, P.; Tait, S. Characterising and modelling free ammonia and ammonium inhibition in anaerobic systems. *Water Res.* **2018**, *143*, 127–135. [CrossRef] [PubMed]
36. Shen, J.; Yan, H.; Zhang, R.; Liu, G.; Chen, C. Characterization and methane production of different nut residue wastes in anaerobic digestion. *Renew. Energy* **2018**, *116*, 835–841. [CrossRef]
37. Pan, X.; Wang, L.; Lv, N.; Ning, J.; Zhou, M.; Wang, T.; Li, C.; Zhu, G. Impact of physical structure of granular sludge on methanogenesis and methanogenic community structure. *RSC Adv.* **2019**, *9*, 29570–29578. [CrossRef]
38. Poggio, D.; Walker, M.; Nimmo, W.; Ma, L.; Pourkashanian, M. Modelling the anaerobic digestion of solid organic waste–Substrate characterisation method for ADM1 using a combined biochemical and kinetic parameter estimation approach. *Waste Manag.* **2016**, *53*, 40–54. [CrossRef]
39. Molano, L.C.; Escalante, H.; Lambis-Benítez, L.E.; Marín-Batista, J.D. Synergistic effects in anaerobic codigestion of chicken manure with industrial wastes. *DYNA* **2018**, *85*, 135–141. [CrossRef]
40. Escalante, H.; Castro, L.; Amaya, M.; Jaimes, L.; Jaimes-Estévez, J. Anaerobic digestion of cheese whey: Energetic and nutritional potential for the dairy sector in developing countries. *Waste Manag.* **2018**, *71*, 711–718. [CrossRef]
41. Ali, M.M.; Ndongo, M.; Bilal, B.; Yetilmezsoy, K.; Youm, I.; Bahramian, M. Mapping of biogas production potential from livestock manures and slaughterhouse waste: A case study for African countries. *J. Clean. Prod.* **2020**, *256*, 120499. [CrossRef]
42. Haitai Power Machinery. Available online: http://www.hitepower.com/index.jsp (accessed on 18 November 2020).
43. González-González, A.; Collares-Pereira, M.; Cuadros, F.; Fartaria, T. Energy self-sufficiency through hybridization of biogas and photovoltaic solar energy: An application for an Iberian pig slaughterhouse. *J. Clean. Prod.* **2014**, *65*, 318–323. [CrossRef]
44. Jin, Y.; Hu, Z.; Wen, Z. Enhancing anaerobic digestibility and phosphorus recovery of dairy manure through microwave-based thermochemical pretreatment. *Water Res.* **2009**, *43*, 3493–3502. [CrossRef]
45. Zheng, Z.; Liu, J.; Yuan, X.; Wang, X.; Zhu, W.; Yang, F.; Cui, Z. Effect of dairy manure to switchgrass co-digestion ratio on methane production and the bacterial community in batch anaerobic digestion. *Appl. Energy* **2015**, *151*, 249–257. [CrossRef]
46. Matthews, C.; Crispie, F.; Lewis, E.; Reid, M.; O'Toole, P.W.; Cotter, P.D. The rumen microbiome: A crucial consideration when optimising milk and meat production and nitrogen utilisation efficiency. *Gut Microbes* **2019**, *10*, 115–132. [CrossRef] [PubMed]
47. RSPCA. How Are Animals Killed for Food? Available online: https://kb.rspca.org.au/knowledge-base/how-are-animals-killed-for-food/ (accessed on 1 October 2020).
48. Ministry of Health and Social Protection. *Resolution No. 240/2013*; Ministry of Health and Social Protection: Bogota, Colombia, 2013.
49. Harris, P.W.; Schmidt, T.; McCabe, B.K. Bovine bile as a bio-surfactant pre-treatment option for anaerobic digestion of high-fat cattle slaughterhouse waste. *J. Environ. Chem. Eng.* **2018**, *6*, 444–450. [CrossRef]
50. Nielsen, H.B.; Angelidaki, I. Strategies for optimizing recovery of the biogas process following ammonia inhibition. *Bioresour. Technol.* **2008**, *99*, 7995–8001. [CrossRef] [PubMed]
51. Cuetos, M.J.; Martinez, E.J.; Moreno, R.; Gonzalez, R.; Otero, M.; Gómez, X. Enhancing anaerobic digestion of poultry blood using activated carbon. *J. Adv. Res.* **2017**, *8*, 297–307. [CrossRef]
52. Jensen, P.; Yap, S.; Boyle-Gotla, A.; Janoschka, J.; Carney, C.; Pidou, M.; Batstone, D.J. Anaerobic membrane bioreactors enable high rate treatment of slaughterhouse wastewater. *Biochem. Eng. J.* **2015**, *97*, 132–141. [CrossRef]

53. Neshat, S.A.; Mohammadi, M.; Darzi, G.N.; Lahijani, P. Anaerobic co-digestion of animal manures and lignocellulosic residues as a potent approach for sustainable biogas production. *Renew. Sustain. Energy Rev.* **2017**, *79*, 308–322. [CrossRef]
54. López, I.; Passeggi, M.; Borzacconi, L. Co-digestion of ruminal content and blood from slaughterhouse industries: Influence of solid concentration and ammonium generation. *Water Sci. Technol.* **2006**, *54*, 231–236. [CrossRef]
55. Cuetos, M.; Gómez, X.; Martínez, E.J.; Fierro, J.; Otero, M. Feasibility of anaerobic co-digestion of poultry blood with maize residues. *Bioresour. Technol.* **2013**, *144*, 513–520. [CrossRef]
56. Zhang, Y.; Banks, C. Co-digestion of the mechanically recovered organic fraction of municipal solid waste with slaughterhouse wastes. *Biochem. Eng. J.* **2012**, *68*, 129–137. [CrossRef]
57. Hernández-Fydrych, V.C.; Benítez-Olivares, G.; Meraz-Rodríguez, M.A.; Salazar-Peláez, M.L.; Fajardo-Ortiz, M.C. Methane production kinetics of pretreated slaughterhouse wastewater. *Biomass Bioenergy* **2019**, *130*, 105385. [CrossRef]
58. Koch, K.; Hafner, S.D.; Weinrich, S.; Astals, S.; Holliger, C. Power and Limitations of Biochemical Methane Potential (BMP) Tests. *Front. Energy Res.* **2020**, *8*, 8. [CrossRef]
59. Feleban; Eco Business Fund; IFC. *What Is the Latin American Banking Sector Doing to Mitigate Climate Change?* Report 2017; Feleban; Eco Business Fund; IFC: Washington, DC, USA, 2017.
60. Portafolio. Colombia Emitirá Bonos Verdes Desde El Próximo Año. 2020. Available online: https://www.portafolio.co/economia/finanzas/colombia-emitira-bonos-verdes-desde-el-proximo-ano-538462 (accessed on 17 November 2020).

Article

Psychrophilic Full Scale Tubular Digester Operating over Eight Years: Complete Performance Evaluation and Microbiological Population

Jaime Jaimes-Estévez [1], German Zafra [2], Jaime Martí-Herrero [3,4,*], Guillermo Pelaz [5], Antonio Morán [5], Alejandra Puentes [1], Christian Gomez [1], Liliana del Pilar Castro [1] and Humberto Escalante Hernández [1]

1 Centro de Estudios en Investigaciones Ambientales (CEIAM) Research Group, Universidad Industrial de Santander, Bucaramanga, Colombia; jamestev7@gmail.com (J.J.-E.); alepuentesl@saber.uis.edu.co (A.P.); chrisgo@saber.uis.edu.co (C.G.); licasmol@saber.uis.edu.co (L.d.P.C.); escala@uis.edu.co (H.E.H.)

2 Grupo de Investigación en Bioquímica y Microbiología (GIBIM), Universidad Industrial de Santander, Bucaramanga, Colombia; geralzaf@uis.edu.co

3 Biomass to Resources Group, Universidad Regional Amazonica Ikiam, Via Tena-Muyuna, Km.7, Tena, Napo, Ecuador

4 Building Energy and Environment Group, Centre Internacional de Mètodes Numérics en Enginyeria (CIMNE), Terrassa, 08034 Barcelona, Spain

5 Chemical, Environmental and Bioprocess Engineering Group, Universidad de León, Av. Portugal 41, 24009 León, Spain; gpelg@unileon.es (G.P.); amorp@unileon.es (A.M.)

* Correspondence: jaime.marti@ikiam.edu.ec

Abstract: Most biogas plants in the world run under psychrophilic conditions and are operated by small and medium farmers. There is a gap of knowledge on the performance of these systems after several years of operation. The aim of this research is to provide a complete evaluation of a psychrophilic, low-cost, tubular digester operated for eight years. The thermal performance was monitored for 50 days, and parameters such as pH, total volatile fatty acid (tVFA), chemical oxygen demand (COD) and volatile solids (VS) were measured every week for the influent and effluent. The digester operated at a stabilized slurry temperature of around 17.7 °C, with a mean organic load rate (OLR) equal to 0.52 kg $VS/m^3{}_{digester}$ *d and an estimated hydraulic retention time (HRT) of 25 days. The VS reduction in the digester was around 77.58% and the COD reduction was 67 ± 3%, with a mean value for the effluent of 3.31 ± 1.20 g COD/Lt, while the tVFA decreased by 83.6 ± 15.5% and the presence of coliforms decreased 10.5%. A BioMethane potential test (BMP) for the influent and effluent showed that the digester reached a specific methane production of 0.40 Nm^3CH_4/kg VS and a 0.21 $Nm^3CH_4/m^3{}_{digester}$ d with 63.1% CH_4 in the biogas. These results, together with a microbiological analysis, show stabilized anaerobic digestion and a biogas production that was higher than expected for the psychrophilic range and the short HRT; this may have been due to the presence of an anaerobic digestion microorganism consortium which was extremely well-adapted to psychrophilic conditions over the eight-year study period.

Keywords: low cost digester; psychrophilic anaerobic digestion; thermal behavior

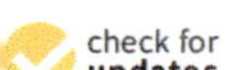

Citation: Jaimes-Estévez, J.; Zafra, G.; Martí-Herrero, J.; Pelaz, G.; Morán, A.; Puentes, A.; Gomez, C.; Castro, L.d.P.; Escalante Hernández, H. Psychrophilic Full Scale Tubular Digester Operating over Eight Years: Complete Performance Evaluation and Microbiological Population. *Energies* **2021**, *14*, 151. https://doi.org/10.3390/en14010151

Received: 27 November 2020
Accepted: 25 December 2020
Published: 30 December 2020

Publisher's Note: MDPI stays neutral with regard to jurisdictional claims in published maps and institutional affiliations.

1. Introduction

In Latin America, the low-cost tubular digester model (also known as the flexible, balloon or plastic model) is the most popular digester for biogas and digestate production from animal waste [1]. The most common livestock wastes used are cattle and pig manures [2,3]. Low cost digesters are characterized by the absence of active mixing devices and/or active heating systems [4]. The controlled use of biodigesters is a sustainable technology for the treatment of animal manure because it produces (i) energy: the biogas produced is often used as fuel for cooking, heating water, and generating electricity for on-site use; (ii) Agricultural benefit: the agronomic use of the effluent from anaerobic digestion, due to the presence of primary nutrients (nitrogen, phosphorous, potassium), is used as a soil amendment to improve plant growth [5,6]; (iii) Environmental quality: organic matter in waste manure is reduced and manure is stabilized (permanent odor and pathogen content reduction); and (iv) Social benefit: digesters improve health (reduction of exposure to wood smoke and volatile organic compounds) and quality of life, especially for women and children (who are able to spend significantly less time cooking) in rural zones [7]. Most of these benefits translate directly to energy and fertilizer cost savings for families living in rural areas in Latin America (on average, savings of USD $600/year for propane and around USD $50/year by using digestate as fertilizer) [8,9].

The anaerobic digestion (AD) process in low cost digesters is strongly influenced by local conditions such as solar radiation and temperature. For example, Castro et al. [5] reported a specific biogas production of 0.15 m^3_{biogas}/kg VS for a 9.5 m^3 (7.1 m^3 operational volume) low cost tubular digester under mesophilic conditions (25 ± 2 °C). In contrast, a tubular reactor with similar characteristics (volume and amount fed) in Peru at 2800 m.a.s.l. (16–20 °C) reached a specific biogas yield of 0.10 m^3_{biogas}/kg VS [10]. Therefore, biodigesters operated under psychrophilic conditions may present limitations due to the facts that: (i) microbial activity is slowed because the optimum growth temperature of bacteria and archaea is 37 °C; (ii) the removal of organic matter decreases, as does the concentration of methane in the biogas, and a fraction of this biogas is solubilized in the digestate; and (iii) in view of the above, the digestate contains organic matter that is converted to ammonia (NH_3) and methane (CH_4) during storage and soil usage [11]. Thus, this can be translated into a loss of energy efficiency and a larger environmental impact due to the aforementioned gaseous emissions [12]. Hence, the search of new strategies that can solve these problems is still in progress.

In order to improve biogas yield under psychrophilic operation conditions, the increase of HRT could be a favorable strategy, but it implies a larger biodigester size. A diminution of around 5 °C in operation temperature requires an increase in HRT from 30 to 50 days in order to maintain a similar biogas production and substrate volume [13] Another alternative to improve biodigester temperature conditions is the implementation of a passive solar heating design (solar radiation gain, insulation and greenhouse) [14] The greenhouse and tank insulation effect allow for the absorption and preservation of heat, which reduce the heat losses from the digester to the environment and to the ground, respectively [1,2]. Additionally, biogas yield can be enhanced with microorganisms which are adapted to low temperatures (<20 °C). According to Feller [15], only microorganisms adapted to psychrophilic conditions can deal with the limitations that occur with temperatures below 20 °C. These adapted microorganisms experience good physiological and ecological conditions in cold environments due to the unique characteristics of their membrane proteins, lipids, and genetic responses to thermal changes. In this sense, psychrotolerant microorganisms make AD possible in cold regions [16]. The AD microbiome comprises several distinct microbial trophic groups from the two evolutionarily distinct domains of bacteria and archaea. High throughput sequencing technology in AD microbiology makes it possible to determine the extensive and complex interactions of microbial communities within their environments and hosts. This procedure has shown that the main microorganism families present in stabilized sludge that proliferate in psychrophilic conditions belong to the families *Pseudomonadaceae, Methylophilaceae, Sphingobacteriaceae, Coriobacteriaceae*; among others [17].

Latin America presents a diversity of geographical and meteorological conditions which lead to a wide range of temperatures, from psychrophilic (<20 °C) to mesophilic (20–45 °C). Biodigesters that operate in tropical and subtropical regions work in mesophilic conditions, which enables the use of a relatively small digester size, high biogas production and good quality sludge [5]. Mesophilic biodigesters operate mainly with cattle manure as a substrate, obtaining biogas yields ranging from 0.15 m^3_{biogas}/kgVS to 0.4 m^3_{biogas}/kgVS, with an average methane quality of 62.6% [5,18]. In those regions, household digesters have been shown to generate the biogas required to satisfy user requirements [18].

Also, there are biodigesters installed at more than 3800 m above sea level [19], and in cold regions [6]. Due to the environmental conditions (mainly temperature), anaerobic digestion performance is affected. Under such conditions, the biogas digester yields ranged from 0.03 to 0.44 $m^3{}_{biogas}$/KgVS d with a reduced quality (60% and 49.6% of CH_4, respectively) [10,19]. Garfi el al. [20] reported that the biogas production in high altitude regions covers just around 60% of fuel needs for cooking; this could be improved by enhancing the digester design and using biofilm carriers [1], despite the fact that the use of digester effluent, known as bioslurry or biol, is, in many cases, more important to small- and medium-scale farmers than biogas [19]. Moreover, after a long period of biodigester operation, some operative problems such as clogging in the inputs and outputs, stagnation in the digester, shortcomings in feeding, changes in diet and solid accumulation may appear [21,22]. Therefore, it is necessary to accomplish a diagnosis of household digester performance after long periods of operation. On the other hand, understanding the microbial communities (through taxonomic analyses) in household digesters in cold climates could help to fundamentally improve the AD process and encourage its widespread application [23]. Unfortunately, most previous investigations have only focused on biogas yield and general monitoring during digester start up, and have not assessed the anaerobic digestion performance after several years of operation. The present research attempts to fill this gap by assessing the performance of psychrophilic rural digesters after several years of operation in continuous mode by examining the following: (i) thermal performance, (ii) bioprocess stability, (iii) microbiological analysis, and (iv) digestate quality. In this study, the performance of a medium-sized biogas plant in a rural area, which had been operating continuously for eight years under psychrophilic conditions, was diagnosed.

2. Materials and Methods

2.1. Site Description

Research was conducted at a Colombian pig farm at an altitude of 2963 m above sea level (m.a.s.l.) and a latitude of N 6°27′45.0″ W 72°24′43.0″. According to the Colombian Institute of Hydrology, Meteorology and Environmental Studies (IDEAM), the environmental temperature varied throughout the year between 12 ± 3 °C [24]. This farm had 456 pigs fed with cheese whey and water. The digester was feed with the excrete produced by 255 animals. The farm is 16.5 km from the Cocuy National Natural Park. The digester had been operated for the last 8 years, and the monitoring period was 50 days.

2.2. Description of Rural Biogas Plant

For the management of the pig manure, the farm used a double layer tubular polyethylene (caliber 8 and UV protection, common greenhouse plastic) digester for over 8 years. The digester dimensions were 30 m in length, 2.5 m in diameter and 147.3 m^3 total volume (operational volume is 103.1 m^3). The digester was covered by a polyethylene greenhouse that provided environmental protection. The daily excrete was composed of a pig manure and urine blend which represented 0.60 m^3 of total fed. This digester was fed daily with 4.16 m^3/d of a mixture of excrete and free-range wash water in a 1:6 ratio. The digester HRT was around 25 days. The biogas produced in the biodigester was used to heat an enclosure containing about 160 piglets. The installed biodigester installed did not have a biogas measurement system.

2.3. Monitoring Temperature in the Biodigester

To carry out the temperature profiles analysis, four datalogger sensors were installed with the objective of monitoring temperature (ambient, interior of the greenhouse, slurry and ground) and luminosity. The datalogger location and specifications are reported in Figure 1 and Table 1.

The sensors were set up to record the temperature every hour. The description of the location and method of installation of the sensors was as follows: (a) Tamt (sensor 1 monitored the ambient temperature around the greenhouse. (b) Tga (sensor 2) was the device by which variations in the temperature of the air inside the greenhouse were recorded; this device was oriented south-north to also monitor the solar luminosity which affected the performance of digester. (c) Ts (sensor 3) recorded variations in the slurry temperature data; this sensor was located one meter inside the biodigester bag. (d) Tgr (sensor 4) monitored soil temperature; this device was located one meter underground.

Meteorological data included solar radiation and ambient temperature, which were measured and collected every hour for 50 days.

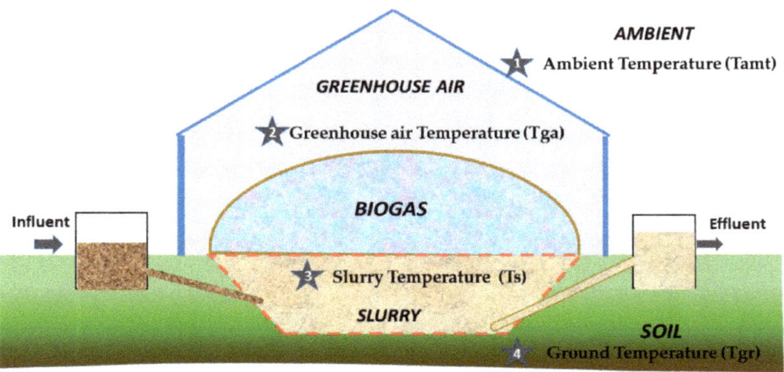

Figure 1. Digester scheme and location of temperature sensors.

Table 1. Location and characteristics of temperature sensors.

Name	Data	Location	Equipment	Resolution	Accuracy
Tamt (sensor 1)	ambient temperature	around the greenhouse	HOBO UA-001-08 Pendant® Waterproof Data Logger	0.14°	±0.53 °C from 0 °C to 50 °C
Tga (sensor 2)	air temperature/solar luminosity	inside greenhouse	HOBO UA-002-64 Pendant® Temperature/Light 64K Data Logger	0.14°/Designed for relative light levels	
Ts (sensor 3)	Slurry temperature	one meter into the biodigester bag	HOBO UA-001-08 Pendant® Waterproof Data Logger	0.14°	
Tgr (sensor 4)	Soil temperature	one meter underground	HOBO UA-001-08 Pendant® Waterproof Data Logger	0.14°	

2.4. Diagnosis of Anaerobic Digestion in the Pig Farm Digester

A diagnosis (performance and stability) of the pig biogas plant was performed by studying the biochemical and microbiological behavior. The biodigester was monitored for 50 days by taking a weekly sample of the influent and effluent. Regular operational conditions remained unaltered. The samples were stored and refrigerated before analysis. All experiments were performed in triplicate. The CH_4 biogas content was determined by gas chromatography using a TCD detector on a GC-Agilent 7890ª brand chromatograph using Argon as a drag gas and a 1010 plot Carboxen capillary column (length 30 m, internal diameter 0.32 mm, 25 µm stationary phase internal layer). The CO_2 content was determined by balance (assuming biogas to be a mixture of CH_4 and CO_2).

2.4.1. Biochemical Assays

The organic matter content was measured in terms of volatile solids (VS) and total chemical oxygen demand (COD), according to standard procedures 2540 G and 5220 D, respectively [25]. Measuring the digester pH, total carbonate alkalinity (TA) and total volatile fatty acids (tVFA) concentrations indicated the process stability. pH was measured with a Metrohm 691 pH Meter. TA and tVFA were measured by pH titration to 4.3, according to the method described by Purser et al. [26]. Individual VFAs (C2–C6) were measured using a 7820A gas chromatograph (Agilent, Santa Clara, CA, USA) equipped with a flame ionization detector and an Innowax column (Agilent, USA).

A biomethane potential (BMP) assay was developed following the methodology proposed by Holliger et al. (2016), but under psychrophilic temperature conditions (15 ± 2 °C). An experimental setup was constructed using 120 mL glass bottles with a 50% working volume. Digestate from the pig farm digester was used as the inoculum, keeping the digester working temperature. The substrate-to-inoculum ratio was 1:1 (VS basis). To measure endogenous methane production, a blank assay (inoculum without substrate) was included. Additionally, a positive control test with crystalline cellulose (97%) was conducted. As BMP, the residual biomethane potential of the digestate was measured in batch experiments (60 mL of inoculum working volume of at 15 ± 2 °C). To guarantee an anaerobic atmosphere, the bottles were flushed with a 80/20% N_2/CO_2 mixture and sealed with aluminum caps and butyl rubber stoppers. Methane production was quantified daily by volume displacement of a sodium hydroxide solution (2 N), and normalized to standard conditions (0 °C and 1 atmosphere). The BMP and residual biomethane potential tests were concluded when the volume of methane accumulated increased by less than 1% for three consecutive days.

2.4.2. Microbiological Analysis of Pig Farm Digester

Microbiological behavior was evaluated as a function of:

(i) Specific Methane Activity (SMA): the inoculum (digestate from pig farm biodigester) specific methanogenic activity at 15 ± 2 °C (local conditions) and 35 ± 2 °C (optimal condition). A SMA test was performed, in accordance with Astals et al. [27]. SMA experiments were performed in triplicate in 120 mL serum bottles, with a working volume of 60 mL. Sodium acetate was used as model substrate. A substrate-free blank was included to measure endogenous methane production from the inoculum. The methane produced during the SMA assay was quantified by measuring the volumetric displacement of an alkaline solution (2 N). The measured methane was normalized and expressed in terms of COD equivalents.

(ii) The taxonomic classification of bacterial and archaeal communities: first, genomic DNA was extracted from the digester influent and effluent using the PowerSoil® DNA Isolation Kit (MoBio Laboratories Inc., Carlsbad, CA, USA), according to the producer's recommendations. PCR reactions were carried out in an Eppendorf Mastercycler and PCR samples were checked for product size on a 1% agarose gel. Then, the entire DNA extract was used for high throughput sequencing of 16S rRNA gene-based massive libraries for eubacterial and archaeal communities. The primer set used for the eubacterial population analysis was 27Fmod (5′-AGRGTTTGATCMTGGCTCAG-3′)/519R modBio (5′-GTNTTACNGCGGCKGCTG-3′) [28]. For the archaeal population analysis, the primer set was Arch 349F (5′-GYGCASCAGKCGMGAAW-3′)/Arch 806R (5′GGACTACVSGGGTATCTAAT-3′) [29]. The obtained DNA reads were compiled in FASTq files for further bioinformatic processing [30]. Finally, operational taxonomic units were taxonomically classified using the Ribosomal Database Project, available at https://rdp.cme.msu.edu/.

(iii) The quantitative analysis of Bacteria and Archaea: bacteria an archaea populations were analyzed by means of quantitative-PCR reaction (qPCR) using PowerUp SYBR Green Master Mix (Applied Biosystems) in a StepOne plus Real Time PCR System (Applied

Biosstems) [31]. The primer sets were 341F and 518R for bacteria and mcrF and mcrR for archaea.

(iv) The pathogen content: fecal and total coliform in the influent and effluent were determined using a serial dilution, deep-plating technique in chromogenic culture media.

The global methodology of this study is presented in Figure 2.

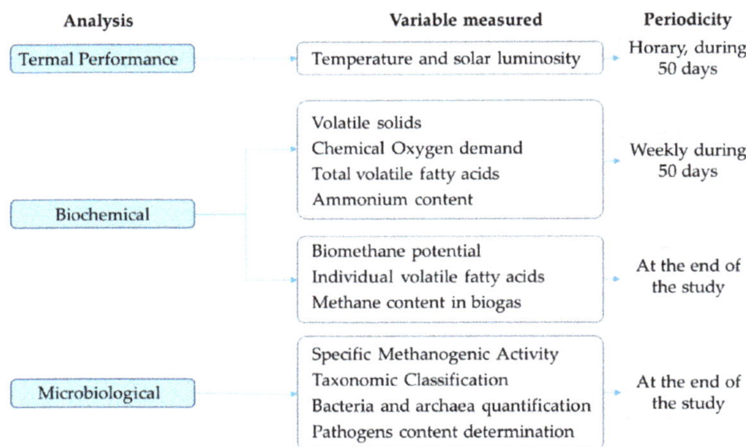

Figure 2. Methodology summary: analysis, variables measured and periodicity during biodigester monitoring.

3. Results and Discussion

3.1. Thermal Behavior of the Digester

Figure 3a shows the typical dynamical daily performance of the temperatures in the digester, while Figure 3b,c show the thermal performance for 5 days (as a tendency example during the monitoring time), and 50 consecutive days, respectively. The ambient temperature (Tamt) shows a typical daily bell shape that the green house temperature follows with higher amplitude. The maximum ambient temperature was 27.9 °C and the minimum was 9.5 °C, while the greenhouse temperature ranged between 35.5 °C and 6.8 °C. These data show that the greenhouse warmed up during the day but cooled during the night due to radiative cooling. In other studies, such as those by Perrigault et al. [14] and Martí-Herrero et al. [32], the greenhouses were built with adobe walls and were airtight, thereby achieving thermal inertia and keeping the greenhouse warm during the night However, in the pig farm digester, the greenhouse was made only of plastic, without thermal inertia and allowing more air exchange, producing lower internal temperatures than those found outside. This night cooling effect can be avoided through the selection of proper plastic; or adding thermal inertia and airtight to the walls of the greenhouse.

The slurry temperature (Ts) showed a flat performance with a mean temperature 17.7 °C, independent of daily variations (Figure 3b). The mean ground temperature (Tgr) was 16.3 °C, which was very close to the mean ambient temperature of 16.6 °C (Figure 3c) This means that the Ts was only 1.1 °C over the ambient temperature, while an improved solar heating design in the digester led to increments of Ts over 10 °C with respect to ambient temperatures [1]. Martí-Herrero et al. [16] showed the performance of a tubular digester under similar weather conditions, where the slurry temperature followed the values of the maximum ambient temperature using just black plastic and 1 cm insulation in the trench, without a greenhouse. This means that a greenhouse without insulation or black plastic is not enough for the solar heating of tubular digesters.

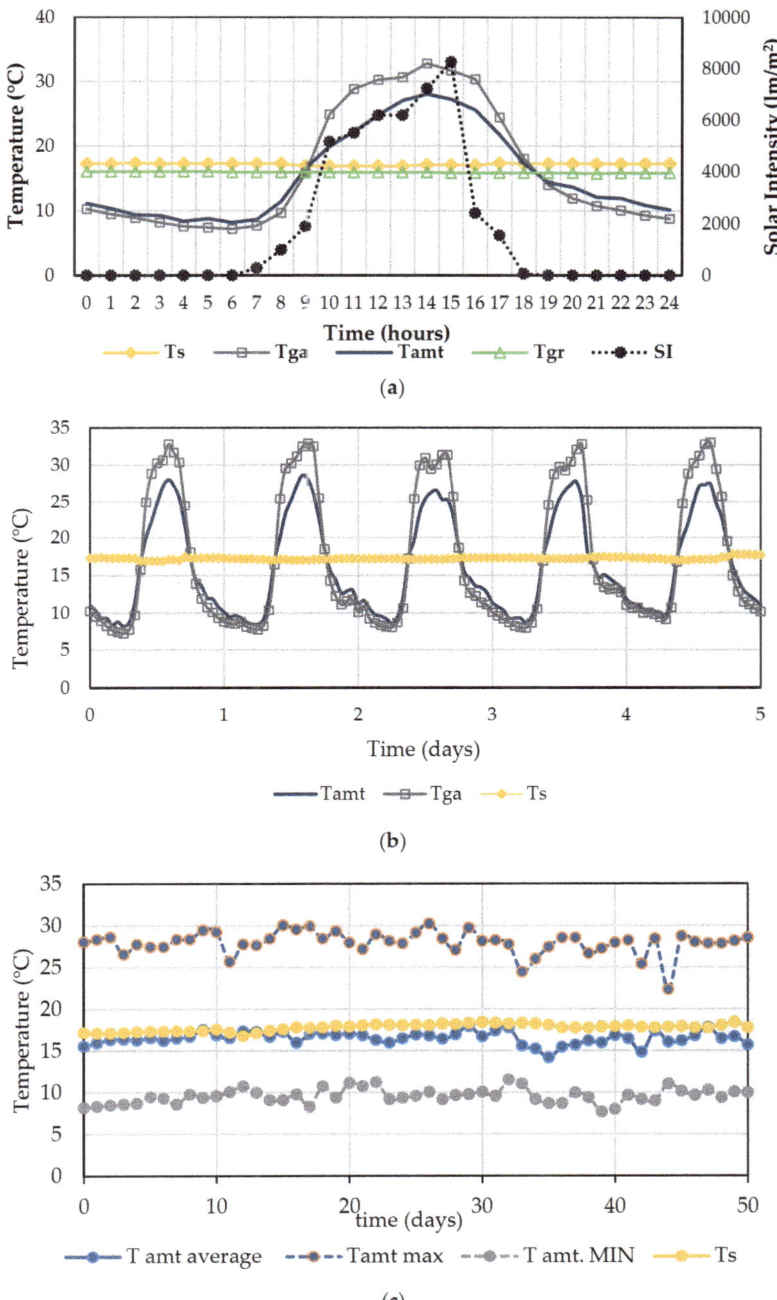

Figure 3. Thermal behavior of the digester. (**a**) Profiles of Ts, Tga, Tamt, Tgr, Tgh, Ts with solar luminosity for a day. (**b**) Profiles of Tamt, Tga, Ts for 5 days, (**c**) Behavior for 50 days of Tamt, Tamt max, Tamt min, Ts average.

3.2. Changes in Control Parameters in Pig Farm Digester

Plug flow digesters regularly operate with OLR from 0.5 kg VS/$m^3{}_{digester}$ *d to 1.0 kg VS/$m^3{}_{digester}$ *d [2]. During monitoring, the OLR ranged from 0.34 kg VS/$m^3{}_{digester}$ *d to 0.76 kg VS/$m^3{}_{digester}$ *d (average HRT = 25 d for a slurry temperature of 17.7 °C). This was because the manure was diluted during the cleaning of the pig shed with no wash water volume regulation. To achieve an adequate functioning of the digester, a previous dilution of the substrates was required which avoided clogging in the load and scum formation on its surface, and ensured continuous flow operation. The operational parameters of the digester are summarized in Table 2. Previous studies carried out with porcine manure reported that a 1:7 dilution favors the hydrolytic and methanogenic activities and the biomethane potential of the process [33]. Regarding the organic matter the average VS of the influent decreased from 12.74 ± 3.52 g VS/kg to a mean value of 2.86 g ± 1.20 g VS/kg, which means an organic matter removal around 77.58%. In comparison with a tubular system fed with swine manure and operating at an average temperature range of 25–30 °C, VS removal was 83% [18]. This comparation allowed us to infer that temperature affects the removal of volatile solids. A similar conclusion was reached in recent research focused on AD in cold regions [34].

The organic matter content (g COD/L) variation with respect to OLR is presented in Figure 4. The influent and effluent average COD were 9.94 ± 3.25 g COD/L and 3.31 ± 1.20 g COD/L, respectively. In the AD process, an increase in OLR caused a decrease in COD removal efficiency. In this study, the results showed a diminution in COD from 70% to 62.5% for OLR of 0.34 kg VS/m^3d and 0.76 kg VS/m^3d, respectively. On average the COD bioconversion achieved in the present study was 67 ± 3%. Previous studies have achieved higher organic matter removal in domestics plug-flow digester systems treating pig manure. Digesters operating at 23 ± 2 °C, 24.5 ± 1.5 °C and 26 ± 1.5 °C reported COD removal rates of around 88.5% [35], 78.5% [36] and 92% [18]. So, digesters working at psychrophilic temperature over 20 °C have been shown to achieve significant organic matter removal (around 28% more).

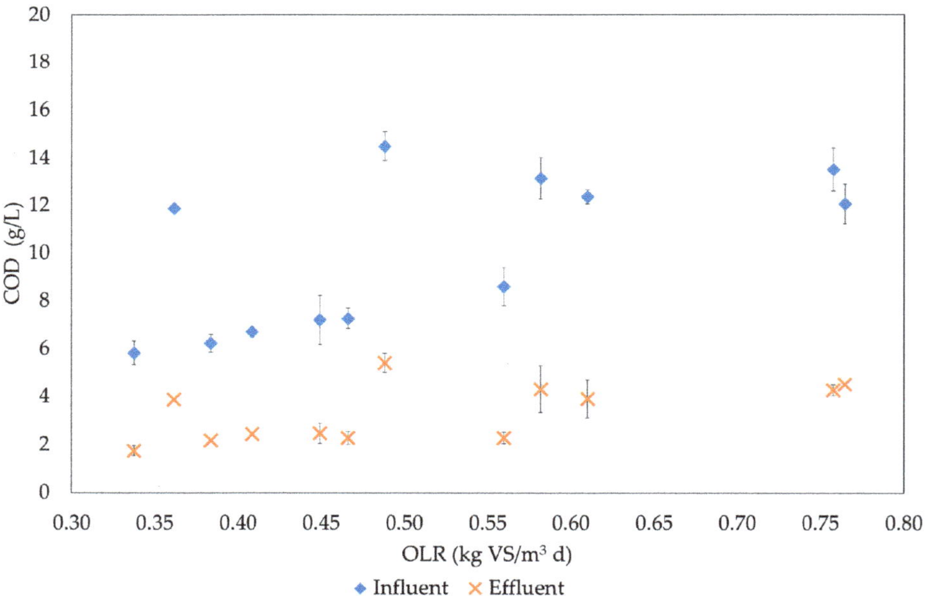

Figure 4. Chemical oxygen demand concentration for influent (blue rhombuses) and effluent (orange cruxes) during OLR changes.

Effluent tVFA concentration represents the easily biodegradable organic matter that was not metabolized in anaerobic processed. Figure 5 shows that the effluent tVFA concentrations were around 0.30 ± 0.08 g COD VFA/L. On average, the tVFA decrease was around 2.6 ± 1.4 g COD $_{VFA}$/L, which represents a bioconversion of 83.6 ± 15.5%. A rural biodigester with one year of continuous operation at 34 °C showed 63% of tVFA bioconversion [5]. Thus, with a longer operational period, it is possible to achieved higher conversion rates of soluble organic matter, even under psychrophilic conditions. These results demonstrated that the pig farm biodigester was operating efficiently, even after 8 years of continuous operation without maintenance.

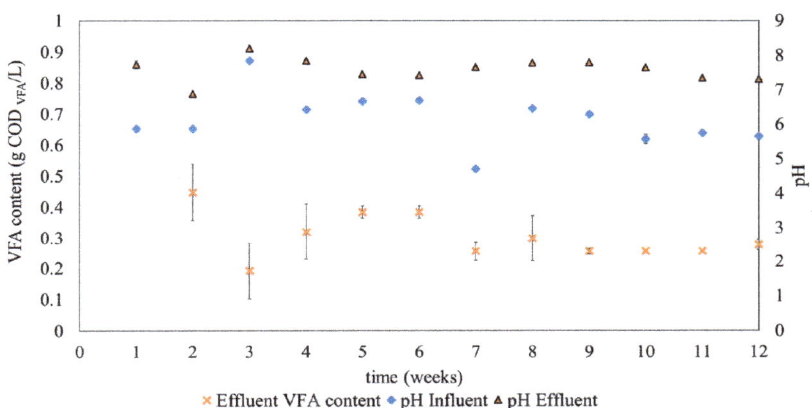

Figure 5. Effluent total volatile fatty acids concentration (orange cruxes) and influent and effluent pH (blue rhombus and orange triangles, respectively) during monitoring.

A diminution in temperature could affect the stability of the fermenting microorganisms. This change in stability may cause pH changes and decrease methane yield [37]. In the present study, the pH value for both the influent and effluent was 6.1 ± 0.8 and 7.6 ± 0.3 at psychrophilic temperature. The pH range for a healthy and continuous AD process is 6.8–8.2 [11]. The FOS/TAC ratio ranged between 0.72 ± 0.2–0.17 ± 0.1 mg of acetic acid/mg of $CaCO_3$ for the affluent and effluent, respectively. This demonstrates the high buffer capacity of pig manure. FOS/TAC values below 0.8 mg acetic acid/mg $CaCO_3$ are adequate for process stability [5]. This confirmed that the pig farm digester was operating properly without inhibition risk. As such, variations in OLR and temperature did not affect the anaerobic processes.

Individual VFA of the affluent and effluent (Figure 6) showed that acetic acid is most prevalent in the affluent (71% in relation to the other acids), indicating stable anaerobic fermentation. Butyric acid showed the best conversion (98%) compared to the other acids, and was the second most prevalent. These results can be compared with those in previous studies, where it has been shown that butyric acid fermentation plays a significant role during low temperature anaerobic degradation [38].

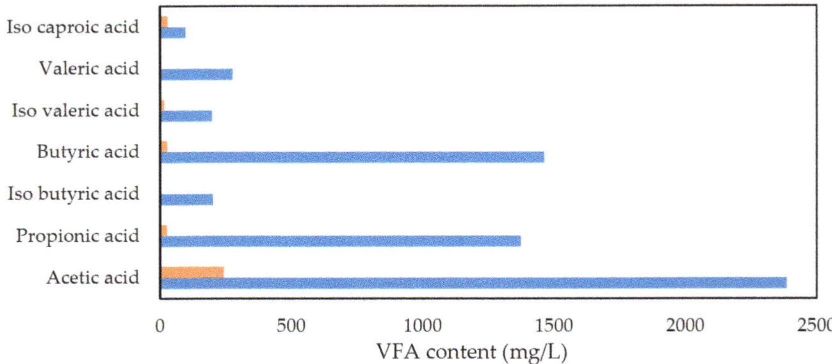

Figure 6. Individual VFAs (C2–C6) concentration in fed (affluent -blue bars-) and digestate (effluent -orange bars-).

According to the literature, a healthy digester has ratios of Pr/Ac below 1.4 with an acetic acid concentration under 800 mg/L. This value indicates that there is a propionate accumulation which represents a reduction in methane content due to hydrogenogenic bacteria inhibition. [39]. Although the VFA content varies along the digester [40], the outlet Pr/Ac ratio was 0.11 (acetic acid concentration of 244 mg/L). Up to now, there are no reports of individual volatile fatty acid values for low cost tubular digesters. As stated above, it can be affirmed that after a significant period of adaptation (8 years), the low-cost digester adapts to the temperature conditions and operates satisfactorily.

Ammonium is attributed to the mineralization of organic matter and is an indicator of bioprocess stability. The changes of NH_4-N during anaerobic fermentation in the low cost tubular digester are shown in Figure 7.

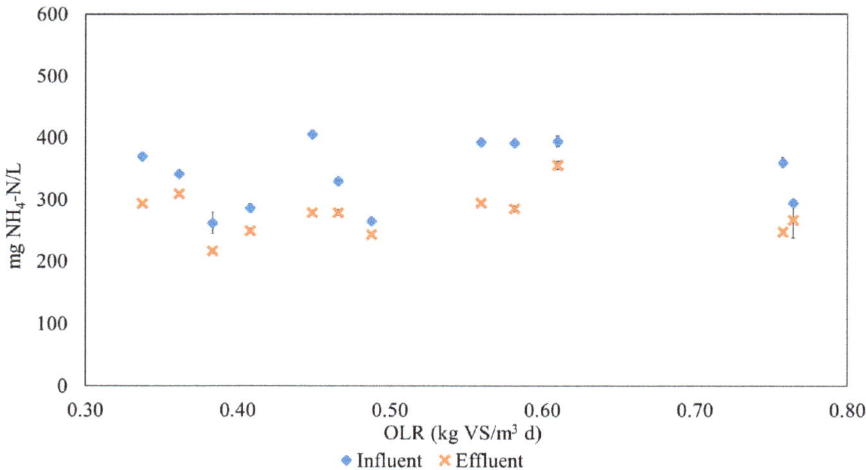

Figure 7. Ammonium changes in affluent (blue rhombus) and effluent (orange cruxes).

Operating under the local environmental conditions (Psychrophilic AD), there were smaller variation of ammonium. Previous studies revealed an ammonium increase under mesophilic and thermophilic conditions [41]. In this study, average NH_4-N decreased from 341.1 ± 52 to 276.8 ± 36 mg/L. The ammonium concentration did not exceed 355.7 mg/L during the whole AD process at 17 °C. It is well known that high concentrations of ammonia (≥3000 mg/L) are toxic to microorganisms [42]. The toxicity of ammonia may have been insignificant in this study. This is because the low-cost tubular digester had low organic loads, and operated at a low temperature, so the accumulation of inhibitors/toxins was likely negligible. Our results agree with those of Massé et al. [43] and Wei and Guo [44], who stated that at low temperatures, ammonium concentrations do not cause failures in digesters.

Process Efficiency and Biogas Quality

Figure 8 presents the biomethane potential kinetic for influent (BMP) and effluent (residual methane potential). The BMP test at 15 °C lasted 40 days, i.e., until methane production was less than 1%. The BMP for the influent at a temperature of 15 °C was 0.49 ± 0.053 Nm^3CH_4/kg VS. Significant differences were not found between the BMP values obtained with a 95% confidence level (p-value = 0.2861). The methane potential of the affluent at 15 °C using the local adapted microorganism consortia was higher than other data reported at 36.5 °C by Kafle and Chen [45], who observed a maximum of 0.33 Nm^3CH_4/kg VS. The residual methane potential at 15 °C of the effluent was 0.09 ± 0.005 Nm^3CH_4/kg VS. Residual methane potential values did not present significant differences (p-value = 0.17 with 95% of confidence level). The biogas composition generated by the biodigester showed a favorable value of 63.1 ± 5.3% for the CH_4 content. The quality of the biogas produced under psychrophilic conditions demonstrated the good performance of the biogas digester. The positive control (crystalline cellulose) test demonstrated the ability of the inoculum to degrade a specific substrate and the quality of BMP assay. BMP from cellulose at 15 °C was 0.40 ± 0.01 Nm^3CH_4/kg VS.

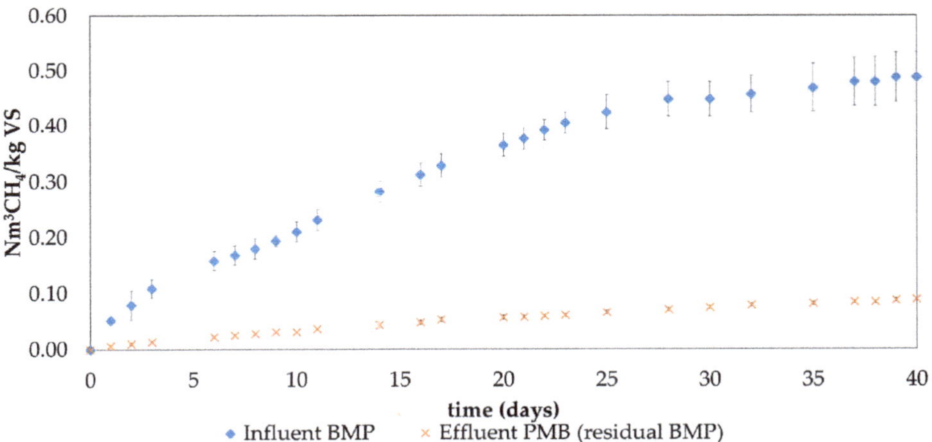

Figure 8. Influent (blue rhombus) and effluent (orange cruxes) biomethane potential kinetic.

Considering the difference of the methane potential between the affluent and effluent, the specific methane production (SMP) of the digester was estimated to be 0.40 Nm^3CH_4/kg VS, which was higher than previously reported values, e.g., Lansing et al. [36] reported 0.29 m^3CH_4/kg VS. The hydraulic retention time of the digester, considering a mean inflow of 4.16 m^3/d and a liquid volume of 103.1 m^3, was around 25 d. This retention time was very low for a 17.7 °C slurry temperature, if compared with other psychrophilic swine manure fed digesters. Martí-Herrero et al. [19] reported the SMP of a 1.5-year old low cost tubular digester divided into two stages, obtaining 0.119 Nm^3CH_4/KgS V for 68.21 d and 0.093 Nm^3CH_4/Kg VS for 34.11 d, and 21.6 °C of slurry temperature in both cases. So, the current digester, even with a short retention time, achieved a good SMP compared with similar digesters, despite working at higher temperatures.

The methane production rate (MPR), considering that the mean OLR was 0.52 kg VS/m$^3_{digester}$ d, was 0.21 Nm3 CH_4/m$^3_{digester}$ d, which was again higher that the values reported by Lansing et al. [36] and Marti-Herrero et al. [19] for low cost tubular digesters fed with swine manure.

Therefore, the digester showed a biogas yield that was higher than expected if compared with similar digesters. The main difference was that in our study, the digester had been working for 8 consecutive years (compared to 1.5 years in the study by Marti-Herrero et al. [19]). This long working period allowed an anaerobic digestion microorganism consortium to develop which was extremely well-adapted to local temperature, operation and influent properties. This phenomenon requires further research. The performance characterization of the digester, is shown in Table 2.

Table 2. Operational conditions, parameter measured and performance characterization of the full-scale low-cost tubular digester.

Operational Conditions	**Units**	**Value**	
Working years	years	8	
Volume	m^3	103.1	
Daily mean load	m^3/d	4.16	
Mean slurry temperature	°C	17.7	
Mean ambient temperature	°C	16.6	
ORL	kg VS/m$^3_{digester}$ d	0.34 to 0.76 (mean 0.52)	
HRT	d	25	
Parameters	Units	Influent value	Effluent value
COD	g COD/L	9.94 ± 3.25	3.31 ± 1.20
VS	gVS/kg	12.74 ± 3.52	2.86 ± 1.2
pH	—	6.15 ± 0.77	7.6 ± 0.3
tVFA	g COD_{VFA}/L	2.9 ± 1.3	0.3 ± 0.08
TA	g $CaCO_3$/L	3.72 ± 1.3	1.95 ± 0.25
Ammonium	g NH_4-N/L	0.34 ± 0.05	0.28 ± 0.03
BMP	Nm^3 CH_4/kgVS	0.46 ± 0.017	0.13 ± 0.06
Coliforms	× 10^6 CFU/mL	3.99	3.57
Performance characterization			
CH_4	%	63.1 ± 5.3	
SMP	Nm^3 CH_4/kg VS	0.40	
MPR	Nm^3 CH_4/m$^3_{digester}$ d	0.21	
COD reduction	%	66.7%	
VS reduction	%	77.6%	
Coliforms reduction	%	10.5%	

3.3. Microbiological Analysis

At low temperatures, methane formation occurred mainly by the acetoclastic route. An acetoclastic methanogenic activity test may be used to delineate the operating conditions for anaerobic systems and a parameter to assess the system performance by giving a better sense of the system and its stability [27]. In the present study, effluent acetoclastic SMA (0.06 g COD CH_4/gVS d) was considerably higher than in the influent (0.01 g COD CH_4/g VS d). The SMA for a three-year operating cattle manure digester was 0.01 g COD CH_4/gVS and 0.04 g COD CH_4/g VS d, for influent and effluent, respectively (previous study, data not shown). This behavior was because the tubular digester design highlighted the separation phases in the axial direction: acid phase (at the beginning of the digester and methane phase (in the final digester). Therefore, the highest number of archaea was at the end of the digester, and consequently, the SMA increased [40]. Moreover, substrate type and temperature were the primary factors influencing microbial activity. From the effluent SMA, it was possible to infer that the bacteria and archaea had adapted to low temperatures after 8 years.

Microbial relative abundances data at the family taxonomic level reflected a remarkable differentiation between inlet and outlet samples (Figure 9).

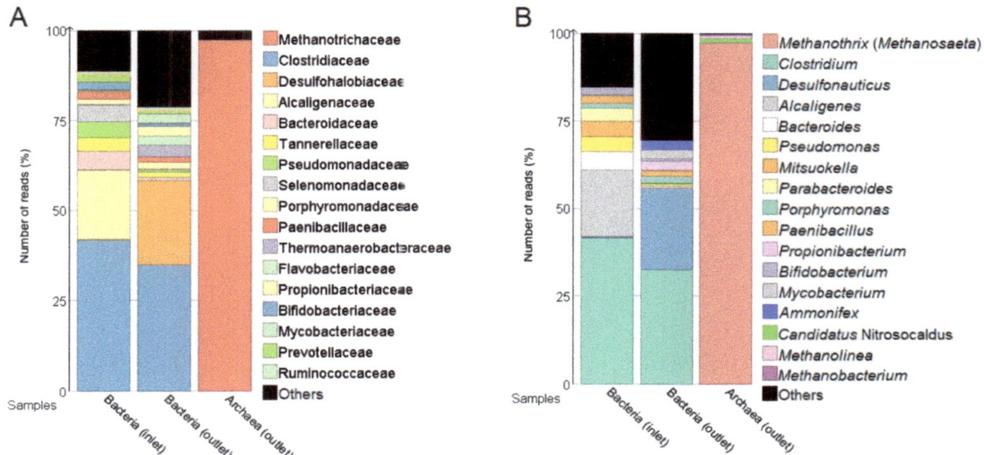

Figure 9. Abundance of bacterial and archaeal taxa at (**A**) family, and (**B**) genus taxonomic level in inlet and outlet samples.

The bacterial populations were dominated by hydrolytic and fermentative organisms which were capable of metabolizing the compounds present in the environment of the digester, such as members of Clostridiaceae and the *Clostridium* genus, whose proportions remained relatively stable throughout the digestion process and showed only a slight decrease in community composition in the outlet sample. Anaerobic and fermentative bacteria typically found in anaerobic digestion systems, such as members of Bacteroidaceae, Propionibacteriaceae, Syntrophaceae, Anaerolineaceae, or Geobacteraeace families, were present in the outlet sample. Propionibacteriaceae members are able to carry out fermentation of sugars to propionic acid [46], present in high quantities in the digester. Syntrophaceae can have a fermentative metabolism or grow in the exclusive presence of H_2 (specifically, genus *Syntrophus* is able to degrade fatty acid chains in a symbiotic relationship with methanogens) [47], whereas Anaerolineaceae can use carbohydrates and Geobacteraceae can oxidize acetic acid and mainly use organic acids and alcohols [48]. Anaerobic conditions also produced a marked increase of *Desulfonauticus* (Desulfohalobiaceae), Firmicutes, and Actinobacteria taxa in outlet samples in comparison to inlet samples. In contrast, the anaerobic conditions created an unfavorable selective pressure for those organisms that were fundamentally aerobic, and resulted in appreciable shifts of most of the remaining bacterial populations. For example, the *Alcaligenes* (Alcaligenaceae) genus was not detectable in the outlet sample. The proportions of other bacterial genera such as *Bacterioides* (Bacteroidaceae), *Parabacteroides* (Porphyromonadaceae), *Pseudomonas* (Pseudomonadaceae) and *Mitsukoella* (Veillonellaceae), among others, drastically reduced between inlet and outlet samples, suggesting a poor adaptation and a displacement in favor of other microorganisms.

Regarding archaeal populations, even though the cell abundance in the inlet sample was not high enough to obtain a 16S rRNA amplicon sequencing dataset, the archaeal populations in the outlet sample were dominated almost exclusively by the methanogenic genus *Methanothrix* (formerly *Methanosaeta*) [49], followed by *Methanobacterium* and *Methanolinea* in lesser proportions. *Methanothrix* species are obligately anaerobic, using acetic acid as their sole source of energy; its metabolism results in the production of CH_4 and CO_2 [50]. The above suggests that the anaerobic process is ongoing, and that these archaeal populations are responsible for the methane found in the outlet sample. Overall, the anaerobic conditions in the reactor produced unfavorable shifts in the composition of the microbial community, favoring taxa which are typically found in anaerobic digestion systems, such as the hydrolytic populations of Clostridiaceae, the fermentative populations of Bacteroidaceae, Propionibacteriaceae, Syntrophaceae, Anaerobilneacea, or *Geobacteraeace*, and the methanogenic populations of Methanotrichaceae, Methanoregulaceae and Nitrosocaldeaceae families.

qPCR data for bacteria and archaea from the inlet and outlet reflected a clear differentiation between samples (Figure 10).

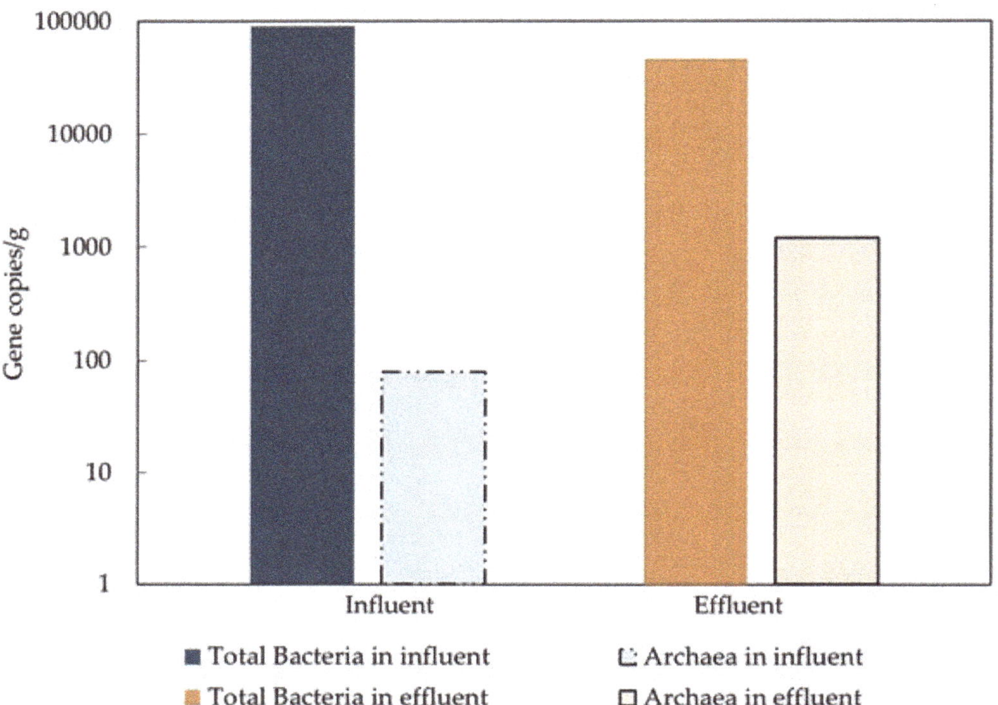

Figure 10. qPCR results for bacteria and archaea from inlet and outlet samples.

The qPCR results seem to indicate that the process provided a selective environment which was favorable to archaea communities with an approximate 15-fold increase from the inlet, which was consistent with the development of a community specialized in methanogenesis. In contrast, the bacteria communities experienced a decrease of approximately 48%. The low values of archaea DNA obtained from the input sample made it impossible to study their relative abundance.

3.4. Pathogen Reduction

The digestate produced in the biodigester was characterized to evaluate its microbiological and nutritional quality. In the affluent and effluent, a concentration of 3.99×10^6 CFU/mL and 3.57×10^6 CFU/mL, respectively, of total coliforms was quantified. This indicated that the digestate was Class B and needed to be stabilized before being dumped into arable soil [51]. Pathogen reduction was 0.42×10^6 CFU/mL; this removal was low compared to that achieved for mesophilic anaerobic processes [52]. This is because the low temperatures did not affect the pathogenic microorganisms.

4. Conclusions

A tubular plastic digester working under psychrophilic conditions for 8 years showed better biogas production than expected when compared with other low-cost tubular digesters A greenhouse over the tubular digester is not enough to heat the digester; therefore, other factors (the color of the reactor and the insulation used in the trench) should be considered for a passive solar heating design. The digester worked properly for organic matter removal and COD reduction, but the effluent still had a large number of coliforms in need of post-treatment (Class B). Performance during the long-term operation of these systems in psychrophilic conditions tends to improve, an aspect not previously considered nor evaluated. The digester had a very short retention time (25 days) for a psychrophilic condition (17.7 °C), indicating that the high methane production (0.40 Nm^3CH_4/kg VS) could be related to the acclimatization and adaptation of the microorganism consortium to the local psychrophilic conditions. A microbiological analysis showed a diverse population adapted to anaerobic digestion conditions, with an increase of methanogenic archaea and a diminution of bacteria populations, resulting in a population that was specialized in hydrolytic and fermentative processes.

Author Contributions J.J.-E., L.d.P.C., H.E.H., A.M. and J.M.-H. developed the conceptualization, wrote the manuscript; C.G. and A.P. conducted the experiments. G.P. and G.Z. realized the microbiological analysis. All authors have read and agreed to the published version of the manuscript.

Funding: This research was funded by Universidad Industrial de Santander. Colombia.

Institutional Review Board Statement: Not applicable.

Informed Consent Statement: Not applicable.

Data Availability Statement: The data presented in this study are available on request from the corresponding author.

Acknowledgments: The authors are grateful to Granja Porcícola La Loma, located in Güicán town, Colombia.

Conflicts of Interest: The authors declare no conflict of interest.

References

1. Martí-Herrero, J.; Alvarez, R.; Rojas, M.; Aliaga, L.; Céspedes, R.; Carbonell, J. Improvement through low cost biofilm carrier in anaerobic tubular digestion in cold climate regions. *Bioresour. Technol.* **2014**, *167*, 87–93. [CrossRef]
2. Garfí, M.; Martí-Herrero, J.; Garwood, A.; Ferrer, I. Household anaerobic digesters for biogas production in Latin America: A review. *Renew. Sustain. Energy Rev.* **2016**, *60*, 599–614. [CrossRef]
3. Sambusiti, C.; Monlau, F.; Ficara, E.; Musatti, A.; Rollini, M.; Barakat, A.; Malpei, F. Comparison of various post-treatments for recovering methane from agricultural digestate. *Fuel Process. Technol.* **2015**, *137*, 359–365. [CrossRef]
4. Guilayn, F.; Jimenez, J.; Martel, J.-L.; Rouez, M.; Crest, M.; Patureau, D. First fertilizing-value typology of digestates: A decision-making tool for regulation. *Waste Manag.* **2019**, *86*, 67–79. [CrossRef]
5. Castro, L.; Escalante, H.; Jaimes-Estévez, J.; Díaz, L.; Vecino, K.; Rojas, G.; Mantilla, L. Low cost digester monitoring under realistic conditions: Rural use of biogas and digestate quality. *Bioresour. Technol.* **2017**, *239*, 311–317. [CrossRef] [PubMed]
6. Dev, S.; Saha, S.; Kurade, M.B.; Salama, E.-S.; El-Dalatony, M.M.; Ha, G.-S.; Chang, S.W.; Jeon, B.-H. Perspective on anaerobic digestion for biomethanation in cold environments. *Renew. Sustain. Energy Rev.* **2019**, *103*, 85–95. [CrossRef]
7. Dohoo, C.; Guernsey, J.R.; Gibson, M.D.; Van Leeuwen, J. Impact of biogas digesters on cookhouse volatile organic compound exposure for rural Kenyan farmwomen. *J. Expo. Sci. Environ. Epidemiol.* **2013**, *25*, 167–174. [CrossRef] [PubMed]
8. Anderson, G.K.; Yang, G. Determination of bicarbonate and total volatile acid concentration in anaerobic digesters using a simple titration. *Water Environ. Res.* **1992**, *64*, 53–59. [CrossRef]
9. Raposo, F.; Borja, R.; Cacho, J.; Mumme, J.; Orupõld, K.; Esteves, S.; Noguerol-Arias, J.; Picard, S.; Nielfa, A.; Scherer, P.; et al. First international comparative study of volatile fatty acids in aqueous samples by chromatographic techniques: Evaluating sources of error. *TrAC Trends Anal. Chem.* **2013**, *51*, 127–143. [CrossRef]
10. Garfí, M.; Ferrer-Martí, L.; Pérez, I.; Flotats, X.; Ferrer, I. Codigestion of cow and guinea pig manure in low-cost tubular digesters at high altitude. *Ecol. Eng.* **2011**, *37*, 2066–2070. [CrossRef]
11. Fotidis, I.A.; Laranjeiro, T.F.V.C.; Angelidaki, I. Alternative co-digestion scenarios for efficient fixed-dome reactor biomethanation processes. *J. Clean. Prod.* **2016**, *127*, 610–617. [CrossRef]
12. Holliger, C.; Alves, M.; Andrade, D.; Angelidaki, I.; Astals, S.; Baier, U.; Bougrier, C.; Buffière, P.; Carballa, M.; De Wilde, V.; et al. Towards a standardization of biomethane potential tests. *Water Sci. Technol.* **2016**, *74*, 2515–2522. [CrossRef] [PubMed]

13. Martí-Herrero, J.; Flores, T.; Alvarez, R.; Pérez, D. How to report biogas production when monitoring small-scale digesters in field. *Biomass- Bioenergy* **2016**, *84*, 31–36. [CrossRef]
14. Perrigault, T.; Weatherford, V.; Martí-Herrero, J.; Poggio, D. Towards thermal design optimization of tubular digesters in cold climates: A heat transfer model. *Bioresour. Technol.* **2012**, *124*, 259–268. [CrossRef] [PubMed]
15. Feller, G. Cryosphere and Psychrophiles: Insights into a Cold Origin of Life? *Life* **2017**, *7*, 25. [CrossRef] [PubMed]
16. Martí-Herrero, J.; Alvarez, R.; Flores, T. Evaluation of the low technology tubular digesters in the production of biogas from slaughterhouse wastewater treatment. *J. Clean. Prod.* **2018**, *199*, 633–642. [CrossRef]
17. Park, J.-H.; Park, J.-H.; Lee, S.-H.; Jung, S.P.; Kim, S.-H. Enhancing anaerobic digestion for rural wastewater treatment with granular activated carbon (GAC) supplementation. *Bioresour. Technol.* **2020**, *315*, 123890. [CrossRef]
18. Lansing, S.; Víquez, J.; Martínez, H.; Botero, R.; Martin, J. Quantifying electricity generation and waste transformations in a low-cost, plug-flow anaerobic digestion system. *Ecol. Eng.* **2008**, *34*, 332–348. [CrossRef]
19. Martí-Herrero, J.; Ceron, M.; Garcia, R.; Pracejus, L.; Alvarez, R.; Cipriano, X. The influence of users' behavior on biogas production from low cost tubular digesters: A technical and socio-cultural field analysis. *Energy Sustain. Dev.* **2015**, *27*, 73–83. [CrossRef]
20. Garfí, M.; Ferrer-Martí, L.; Velo, E.; Ferrer, I. Evaluating benefits of low-cost household digesters for rural Andean communities. *Renew. Sustain. Energy Rev.* **2012**, *16*, 575–581. [CrossRef]
21. Gulhane, M.; Khardenavis, A.; Karia, S.; Pandit, P.; Kanade, G.S.; Lokhande, S.; Vaidya, A.N.; Purohit, H.J. Biomethanation of vegetable market waste in an anaerobic baffled reactor: Effect of effluent recirculation and carbon mass balance analysis. *Bioresour. Technol.* **2016**, *215*, 100–109. [CrossRef] [PubMed]
22. Martí-Herrero, J.; Soria-Castellón, G.; Diaz-De-Basurto, A.; Alvarez, R.; Chemisana, D. Biogas from a full scale digester operated in psychrophilic conditions and fed only with fruit and vegetable waste. *Renew. Energy* **2019**, *133*, 676–684. [CrossRef]
23. McAteer, P.G.; Trego, A.C.; Thorn, C.; Mahony, T.; Abram, F.; O'Flaherty, V. Reactor configuration influences microbial community structure during high-rate, low-temperature anaerobic treatment of dairy wastewater. *Bioresour. Technol.* **2020**, *307*, 123221. [CrossRef] [PubMed]
24. IDEAM. GLACIARES. 2020. Available online: http://www.ideam.gov.co/web/otros-ecosistemas/investigacion-y-publicaciones?p_p_id=31_INSTANCE_x5l2d9DnqhhZ&p_p_lifecycle=0&p_p_state=normal&p_p_mode=view&p_p_col_id=column-1&p_p_col_pos=1&p_p_col_count=3&_31_INSTANCE_x5l2d9DnqhhZ_struts_action=%2Fimage_g (accessed on 29 December 2020).
25. APHA. *Standard Methods for the Examination of Water & Wastewater*; Eaton, D.A., Franson, H.M.A., Eds.; American Public Health Association (APHA): Leamington, ON, Canada, 2005.
26. Purser, B.J.; Thai, S.-M.; Fritz, T.; Esteves, S.; Dinsdale, R.; Guwy, A. An improved titration model reducing over estimation of total volatile fatty acids in anaerobic digestion of energy crop, animal slurry and food waste. *Water Res.* **2014**, *61*, 162–170. [CrossRef]
27. Astals, S.; Batstone, D.; Tait, S.; Jensen, P. Development and validation of a rapid test for anaerobic inhibition and toxicity. *Water Res.* **2015**, *81*, 208–215. [CrossRef]
28. Callaway, T.R.; Dowd, S.E.; Wolcott, R.D.; Sun, Y.; McReynolds, J.; Edrington, T.; Byrd, J.A.; Anderson, R.C.; Krueger, N.; Nisbet, D.J. Evaluation of the bacterial diversity in cecal contents of laying hens fed various molting diets by using bacterial tag-encoded FLX amplicon pyrosequencing. *Poult. Sci.* **2009**, *88*, 298–302. [CrossRef]
29. Takai, K.; Horikoshi, K. Rapid Detection and Quantification of Members of the Archaeal Community by Quantitative PCR Using Fluorogenic Probes. *Appl. Environ. Microbiol.* **2000**, *66*, 5066–5072. [CrossRef]
30. Sotres, A.; Tey, L.; Bonmatí, A.; Viñas, M. Microbial community dynamics in continuous microbial fuel cells fed with synthetic wastewater and pig slurry. *Bioelectrochemistry* **2016**, *111*, 70–82. [CrossRef]
31. San-Martín, M.I.; Sotres, A.; Alonso, R.M.; Díaz-Marcos, J.; Morán, A.; Escapa, A. Assessing anodic microbial populations and membrane ageing in a pilot microbial electrolysis cell. *Int. J. Hydrogen Energy* **2019**, *44*, 17304–17315. [CrossRef]
32. Martí-Herrero, J.; Chipana, M.; Cuevas, C.; Paco, G.; Serrano, V.; Zymla, B.; Heising, K.; Sologuren, J.; Gamarra, A. Low cost tubular digesters as appropriate technology for widespread application: Results and lessons learned from Bolivia. *Renew. Energy* **2014**, *71*, 156–165. [CrossRef]
33. Castro-Molano, L.D.P.; Parrales-Ramírez, Y.A.; Escalante-Hernández, H. Co-digestión anaerobia de estiércoles bovino, porcino y equino como alternativa para mejorar el potencial energético en digestores domésticos. *Rev. ION* **2019**, *32*, 29–39. [CrossRef]
34. Yao, Y.; Huang, G.; An, C.; Chen, X.; Zhang, P.; Xin, X.; Shen, J.; Agnew, J. Anaerobic digestion of livestock manure in cold regions: Technological advancements and global impacts. *Renew. Sustain. Energy Rev.* **2020**, *119*, 109494. [CrossRef]
35. Pedraza, G.; Chará, J.; Conde, N.; Giraldo, S.; Giraldo, L. Evaluation of polyethylene and PVC tubular biodigesters in the treatment of swine wastewater. *Livest. Res. Rural Dev.* **2002**, *14*. Available online: http://www.lrrd.org/lrrd14/1/Pedr141.htm (accessed on 29 December 2020).
36. Lansing, S.; Martin, J.F.; Botero, R.B.; Da Silva, T.N.; Da Silva, E.D. Methane production in low-cost, unheated, plug-flow digesters treating swine manure and used cooking grease. *Bioresour. Technol.* **2010**, *101*, 4362–4370. [CrossRef]
37. Gerardi, M.H. Temperature. In *The Microbiology of Anaerobic Digesters*; John Wiley & Sons, Inc.: Hoboken, NJ, USA, 2003; pp. 89–92.
38. Nozhevnikova, A.N.; Rebak, S.; Kotsyurbenko, O.R.; Parshina, S.; Holliger, C.; Lettinga, G. Anaerobic production and degradation of volatile fatty acids in low temperature environments. *Water Sci. Technol.* **2000**, *41*, 39–46. [CrossRef]

39. Hill, D.T.; Cobb, S.A.; Bolte, J.P. Using Volatile Fatty Acid Relationships to Predict Anaerobic Digester Failure. *Trans. ASAE* **1987**, *30*, 0496–0501. [CrossRef]
40. Jaimes-Estévez, J.; Castro, L.; Escalante, H.; Carrillo, D.; Portillo, S.; Sotres, A.; Morán, A. Cheese whey co-digestion treatment in a tubular system: Microbiological behaviour along the axial axis. *Biomass Convers Biorefin.* **2020**. [CrossRef]
41. Yenigun, O.; Demirel, B. Ammonia inhibition in anaerobic digestion: A review. *Process. Biochem.* **2013**, *48*, 901–911. [CrossRef]
42. Hobson, P.; Shaw, B. Inhibition of methane production by Methanobacterium formicicum. *Water Res.* **1976**, *10*, 849–852. [CrossRef]
43. Massé, D.I.; Rajagopal, R.; Singh, G. Technical and operational feasibility of psychrophilic anaerobic digestion biotechnology for processing ammonia-rich waste. *Appl. Energy* **2014**, *120*, 49–55. [CrossRef]
44. Wei, S.; Guo, Y. Comparative study of reactor performance and microbial community in psychrophilic and mesophilic biogas digesters under solid state condition. *J. Biosci. Bioeng.* **2018**, *125*, 543–551. [CrossRef]
45. Kafle, G.K.; Chen, L. Comparison on batch anaerobic digestion of five different livestock manures and prediction of biochemical methane potential (BMP) using different statistical models. *Waste Manag.* **2016**, *48*, 492–502. [CrossRef] [PubMed]
46. Stackebrandt, E.; Cummins, C.S.; Johnson, J.L. Family Propionibacteriaceae: The Genus Propionibacterium. *Prokaryotes* **2006**, *2006*, 400–418. [CrossRef]
47. Kuever, J. The Family Syntrophaceae. In *The Prokaryotes: Deltaproteobacteria and Epsilonproteobacteria*; Springer: Heidelberg, Germany, 2014; Volume 9783642390, pp. 1–413. [CrossRef]
48. Zhuang, L.; Tang, Z.; Yu, Z.; Li, J.; Tang, J. Methanogenic Activity and Microbial Community Structure in Response to Different Mineralization Pathways of Ferrihydrite in Paddy Soil. *Front. Earth Sci.* **2019**, *7*, 1–11. [CrossRef]
49. The Prokaryotes. *The Prokaryotes*; Springer Science and Business Media LLC: Berlin/Heidelberg, Germany, 2014; Volume 2014, pp. 298–305.
50. McIlroy, S.J.; Kirkegaard, R.H.; Dueholm, M.S.; Fernando, E.; Karst, S.M.; Albertsen, M.; Nielsen, P.H. Culture-Independent Analyses Reveal Novel Anaerolineaceae as Abundant Primary Fermenters in Anaerobic Digesters Treating Waste Activated Sludge. *Front. Microbiol.* **2017**, *8*, 1134. [CrossRef]
51. USEPA. *Enviromental Regulations and Technology. Control of Pathogens and Vector Attraction in Sewage Sludge*; United States Environment Protection Agency: Washington, DC, USA, 2003.
52. Ju, F.; Liping, M.; Ma, L.; Wang, Y.; Huang, D.; Zhang, T. Antibiotic resistance genes and human bacterial pathogens: Co-occurrence, removal, and enrichment in municipal sewage sludge digesters. *Water Res.* **2016**, *91*, 1–10. [CrossRef]

Article

Scaling-Up the Anaerobic Digestion of Pretreated Microalgal Biomass within a Water Resource Recovery Facility

Rubén Díez-Montero [1], Lucas Vassalle [1,2], Fabiana Passos [2], Antonio Ortiz [1], María Jesús García-Galán [1], Joan García [1] and Ivet Ferrer [1,*]

1 GEMMA—Group of Environmental Engineering and Microbiology, Department of Civil and Environmental Engineering, Universitat Politècnica de Catalunya·BarcelonaTech, c/Jordi Girona 1-3, Building D1, 08034 Barcelona, Spain; ruben.diez.montero@upc.edu (R.D.-M.); lucas.vassalle@upc.edu or lvassalle@ufmg.br (L.V.); antonio.ortiz.ruiz@upc.edu (A.O.); chus.garcia@upc.edu (M.J.G.-G.); joan.garcia@upc.edu (J.G.)

2 Department of Sanitary and Environmental Engineering, Universidade Federal de Minas Gerais, Av. Antônio Carlos, 6627, Belo Horizonte 31270-901, Brazil; fabiana@desa.ufmg.br

* Correspondence: ivet.ferrer@upc.edu; Tel.: +34-93-401-6463

Received: 23 August 2020; Accepted: 14 October 2020; Published: 20 October 2020

Abstract: Microalgae-based wastewater treatment plants are low-cost alternatives for recovering nutrients from contaminated effluents through microalgal biomass, which may be subsequently processed into valuable bioproducts and bioenergy. Anaerobic digestion for biogas and biomethane production is the most straightforward and applicable technology for bioenergy recovery. However, pretreatment techniques may be needed to enhance the anaerobic biodegradability of microalgae. To date, very few full-scale systems have been put through, due to acknowledged bottlenecks such as low biomass concentration after conventional harvesting and inefficient processing into valuable products. The aim of this study was to evaluate the anaerobic digestion of pretreated microalgal biomass in a demonstration-scale microalgae biorefinery, and to compare the results obtained with previous research conducted at lab-scale, in order to assess the scalability of this bioprocess. In the lab-scale experiments, real municipal wastewater was treated in high rate algal ponds (2 × 0.47 m^3), and harvested microalgal biomass was thickened and digested to produce biogas. It was observed how the methane yield increased by 67% after implementing a thermal pretreatment step (at 75 °C for 10 h), and therefore the very same pretreatment was applied in the demonstration-scale study. In this case, agricultural runoff was treated in semi-closed tubular photobioreactors (3 × 11.7 m^3), and harvested microalgal biomass was thickened and thermally pretreated before undergoing the anaerobic digestion to produce biogas. The results showed a VS removal of 70% in the reactor and a methane yield up to 0.24 L CH_4/g VS, which were similar to the lab-scale results. Furthermore, photosynthetic biogas upgrading led to the production of biomethane, while the digestate was treated in a constructed wetland to obtain a biofertilizer. In this way, the demonstration-scale plant evidenced the feasibility of recovering resources (biomethane and biofertilizer) from agricultural runoff using microalgae-based systems coupled with anaerobic digestion of the microalgal biomass.

Keywords: agricultural runoff; anaerobic digestion; biogas; biomethane; biorefinery; microalgae; photobioreactor; pretreatment; wastewater

1. Introduction

The treatment of wastewater is fundamental for ensuring public health and environmental quality. European regulations such as the Urban Waste Water Treatment Directive (91/271/EEC) [1] and the

Water Framework Directive (2000/60/EC) [2] aim at protecting surface waters from the adverse effects of wastewater discharges, such as organic pollution and oxygen depletion, which degrade aquatic life. This has been partially achieved through the collection and treatment of wastewater in urban settlements. In most of these cases, wastewater is subject to biological treatment (secondary treatment) for the removal of organic matter and suspended solids, but in cases where the receiving water bodies are considered sensitive to eutrophication, more stringent tertiary treatment may be required to reduce nitrogen and phosphorus pollution. In 2015, the percentage of population connected to wastewater treatment facilities ranged from 75% in Eastern Europe to 97% in central Europe, while the percentage connected to wastewater treatment plants that implement tertiary treatment ranged from 21% in south Eastern Europe to 80% in central Europe [3]. The percentage not connected to wastewater treatment facilities mostly corresponds to population living in scattered communities outside agglomerations, usually in rural areas.

Nature-based sanitation systems, such as constructed wetlands and microalgae-based systems, may be the most feasible solution for rural areas, since they have lower costs and less sophisticated operation and maintenance requirements [4,5]. Moreover, these systems can provide treatment efficiencies similar to those of activated sludge wastewater treatment plants (WWTPs) including tertiary treatment. The main disadvantages of natural systems are that they are susceptible to seasonality and require larger land areas compared to conventional treatment systems [6]. The effects of seasonality can be lessened by a proper design under the most adverse conditions. Regarding land availability, it may not be an issue in rural areas as compared to urban agglomerations. In addition, these systems are suitable for the treatment of agricultural runoff.

In particular, microalgae-based treatment systems have much lower energy input compared to conventional activated sludge units, since oxygen for biological treatment is supplied through microalgae photosynthesis. Moreover, these microorganisms are responsible for nutrient assimilation, allowing nitrogen and phosphorus removal [7,8]. Experimental and demonstration-scale facilities of microalgae-based systems treating municipal wastewater have shown removal efficiencies of 90% for COD, 75–95% for $N\text{-}NH_4$ and 37% of $P\text{-}PO_4$ [9–11]. On the other hand, WWTPs are shifting from being just a sanitation technology towards a bioproduct recovery industry, as biorefineries or water resources recovery facilities (WRRFs). Microalgae-based systems fit in this approach, since the treatment of wastewater is associated with the production of microalgal biomass that could be recovered or reused for further purposes. Thus, microalgae have gained research interest due not only to their great potential and impact applications on wastewater treatment, but also for resource recovery and societal development [12,13]. Harvested microalgal biomass can be processed into protein for animal feed, agricultural fertilizer, pigments and biopolymers, while biogas can be produced by means of anaerobic digestion of the total or residual biomass [14–20]. Biogas production from microalgae is suitable and of special interest for small agglomerations and rural areas, since a positive energy balance can be achieved, producing more energy from the biogas than the energy required for the operation of the whole plant, if environmental conditions (solar radiation, temperature) are appropriate [11,21].

For microalgae-based wastewater treatment, open ponds are normally justified as more economical than closed photobioreactors, which seem to be only recommended for high-value by-products. Nonetheless, closed tubular photobioreactors have interesting advantages, as more independency on weather conditions, lower risk of microbial contamination and lower CO_2 losses [22]. Systems combining open and closed compartments aim at taking advantage of the features and avoiding the main drawbacks of both types of systems, which may encourage the use of semi-closed photobioreactors in microalgae-based WRRFs [23].

Regarding bioenergy production, anaerobic digestion is the most straightforward and applicable technology to date. According to the literature, results on microalgal biomass methane yield at lab-scale range from 0.07 to 0.56 L CH_4 g VS^{-1}, depending on microalgae species, substrate characteristics and operating conditions, among other factors [15]. In any case, for improving biomass biodegradability, pretreatment methods have been tested in order to disrupt the cell wall and enhance the hydrolysis step. Pretreatment techniques that have so far been applied to microalgae include physical, chemical and biological methods, as well as their combinations [24]. Even if they all seem effective in terms of methane production increase, thermal pretreatments at low temperature (<100 °C) seem more feasible to scale-up, since they have led to 70% methane yield increase and positive energy balances in lab scale reactors [25–27]. However, full-scale experience on anaerobic digestion of pretreated microalgal biomass is limited, despite its implementation is increasing according to the number of research projects worldwide [28].

In this context, a demonstration-scale plant including anaerobic digestion of pretreated microalgal biomass was implemented and operated in the framework of the projects INCOVER and AL4BIO. The projects aimed at changing the current wastewater treatment concept towards a bioproduct recovery industry and a reclaimed water supplier. One of the main outcomes of the projects was the evaluation at demonstration-scale processes and technologies that were previously tested only at the lab or pilot-scale. In particular, agricultural runoff and domestic wastewater were treated in demonstration-scale semi-closed photobioreactors, assessing the feasibility of selection of cyanobacteria and accumulation of polyhydroxybutyrate (PHB) and carbohydrates [29,30]. The biomass was harvested in a lamella settling tank and thickened in gravity settlers. Subsequently, the biomass was digested anaerobically for the production of biogas, after undergoing thermal pretreatment. The biogas was upgraded to biomethane in a photosynthetic absorption column [17], while the digestate was further stabilised and dewatered in a sludge treatment wetland for the production of biofertilizer.

This study compiles the data from the anaerobic digestion of pretreated microalgal biomass, with the objective of evaluating the results and comparing the production of biogas with previous research conducted at lab-scale. The discussion regarding the performances obtained at both scales aims at assessing the scalability of this bioprocess.

2. Materials and Methods

2.1. Demonstration-Scale Set-Up

The microalgae-based WRRF was located outdoors in the Agròpolis Campus of the Universitat Politècnica de Catalunya (UPC) in Viladecans (Barcelona, Spain, Figure 1). It treated a mixture of agricultural runoff (90% *v*/*v*) and domestic wastewater from a septic tank (10% *v*/*v*). The agricultural runoff was obtained from a drainage collection channel beside the campus. The system comprised three horizontal tubular semi-closed photobioreactors, a lamellar settler with polymer addition for biomass harvesting, two gravity thickeners, an anaerobic digestion unit for biogas production and upgrading to biomethane, and a constructed wetland for digestate stabilisation and dewatering. The clarified effluent was post-treated in a solar-driven ultrafiltration-disinfection unit and in three adsorption columns for nutrients recovery, and eventually reused for irrigation of rapeseed and sunflower crops by means of a smart irrigation system. Further details on the start-up of the plant may be found in [31].

2.1.1. Microalgal Biomass Production and Harvesting

Agricultural runoff was pumped from the collection channel to a homogenisation tank (10 m^3), where it was mixed with the partially treated domestic wastewater pumped from a septic tank. The influent was conveyed to the three semi-closed tubular photobioreactors. Each photobioreactor (11.7 m^3) was composed by two lateral open tanks (2.5 m^3) equipped with paddle-wheels, connected by sixteen horizontal transparent tubes (9.2 m^3). The paddle-wheels in the lateral open tanks provided a difference in the water level between the two tanks, causing the mix liquor to flow from one tank to the other through eight tubes and returning to the first open tank through the other eight tubes. The liquid velocity inside the tubes was 0.25 m/s, ensuring a turbulent flow and homogeneous mixing. Moreover, the open tanks provided dissolved oxygen release and preserved temperature increase.

The system was operated in a semi-batch mode, with a discharge of 2.3 m^3 of mixed liquor from each photobioreactor, followed by feeding the same volume of influent wastewater, each and every day at 5 a.m. and 7 a.m., respectively. During the experimental period, the operation of the photobioreactors was changed according to the research and innovation objectives and the goals to be attained, e.g., wastewater treatment and biomass production optimisation or PHB accumulation by cyanobacteria. On the whole the plant was operated for 20 months; during the first 12 months the photobioreactors were operated in parallel with a HRT of 5 days, while during the following 8 months they were connected in series with a total HRT of 15 days [29,30].

Figure 1. Global view of the demonstration-scale plant.

Microalgal biomass was harvested in a lamellar settler (700 L), which comprised a flocculation chamber (50 L), which received the influent mixed liquor and an addition of coagulant; a stilling zone (180 L) after the flocculation chamber; a lamellar zone (350 L) which was the main settling volume; an effluent weir and collection channel over the lamella zone; and a sludge hopper at the bottom for collecting the settled biomass (120 L). The total daily volume of mixed liquor discharged from the photobioreactors was pumped to the settling tank at a surface loading rate of 0.135 m/h (including the lamellae's surface), with a HRT of 1.75 h. Biomass coagulation and flocculation was enhanced by dosing aluminium polychloride. The dose of coagulant was modified according to the influent mixed liquor characteristics. The sludge was drawn off from the bottom of the settling tank by means of an electro valve and a timer several times every day, until no more sludge remained in the hopper. Following, harvested microalgal biomass was further thickened in two gravity settlers (200 L each) working in series.

2.1.2. Thermal Pretreatment and Anaerobic Digestion

A diagram and an image of the thermal pretreatment and the anaerobic digestion unit are shown in Figure 2. Thickened microalgal biomass was conveyed to a homogenisation tank (100 L) under constant stirring. The biomass was then fed to the thermal pretreatment unit at a flow rate between 15 and 30 L per day. In order to distribute the load during the day, the microalgal biomass was pumped at a constant flow of 0.5 L/min during one minute every 25–45 min (OEM 520FAM/R2 peristaltic pump, Watson-Marlow®, United Kingdom). The time interval between each consecutive pumping event was adjusted in order to feed the desired total volume of biomass. The thermal pretreatment was carried out in a stainless steel tank (25 L), with constant stirring and an electrical resistance (1.5 kW, Electricfor SA, Barcelona, Spain) for maintaining the temperature at 75 °C. The pretreatment temperature was selected according to previous studies on the increase of microalgae anaerobic biodegradability after evaluating several pretreatment methods and validating the thermal pretreatment in continuous lab-scale reactors [26,27,32]. The pretreatment tank was equipped with an electronic temperature sensor (TD2517, IFM electronic LTD, Essen, Germany), and temperature data were collected and recorded in a datalogger every 20 min. The tank also included an electronic liquid level sensor (PI2789, IFM) to control filling and emptying operations.

Pretreated biomass was pumped to the anaerobic digester (Watson-Marlow OEM 520FAM/R2 peristaltic pump, United Kingdom). The anaerobic digester (1 m^3) was maintained under constant stirring by means of liquid recirculation at 2 m^3/h (BN 2–6 L rotating positive-displacement pump, Seepex, TD2517, IFM electronic LTD, Essen, Germany) and at mesophilic temperature (35 °C) by means of an electrical resistance (CR212II0030 M77 LIR 589, Electricfor SA, Barcelona, Spain). Furthermore, the digester was equipped with electronic liquid level sensors, pressure, temperature and redox (PI2798, PI008A and TD2517—IFM Electronic LTD.—Essen, Germany) and pH (K100, Seko—Santa Rufina, Italy). Data of these parameters were measured online and recorded in a datalogger every 20 min. Biogas production was quantified using a mechanical flowmeter (TG0.5-PVC-PVC, Ritter® Bochum, Germany) and stored in a gasometer. The volume of biogas was recorded manually from the mechanical flowmeter every working day. Therefore, results are expressed as weekly average values of biogas production (L biogas/$L_{reactor}$·day) and methane yield (L CH_4/g VS).

Over an experimental period of 14 months (420 days), the digester was operated with two different HRT: 20 days (Period 1, until day 271) and 32 days (Period 2, days 272 to 420). Previous research had shown how increasing the HRT to 28–30 days could improve the biogas production from microalgae [27,33], and here we wanted to evaluate if it was also the case with pretreated microalgal biomass.

2.2. Analytical Methods

The performance of the photobioreactors and harvesting unit was monitored as described elsewhere [29]. In brief, grab samples from the influent wastewater (homogenization tank), the effluent of each photobioreactor, and the effluent of the lamella settling tank were collected and analysed weekly. The main operational parameters were analysed, among them the nutrients orthophosphate (PO_4^{3-}-P) and ammonium (NH_4^+-N) in the influent and in the photobioreactors, and turbidity, total suspended solids (TSS) and volatile suspended solids (VSS) in the photobioreactors and the harvesting unit. Turbidity was measured using a HI-93703 turbidimeter (Hanna instrumental, Limena, Italy). TSS and VSS were analyzed following Standard Methods for the Examination of Water and Wastewater [34]. NH_4^+-N was analyzed according to the methods described in Solórzano (1969) [35] and PO_4^{3-}-P was measured by means of a DIONEX ICS1000 ion chromatography system (Thermo-Scientific®, Waltham, MA, USA).

Samples of biomass were observed under a bright light microscope (Motic, Kowloon, Hong Kong) equipped with a camera (Fi2, Nikon, Tokyo, Japan) and a fluorescence microscope (Eclipse E200, Nikon, Tokyo, Japan) using the NIS-Element viewer® software, in order to observe the composition of microorganisms during the experimental period. The identification of microalgae and cyanobacteria was based on taxonomic books and databases [36,37].

The performance of the anaerobic digester was monitored as follows. The pH, redox potential, temperature and volume of the digester were continuously monitored on-site and recorded every 20 min, as well as the temperature and volume of the pretreatment unit. The volume of produced biogas was recorded every working day. The CH_4 and CO_2 content were periodically analysed from biogas samples using a GC equipped with a thermal conductivity detector (Trace GC with Hayesep packed column, Thermo Finnigan—Thermo-Scientific®, Waltham, MA, USA), as described by Marín et al. [17]. Samples of the influent biomass, pretreated biomass and digestate were analysed on a weekly basis. The concentration of Total Solids (TS), Volatile Solids (VS), total and soluble Chemical Oxygen Demand (COD and CODs) were determined according to the Standard Methods for the Examination of Water and Wastewater [34]. Total organic carbon (TOC) and total nitrogen (TN) were measured using an automatic analyser (multi N/C® 2100S analyser, Analytik Jena—Jena, Germany). TOC was analysed with an infrared detector (NDIR) according to the combustion-infrared method of the Standard Methods for the Examination of Water and Wastewater [34], by means of catalytic oxidation at 800 °C using CeO_2 as catalyst. Following, a solid-state chemical detector (ChD) was used to quantify TN as NOx.

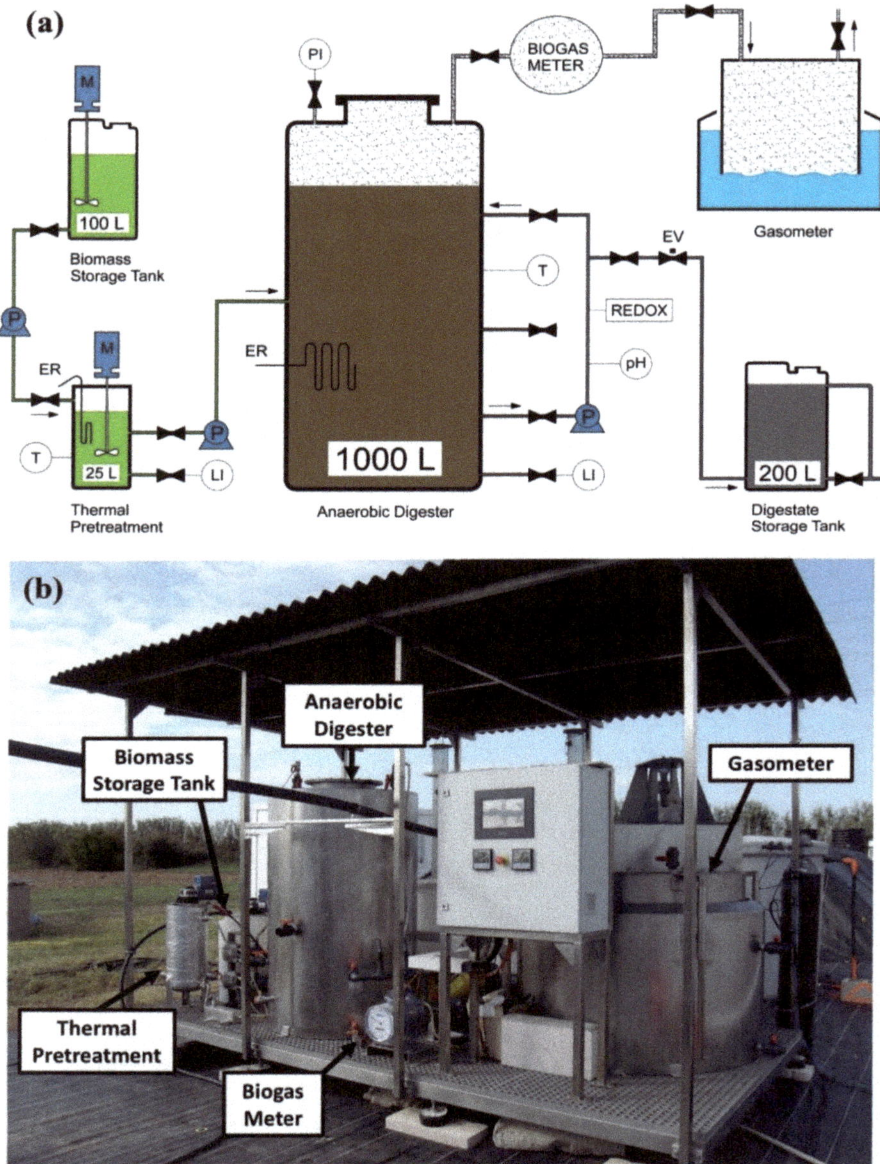

Figure 2. (**a**) Diagram of the anaerobic digestion plant. Mixers (M), pumps (P), electrical resistances (ER) and electrovalves (EV) are indicated in the figure, as well as the temperature (T), liquid level (LI), pressure (PI), redox and pH sensors. (**b**) Image of the anaerobic digestion plant.

2.3. Determination of Parameters

The performances of the thermal pretreatment, anaerobic digestion and biogas production were evaluated by calculating the following parameters.

The degree of solubilisation of microalgal biomass in the thermal pretreatment was calculated according to Equations (1) (S, percentage of solubilisation of the influent particulate COD) and (2) (SR, solubilisation ratio), where COD_{sp} is the soluble COD after pretreatment, COD_{so} is the soluble COD of the influent microalgal biomass, and COD_o is the total COD of the influent microalgal biomass.

$$S = \frac{COD_{sp} - COD_{so}}{COD_o - COD_{so}} \cdot 100 \quad (1)$$

$$SR = \frac{COD_{sp}}{COD_{so}} \quad (2)$$

The removal of VS ($VS_{removed}$, %) in the anaerobic digester was calculated as the difference between the VS concentration in the influent and effluent, with respect to the VS concentration in the influent, according to Equation (3), where VS_{inf} and VS_{eff} are the influent and effluent concentration of VS. VS_{inf} has been estimated as the mobile average of the influent VS concentration during the previous HRT period:

$$VS_{removed} = \frac{VS_{inf} - VS_{eff}}{VS_{inf}} \cdot 100 \quad (3)$$

The organic loading rate (OLR, kg VS/m^3·d) was determined as the amount of organic matter fed to the anaerobic digester per day, referred to the reactor working volume ($V_{reactor}$), according to Equation (4). At this aim, the organic matter concentration in the influent was expressed as the concentration of VS (VS_{fed}):

$$OLR = \frac{Q \cdot VS_{fed}}{V_{reactor}} \quad (4)$$

The methane production rate ($P_{methane}$, L CH_4/L·d) was calculated as the volume of methane produced per day, referred to the reactor working volume, according to Equation (5), where %CH_4 is the methane content in the biogas:

$$P_{methane} = \frac{\text{L of methane per day}}{V_{reactor}} = \frac{\text{L of biogas per day} \cdot \%CH_4}{V_{reactor}} \quad (5)$$

Finally, the methane yield (Y_{CH4}, L CH_4/g VS) or specific methane production, was calculated by referring the methane production rate to the organic loading rate, according to Equation (6):

$$Y_{CH4} = \frac{P_{methane}}{OLR} \quad (6)$$

3. Results

3.1. Microalgal Biomass Production and Harvesting

Microalgal biomass produced in the semi-closed photobioreactors varied throughout the experimental period, as a result of the mode of operation and performance of the photobioreactors and harvesting unit, the weather conditions of the season and the variability of influent wastewater characteristics. Indeed, the biomass (expressed as VSS) concentration fluctuated with the solar radiation and water temperature, attaining low microalgae production during winter (7 g/m^3·day) and early spring, and increasing in summer and early autumn (up to 43 g/m^3·day) [29].

The operational conditions and performance of the harvesting unit also varied during the experiment. The turbidity of the influent mixed liquor to the lamella settling tank ranged between 20 and 500 NTU, and the doses of coagulant ranged between 1 to 12 mg Al/L for achieving an effluent turbidity < 5 NTU. Harvested biomass was further thickened by gravity, reaching a concentration of VS between 2 and 18 g VS/L (Figure 3).

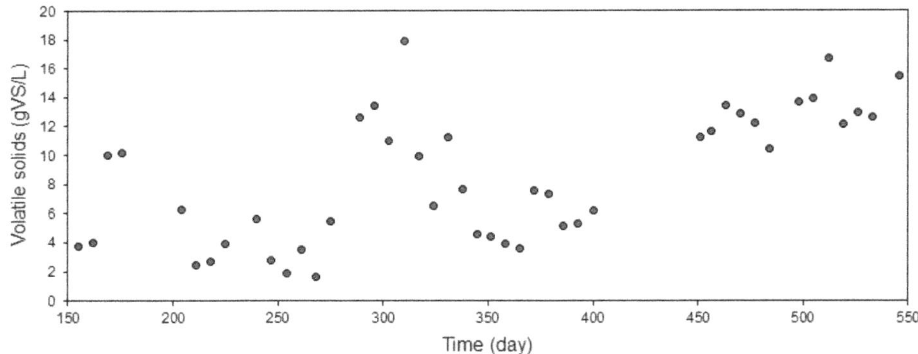

Figure 3. Concentration of volatile solids (VS) of thickened microalgal biomass.

The production of microalgal biomass seemed to be limited by the concentration of nutrients in the influent agricultural runoff, with average seasonal concentrations of N-NH_4 ranging between 1.2 and 3.6 mg/L and of P-PO_4 between 0.32 and 1.84 mg/L [38]. These values are quite low when compared to primary treated domestic wastewater (24–53 mg N-NH_4/L and 8–25 mg P-PO_4/L) [39]. In addition, the modification of the photobioreactors operation mode on day 330, from operation in parallel (5 days of HRT) to operation in series (15 days of HRT), also had an influence on the biomass production. Indeed, in spite of the favourable environmental conditions of springtime, after the modification the biomass production decreased, which was attributed to the lower influent flowrate and nutrients loading during the operation of the photobioreactors in series, with a total HRT of 15 days.

In general, the mixed culture was dominated throughout the whole period by cyanobacteria belonging to a coccoid species resembling *Synechococcus* sp. (especially during the operation in series), along with some filamentous cyanobacteria like *Pseudanabaena* sp. and green microalgae [29,30,38] (Figure 4a–d).

3.2. *Thermal Pretreatment of Microalgal Biomass*

The anaerobic digestion system was operated for 18 months. For the purposes of this study, only periods of stable operation were considered, in order to compare the anaerobic digestion of thermally pretreated microalgal biomass under lab-scale controlled conditions [26,27] and pilot-scale real conditions, and assess the scalability of the process. Thus, results from steady-state operation (days 204 to 455) are shown in Table 1. The temperature of the thermal pretreatment was steadily maintained at about 75 °C during the whole period, and the exposure time was around 20 h.

Table 1. Average values and standard deviation (SD) of thermal pretreatment parameters.

	Average	**SD**
HRT (h)	20.0	4.5
Influent		
Soluble COD before (mg/L)	456	265
VS/TS	0.47	0.08
Effluent		
Soluble COD after (mg/L)	2625	1262
VS/TS	0.45	0.08

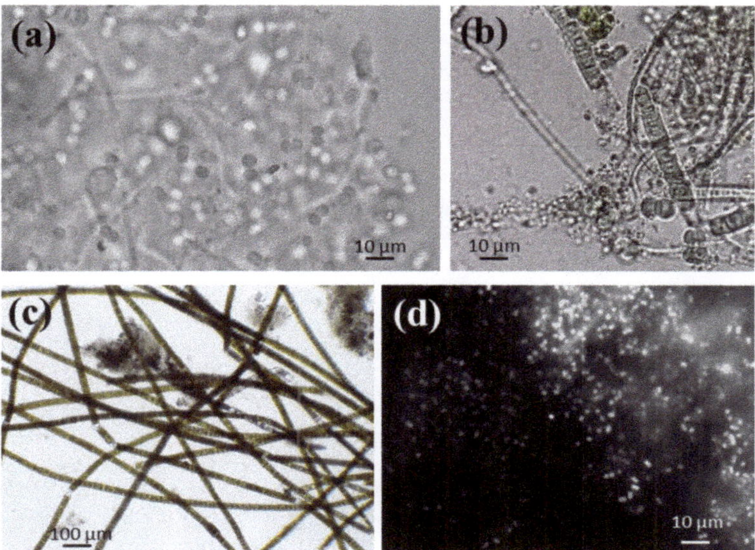

Figure 4. Microscopic images of the mixed liquor of the photobioreactors: (**a**) Coccal Cyanobacteria resembling *Synechococcus* sp. and small filamentous Cyanobacteria, surrounded by green microalgae, (**b**) filamentous Cyanobacteria resembling to *Oscillatoria* sp. and *Leptolyngbya* sp. and coccoid Cyanobacteria resembling to *Chroococcus* sp., *Synechococcus* sp. and *Synechocystis* sp., and (**c**) filamentous green microalgae, observed under bright light microscopy during the operation of the photobioreactors in parallel; and (**d**) higher dominance of *Synechococcus* sp. with some presence of *Pseudanabaena* sp., observed under fluorescence microscopy during the operation of the photobioreactors in series.

One of the most important parameters for evaluating the pretreatment effectiveness is the solubilisation of organic matter. Since microalgae cells are complex and resistant, in particular those grown in wastewater, organic compounds may be retained inside the cell wall, hindering the anaerobic biodegradability. Pretreatment methods aim at disrupting the cell wall and releasing intracellular compounds, enhancing the bioavailability of these compounds for anaerobic bacteria, and ultimately enhancing the anaerobic digestion rate and extent. This is commonly measured by the degree of solubilisation achieved after applying the pretreatment. In this study, the solubilisation degree (calculated from Equation (1)) was on average 45.7%, which means that almost half of the influent particulate COD was converted into soluble COD. When comparing the soluble COD before and after the pretreatment, it was increased from 456 to 2625 mg/L, representing a 5.8-fold solubilisation (calculated from Equation (2)).

These results fall within the range reported in the literature under laboratory conditions. For instance, the thermal pretreatment of mixed microalgal biomass at 75 °C for 10 h reached a 10.6-fold solubilisation [32], while the pretreatment of *Scenedesmus* biomass at 90 °C for 3 h increased soluble organic matter by 4.4-fold [40]. Indeed, the pretreatment effectiveness may vary depending on the microalgae species and growth characteristics, which depend on the culture medium composition [24]. For instance, in this study microalgal biomass was mainly composed of cyanobacteria in the demonstrative-scale plant treating agricultural runoff (with nutrients limitation), while in our lab-scale studies treating municipal wastewater the predominant species were green microalgae such as *Stigeoclonium* sp., *Monorraphidium* sp., or the diatoms *Nitzchia* sp. and *Amphora* sp.; the latter ones with an extremely resistant cell wall composed of silica [26].

According to results obtained, it seems that there was no organic matter loss during the pretreatment at 75 °C for 20 h, as the VS/TS ratio was maintained (Table 1), reproducing what was already observed in the lab-scale [26,27]. This is a matter of concern, since organic matter should not be lost prior to its conversion into biogas in the anaerobic digester.

3.3. Anaerobic Digestion Performance and Biogas Production

The anaerobic digestion performance is shown in Figures 5–7, where two experimental periods are differentiated: Period 1, when the anaerobic digester was operated with a HRT of 20 days (until day 271); and Period 2, when HRT was 32 days (days 272 to 420). Both periods operated under mesophilic conditions (35.8 ± 0.3 °C). The OLR ranged from 0.2 to 0.5 g VS/L·day in Period 1 and from 0.2 to 1.0 g VS/L·day in Period 2 (Figure 5). Indeed, it was more stable during the first period than during the second one, as a result of the VS concentration in thickened microalgal biomass, which follows a similar trend (Figure 3). Despite the variability, the average OLR was higher during the second period (0.5 vs. 0.28 g VS/L·day), even if the HRT was increased from 20 to 32 days. The reason for this is the increase in microalgal biomass production during summer time (Period 2), when microalgae growth was the highest (around 40 g/m^3·day). The correlation between the photosynthetic activity and the weather conditions is widely reported in the literature. In a pilot-scale study carried out at the same location, microalgae growth and biomass production followed the same trend as the solar radiation, reaching the highest values in spring and summer [11].

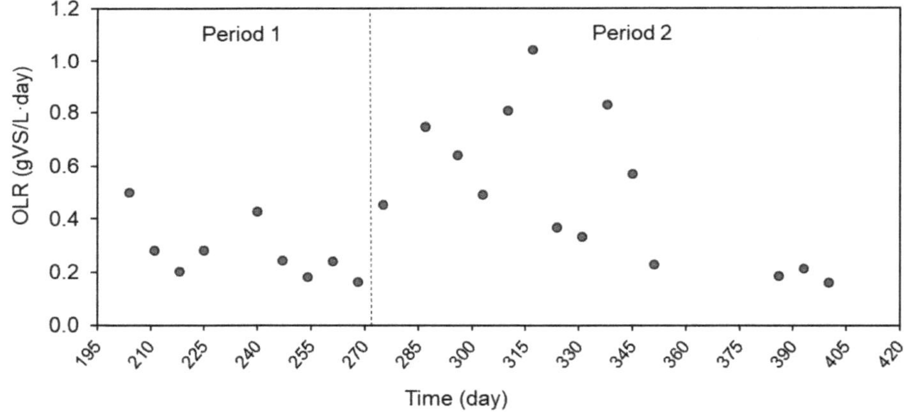

Figure 5. Organic loading rate in the anaerobic digester over the experimental periods 1 (HRT of 20 days) and 2 (HRT of 32 days).

The biogas production rate showed a similar trend as the OLR, with the highest values during summer (days 280–320) (Figure 6a). Indeed, the OLR was fairly low, and therefore increasing the OLR also increased the biogas production resulting from higher organic matter biodegradation. The methane content in biogas was around 76% in both periods, which is considered high upon the anaerobic digestion of particulate organic matter, suggesting an appropriate methanogenic activity. In terms of methane yield (Figure 6b), it ranged between 0.11 and 0.38 L CH_4/g VS during the first period and between 0.07 and 0.28 L CH_4/g VS during the second one, with average values of 0.24 and 0.16 L CH_4/g VS, respectively. It could be speculated that increasing the HRT (from 20 to 32 days) would concomitantly increase the anaerobic biodegradability and methane yield, as previously reported [27,33] and especially upon the anaerobic digestion of particulate organic matter, characterised by a slow hydrolysis step. However, microalgal biomass had already been pretreated with the aim of accelerating the hydrolysis, and in this case no further improvement was observed by increasing the HRT from

20 to 32 days. Most probably, all the soluble organic matter attained after the thermal pretreatment was already digested at 20 days of HRT and no further intracellular, hardly digestible or recalcitrant components were converted into biogas at 32 days of HRT. This indicates that the lower HRT of 20 days was already enough for operating the anaerobic reactor under the conditions of this study.

Another strategy for improving the anaerobic digestion performance would be the co-digestion with carbon-rich substrates, as agricultural biomass, to counter-balance the low C/N ratio of microalgae [41]. Indeed, the C/N ratio of pretreated microalgal biomass was fairly low, ranging from 4 to 10 (Figure 7), as a result of the high protein concentration in cells. This may jeopardize anaerobic digestion when ammonium concentrations arrive at inhibitory or toxic levels. According to the literature, optimal values for microbial growth are around 25–30 [42], which may lead to faster and higher methane production, while promoting the stability of the anaerobic digestion process. Besides, it is a way of increasing the OLR and biogas production potential.

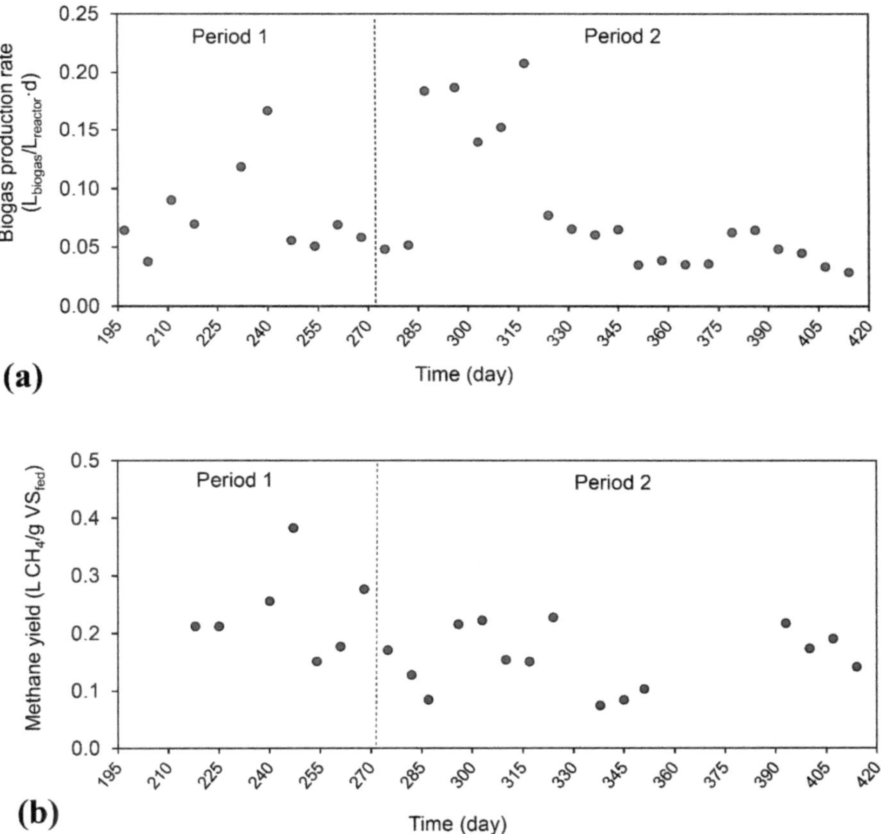

Figure 6. Biogas production rate (**a**) and methane yield (**b**) over the experimental periods 1 (HRT of 20 days) and 2 (HRT of 32 days).

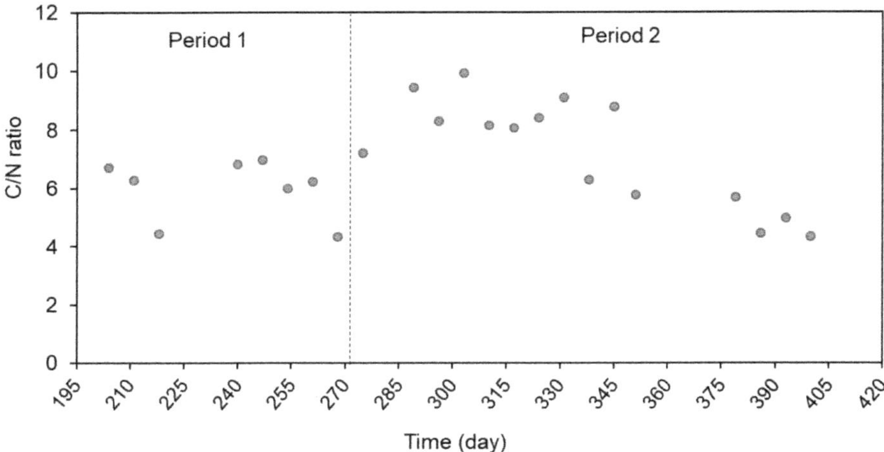

Figure 7. C/N ratio of the anaerobic digester influent (pretreated microalgal biomass) over the experimental periods 1 (HRT of 20 days) and 2 (HRT of 32 days).

4. Discussion

This study was intended to evaluate the anaerobic digestion of thermally pretreated microalgal biomass within a microalgae-based WRRF at demonstration-scale in outdoors conditions. Microalgae-based systems and biomass valorisation technologies have mostly been investigated in lab-scale facilities under controlled conditions. Such experiments are useful to quantify and compare operating conditions, yet do not provide information on the scalability under real conditions, with a strong seasonality and variations in influent wastewater characteristics, which are known to affect the wastewater treatment effectiveness, microalgal biomass production and biomass characteristics (predominant microalgae species and macromolecular composition). In fact, a recent study comparing microalgae-based systems at a lab-scale (5 m^2), pilot-scale (330 m^2) and full-scale (1 ha) revealed that full-scale units showed the lowest values in nutrient removal and microalgal biomass production [43]. The mentioned work indicated that the use of lab-scale data for designing and optimising full-scale plants is still uncertain. On the other hand, literature also suggests that there is an urgent need for more pilot and full-scale studies, since that represents a more realistic approach of the technology in comparison with lab-scale results [44].

In our previous studies, the thermal pretreatment conditions were optimised by comparing the effect of different temperatures (55, 75 and 95 °C) and exposure times (5, 10 and 15 h) on microalgal biomass solubilisation and biochemical methane potential (BMP) [32]. Subsequently, semi-continuous lab-scale reactors (1.5 L) were operated with microalgal biomass pretreated under the optimal conditions (75 °C for 10 h) [26,27]. Both studies were carried out under mesophilic conditions (35 °C) with a HRT of 20 days [26] and 30 days [27]; and in both cases two reactors were run in parallel, the first one receiving pretreated microalgal biomass and the second one raw microalgal biomass (control). In the present study, the same bioprocess was scaled-up in a microalgae-based WRRF, where the biogas produced was upgraded to biomethane and the digestate was post-treated in a constructed wetland to produce a biofertilizer. Thermal pretreatment has been described in the literature as the method to give the best result in microalgae pretreatment [15], however still with very few results in pilot and full-scale systems [7].

The main results obtained in the lab-scale reactors [26,27] and pilot set-up (Periods 1 and 2) are summarised in Table 2. In all cases the anaerobic digesters operated under mesophilic conditions (35–37 °C) with a HRT of 20 or 30–32 days. Microalgal biomass pretreatment was always conducted at

75 °C, with an exposure time of 10 h in the lab-scale experiments and 20 h in the pilot set-up. The OLR was considerably higher in the lab-scale reactors (around 0.7–0.8 g VS/L·day) than in the pilot ones (around 0.3 g VS/L·day in Period 1 and 0.5 g VS/L·day in Period 2), which is attributed to different influent wastewater characteristics, hence biomass production. In the lab-scale experiments, microalgae were grown in high rate algal ponds (HRAPs) treating urban wastewater (without limitation of N and P), manually harvested and thickened reaching higher concentration of VS than in the automated demonstration-scale facility, where microalgae were grown in photobioreactors treating agricultural runoff with some nutrients limitation [33]. In fact, a previous study using microalgae for treating agricultural stormwater showed nutrient limitation, which hampered biomass production, mainly in months with low rainfall events [45].

Table 2. Anaerobic digestion performance for thermally pretreated microalgal biomass in laboratory-scale and pilot-scale reactors. Mean values (standard deviation).

Parameter	Laboratory Scale *	Demonstration-Scale (Period 1)	Laboratory Scale **	Demonstration-Scale (Period 2)
Operational conditions				
Thermal pre-treatment HRT (h)	10	21.3 (0.0)	10	21.7 (5.6)
Anaerobic digester HRT (days)	20	20 (0)	30	32 (10)
OLR (g VS/L·day)	0.68 (0.10)	0.28 (0.11)	0.81 (0.02)	0.50 (0.28)
Influent composition				
VS (g/L)	11.2 (1.40)	6.4 (0.7)	23.7 (1.00)	18.1 (7.2)
TS (g/L)	21.1 (3.10)	16.6 (1.7)	34.2 (2.80)	36.1 (15.4)
COD (g/L)	11.84 (0.71)	9.04 (0.98)	25.2 (1.8)	20.92 (11.98)
N-NH_4 (mg/L)	218 (9.54)	156 (120)	260 (6.00)	312 (300)
Effluent composition				
VS (g/L)	9.50 (1.0)	1.8 (1.2)	14.5 (1.10)	9.9 (5.5)
TS (g/L)	19.80 (2.70)	4.6 (3.1)	26.7 (2.70)	23.4 (13.2)
COD (g/L)	10.6 (0.5)	11.3 (8.9)	25.2 (2.1)	14.7 (10.1)
N-NH_4 (mg/L)	323 (17.15)	458 (250)	8.0 (1.0)	456 (310)
Anaerobic digester pH	7.6 (0.4)	7.0 (0.2)	7.55 (0.08)	7.4 (0.1)
VFA (mg COD/L)	150 (58.6)	-	130 (<596 [1])	-
Anaerobic digestion performance				
VS removal (%)	52.3 (3.8)	70.0 (23.6)	39.5 (3.7)	45.7 (18.0)
Methane production rate (L CH_4/L·day)	0.20 (0.10)	0.072 (0.035)	0.19 (0.07)	0.064 (0.053)
Methane yield (L CH_4/g VS)	0.30 (0.09)	0.24 (0.08)	0.24 (0.07)	0.16 (0.05)
Methane content in biogas (%)	68.1 (0.6)	76.7 (0.0)	69.5 (1.7)	76.8 (2.0)

Note: * Data published by Passos and Ferrer, 2014 [26]; ** Data published by Solé-Bundó et al., 2018 [27]; [1] Maximum value achieved.

Consequently, in our study, the methane production rate was much higher in the lab-scale experiments (Table 2), yet the methane yield was not so different. With a HRT of 20 days, the methane yield was 25% higher in the lab-scale experiment (0.30 vs. 0.24 L CH_4/g VS) but with a HRT of 30–32 days it was 50% higher (0.24 vs. 0.16 L CH_4/g VS). When comparing the results, we should bear in mind that the anaerobic biodegradability depends on the microalgae species, which in systems treating real wastewater keep changing over time, depending on the weather conditions and influent characteristics [28,44]. Furthermore, these experiments were conducted with spontaneous mixed cultures dominated by green microalgae in the HRAPs treating urban wastewater (lab-scale experiments), and by cyanobacteria in the photobioreactors treating agricultural runoff (demonstration-scale facility). In addition, lab-scale experiments were conducted under controlled conditions, and manual microalgae harvesting and digester feeding ensured a constant flow rate of thickened microalgal biomass with a fairly stable OLR. Conversely, the demonstration-scale facility was fully automated, meaning that the operation of a process depended on the success of the previous one and, despite the complexity of operating a microalgae biorefinery like this, with operational issues occurring regularly, the anaerobic digestion stage showed to be quite robust and reproduced reasonably well lab-scale results under real conditions resembling full-scale operation. This was reinforced by the results of stable pH, high methane content in biogas and the similar methane yield when compared to lab-scale results. On the whole, the results suggests that even with a variable microalgal biomass production and composition, and a lower OLR, the anaerobic digestion was a quite robust and straight forward downstream option for microalgal biomass valorisation at demonstration-scale.

In the context of microalgae-based biorefinery or WRRF, the bioproducts obtained in the anaerobic digester were further processed. The produced biogas was subsequently sparged into a 45 L absorption column, fed with mixed liquor from the photobioreactors. The photosynthetic biogas upgrading process was validated at demonstration-scale under outdoors conditions. The continuous operation of the system resulted in the production of biomethane, reducing the content of CO_2 and H_2S and obtaining a concentration of CH_4 between 94.1% and 98.8%, complying with most international regulations for methane injection into natural gas grids [17]. Moreover, the digestate was further stabilised in a sludge treatment wetland with an effective surface area of 6 m^2 and height of 1.5 m. The wetland was planted with common reed (*Phragmites australis*) and the digestate was daily pumped and fed to the wetland through a sludge distribution system consisting in a net of pipes with risers. The digestate was mineralised and dewatered in the wetland, producing a soil like structure with 12.5–12.8% dry matter content. According to the nutrient and heavy metals content (below the limits for reuse of sludge in arable land), the material could be used as soil amendment or biofertilizer.

5. Conclusions

This study assessed the scalability of the anaerobic digestion of pretreated microalgal biomass by comparing the results from a demonstration-scale microalgae biorefinery with those previously obtained at lab-scale. With the thermal pretreatment of microalgal biomass, the degree of solubilisation was on average 45.7%, which means that almost half of the influent particulate COD was converted into soluble COD. When comparing the soluble COD before and after the pretreatment, it was increased from 456 to 2625 mg/L, representing a 5.8-fold solubilisation. In the anaerobic digester, the average VS removal was 70% and the methane yield up to 0.24 L CH_4/g VS, which were similar to the lab-scale results. Overall, the anaerobic digestion step of the microalgae biorefinery showed to be quite robust and reproduced reasonably well lab-scale results under real conditions resembling full-scale operation.

Author Contributions: Conceptualization, I.F.; methodology, I.F. and R.D.-M.; formal analysis, R.D.-M., I.F. and F.P.; investigation, A.O., M.J.G.-G. and R.D.-M.; resources, A.O.; data curation, R.D.-M., F.P. and I.F.; writing—original draft preparation, R.D.-M., F.P., A.O. and L.V.; writing—review and editing, R.D.-M., L.V., F.P., and I.F.; supervision, J.G. and I.F.; project administration, J.G., M.J.G.-G. and R.D.-M.; funding acquisition, J.G. All authors have read and agreed to the published version of the manuscript.

Funding: This research was funded by the European Commission (H2020 project INCOVER, GA 689242); the Spanish Ministry of Science, Innovation and Universities (MCIU), Research National Agency (AEI), and European Regional Development Fund (FEDER) (AL4BIO project, RTI2018-099495-B-C21); and the Generalitat de Catalunya (Consolidated Research Group 2017 SGR 1029).

Acknowledgments: R. Díez-Montero and M.J. García-Galán acknowledge the Spanish Ministry of Industry and Economy for their research grants (FJCI-2016-30997 and IJCI-2017-34601, respectively). L. Vassalle acknowledges the CNPQ for his scholarship 204026/2018-0. The authors would like to thank Katerina Ilijovska for her support during data acquisition and analysis.

Conflicts of Interest: The authors declare no conflict of interest.

References

1. Union. Council Directive 91/271/EEC of 21 May 1991 concerning urban wastewater treatment. *Off. J. Eur. Commun.* **1991**, *135*, 40–52.
2. Union. European Parliament and Council Directive 2000/60/EC of 23 of October 2000 establishing a framework for the Community action in the field of water policy. *Off. J. Eur. Commun.* **2000**, *22*, 35–38.
3. Statistical Office of the European Union (Eurostat). *Resident Population Connected to Wastewater Collection and Treatment Systems Reported to the "OECD/Eurostat Joint Questionnaire—Inland Waters-2012"*; Statistical Office of the European Union (Eurostat): Luxembourg, 2017.
4. Lens, P.N.L.; Zeeman, G.; Lettinga, G. *Decentralised Sanitation and Reuse–Concepts, Systems and Implementation*; IWA Publishing: London, UK, 2001; ISBN 9781900222471.
5. Von Sperling, M. Comparison among the most frequently used systems for wastewater treatment in developing countries. *Water Sci. Technol.* **1996**, *33*, 59–62. [CrossRef]
6. Von Sperling, M. *Wastewater Characteristics, Treatment and Disposal*, 1st ed.; IWA Publishing: London, UK, 2007; Volume 1, ISBN 1 84339 161 9.
7. De Godos, I.; Arbib, Z.; Lara, E.; Rogalla, F. Evaluation of High Rate Algae Ponds for treatment of anaerobically digested wastewater: Effect of CO2 addition and modification of dilution rate. *Bioresour. Technol.* **2016**, *220*, 253–261. [CrossRef] [PubMed]
8. Santiago, A.F.; Calijuri, M.L.; Assemany, P.P.; Calijuri, M.D.C.; Reis, A.J.D. Dos Algal biomass production and wastewater treatment in high rate algal ponds receiving disinfected effluent. *Environ. Technol.* **2017**, *34*, 1877–1885. [CrossRef] [PubMed]
9. Arashiro, L.T.; Ferrer, I.; Rousseau, D.P.L.; Van Hulle, S.W.H.; Garfí, M. The effect of primary treatment of wastewater in high rate algal pond systems: Biomass and bioenergy recovery. *Bioresour. Technol.* **2019**, *280*, 27–36. [CrossRef]
10. Park, J.B.K.; Craggs, R.J. Algal production in wastewater treatment high rate algal ponds for potential biofuel use. *Water Sci. Technol.* **2011**, *63*, 2403–2410. [CrossRef]
11. Passos, F.; Gutiérrez, R.; Uggetti, E.; Garfí, M.; García, J.; Ferrer, I. Towards energy neutral microalgae-based wastewater treatment plants. *Algal Res.* **2017**, *28*, 235–243. [CrossRef]
12. Barkia, I.; Saari, N.; Manning, S.R. Microalgae for high-value products towards human health and nutrition. *Mar. Drugs* **2019**, *17*, 304. [CrossRef]
13. Vernès, L.; Li, Y.; Chemat, F.; Abert-Vian, M. Plant Based "Green Chemistry 2.0." In *Green Chemistry and Sustainable Technology*; Li, Y., Chemat, F., Eds.; Green Chemistry and Sustainable Technology; Springer: Singapore, 2019; pp. 15–50. ISBN 978-981-13-3809-0.
14. Arashiro, L.T.; Ferrer, I.; Pániker, C.C.; Gómez-Pinchetti, J.L.; Rousseau, D.P.L.; Van Hulle, S.W.H.; Garfí, M. Natural pigments and biogas recovery from microalgae grown in wastewater. *ACS Sust. Chem. Eng.* **2020**, *8*, 10691–10701. [CrossRef]
15. Jankowska, E.; Sahu, A.K.; Oleskowicz-Popiel, P. Biogas from microalgae: Review on microalgae' s cultivation, harvesting and pretreatment for anaerobic digestion. *Renew. Sust. Energy Rev.* **2017**, *75*, 692–709. [CrossRef]
16. Khan, M.I.; Shin, J.H.; Kim, J.D. The promising future of microalgae: Current status, challenges, and optimization of a sustainable and renewable industry for biofuels, feed, and other products. *Microb. Cell Fact.* **2018**, *17*, 1–21. [CrossRef]
17. Marín, D.; Ortíz, A.; Díez-Montero, R.; Uggetti, E.; García, J.; Lebrero, R.; Muñoz, R. Influence of liquid-to-biogas ratio and alkalinity on the biogas upgrading performance in a demo scale algal-bacterial photobioreactor. *Bioresour. Technol.* **2019**, *280*, 112–117. [CrossRef] [PubMed]
18. Morales, M.; Collet, P.; Lardon, L. Hélias, A.; Steyer, J.-P.; Bernard, O. *Life-Cycle Assessment of Microalgal-Based Biofuel*, 2nd ed.; Elsevier: Amsterdam, The Netherlands, 2019; ISBN 9780444641922.
19. Wang, M.; Park, C. Investigation of anaerobic digestion of *Chlorella* sp. and *Micractinium* sp. grown in high-nitrogen wastewater and their co-digestion with waste activated sludge. *Biomass Bioenergy* **2015**, *80*, 30–37. [CrossRef]
20. Vassalle, L.; Díez-Montero, R.; Machado, A.T.R.; Moreira, C.; Ferrer, I.; Mota, C.R.; Passos, F. Upflow anaerobic sludge blanket in microalgae-based sewage treatment: Co-digestion for improving biogas production. *Bioresour. Technol.* **2020**, *300*, 9. [CrossRef] [PubMed]

21. Díez-Montero, R.; Solimeno, A.; Uggetti, E.; García-Galán, M.J.; García, J. Feasibility assessment of energy-neutral microalgae-based wastewater treatment plants under Spanish climatic conditions. *Process. Saf. Environ. Prot.* **2018**, *119*, 242–252. [CrossRef]
22. Acien, F.G.; Sevilla, J.M.F.; Grima, E.M. Microalgae: The Basis of Mankind Sustainability. In *Case Study of Innovative Projects–Successful Real Cases*; InTech: London, UK, 2017; Volume 1, p. 17.
23. García-Galán, M.J.; Gutiérrez, R.; Uggetti, E.; Matamoros, V.; García, J.; Ferrer, I. Use of full-scale hybrid horizontal tubular photobioreactors to process agricultural runoff. *Biosyst. Eng.* **2018**, *166*, 138–149. [CrossRef]
24. Passos, F.; Uggetti, E.; Carrère, H.; Ferrer, I. Pretreatment of microalgae to improve biogas production: A review. *Bioresour. Technol.* **2014**, *172*, 403–412. [CrossRef]
25. Ehimen, E.A.; Holm-Nielsen, J.B.; Poulsen, M.; Boelsmand, J.E. Influence of different pre-treatment routes on the anaerobic digestion of a filamentous algae. *Renew. Energy* **2013**, *50*, 476–480. [CrossRef]
26. Passos, F.; Ferrer, I. Microalgae conversion to biogas: Thermal pretreatment contribution on net energy production. *Environ. Sci. Technol.* **2014**, *48*, 7171–7178. [CrossRef] [PubMed]
27. Solé-Bundó, M.; Salvadó, H.; Passos, F.; Garfí, M.; Ferrer, I. Strategies to optimize microalgae conversion to biogas: Co-digestion, pretreatment and hydraulic retention time. *Molecules* **2018**, *23*, 2096. [CrossRef] [PubMed]
28. Cavinato, C.; Ugurlu, A.; de Godos, I.; Kendir, E.; Gonzalez-Fernandez, C. *Biogas Production from Microalgae*; Elsevier Ltd.: Amsterdam, The Netherlands, 2017; ISBN 9780081010273.
29. García, J.; Ortiz, A.; Álvarez, E.; Belohlav, V.; García-Galán, M.J.; Díez-Montero, R.; Álvarez, J.A.; Uggetti, E. Nutrient removal from agricultural run-off in demonstrative full scale tubular photobioreactors for microalgae growth. *Ecol. Eng.* **2018**, *120*, 513–521. [CrossRef]
30. Rueda, E.; García-Galán, M.J.; Díez-Montero, R.; Vila, J.; Grifoll, M.; García, J. Polyhydroxybutyrate and glycogen production in photobioreactors inoculated with wastewater borne cyanobacteria monocultures. *Bioresour. Technol.* **2020**, *295*, 122233. [CrossRef] [PubMed]
31. Uggetti, E.; García, J.; Álvarez, J.A.; García-Galán, M.J. Start-up of a microalgae-based treatment system within the biorefinery concept: From wastewater to bioproducts. *Water Sci. Technol.* **2018**, *78*, 114–124. [CrossRef]
32. Passos, F.; García, J.; Ferrer, I. Impact of low temperature pretreatment on the anaerobic digestion of microalgal biomass. *Bioresour. Technol.* **2013**, *138*, 79–86. [CrossRef]
33. Ras, M.; Lardon, L.; Bruno, S.; Bernet, N.; Steyer, J.P. Experimental study on a coupled process of production and anaerobic digestion of *Chlorella vulgaris. Bioresour. Technol.* **2011**, *102*, 200–206. [CrossRef]
34. APHA-AWWA-WEF. *Standard Methods for the Examination of Water and Wastewater*, 22nd ed.; Rice, E.W., Baird, R.B., Eaton, A.D., Clesceri, L.S., Eds.; The American Water Works Association; The American Public Health Association; The Water Environment Federation: Washington, DC, USA, 2012; ISBN 9780875530130.
35. Solórzano, L. Determination of ammonia in natural seawater by the phenol-hypochlorite method. *Limnol. Oceanogr.* **1969**, *14*, 799–801. [CrossRef]
36. Streble, H.; Krauter, D. *Atlas de los Microorganismos de Agua Dulce, La Vida en Una Gota de Agua*; Omega: Barcelona, Spain, 1987; ISBN 842820800X 9788428208000.
37. Palmer, C.M. *Algas en Abastecimientos de Agua: Manual Ilustrado Acerca de la Identificación, Importancia y Control de las Algas en los Abastecimientos de Agua*; Interamericana: Ciudad de México, Mexico, 1962.
38. Díez-Montero, R.; Belohlav, V.; Ortiz, A.; Uggetti, E.; García-Galán, M.J.; García, J. Evaluation of daily and seasonal variations in a semi-closed photobioreactor for microalgae-based bioremediation of agricultural runoff at full-scale. *Algal Res.* **2020**, *47*, 101859. [CrossRef]
39. Tchobanoglous, G.; Burton, F.; Stensel, H.D. *Wastewater Engineering: Treatment and Reuse*, 4th ed.; Metcalf & Eddy, McGraw-Hill Education: New York, NY, USA, 2003; ISBN 9780070418783.
40. González-Fernández, C.; Sialve, B.; Bernet, N.; Steyer, J.P. Thermal pretreatment to improve methane production of Scenedesmus biomass. *Biomass Bioenergy* **2012**, *40*, 105–111. [CrossRef]
41. Solé-Bundó, M.; Passos, F.; Romero-Güiza, M.S.; Ferrer, I.; Astals, S. Co-digestion strategies to enhance microalgae anaerobic digestion: A review. *Renew. Sust. Energy Rev.* **2019**, *112*, 471–482. [CrossRef]
42. McCarty, P.L. Anaerobic Waste Treatment Fundamentals. *Public Work* **1964**, *95*, 91–94.

43. Sutherland, D.L.; Park, J.; Heubeck, S.; Ralph, P.J.; Craggs, R.J. Size matters—Microalgae production and nutrient removal in wastewater treatment high rate algal ponds of three different sizes. *Algal Res.* **2020**, *45*, 101734. [CrossRef]
44. Zabed, H.M.; Akter, S.; Yun, J.; Zhang, G.; Zhang, Y.; Qi, X. Biogas from microalgae: Technologies, challenges and opportunities. *Renew. Sust. Energy Rev.* **2020**, *117*, 109503. [CrossRef]
45. Bohutskyi, P.; Chow, S.; Ketter, B.; Fung Shek, C.; Yacar, D.; Tang, Y.; Zivojnovich, M.; Betenbaugh, M.J.; Bouwer, E.J. Phytoremediation of agriculture runoff by filamentous algae poly-culture for biomethane production, and nutrient recovery for secondary cultivation of lipid generating microalgae. *Bioresour. Technol.* **2016**, *222*, 294–308. [CrossRef] [PubMed]

Article

Potential Applications of Biogas Produced in Small-Scale UASB-Based Sewage Treatment Plants in Brazil

Fabiana Passos *, Thiago Bressani-Ribeiro, Sonaly Rezende and Carlos A. L. Chernicharo

Department of Sanitary and Environmental Engineering, Universidade Federal de Minas Gerais, Av. Antônio Carlos, 6627, Belo Horizonte 31270-901, Brazil; thiagobressani@ufmg.br (T.B.-R.); srezende@desa.ufmg.br (S.R.); calemos@desa.ufmg.br (C.A.L.C.)

* Correspondence: fabiana@desa.ufmg.br; Tel.: +55-31-34093593

Received: 31 January 2020; Accepted: 20 May 2020; Published: 1 July 2020

Abstract: Rural sanitation is still a challenge in developing countries, such as Brazil, where the majority population live with inadequate services, compromising public health and environmental safety. In this context, this study analyzed the demographic density of these rural agglomerations using secondary data from the Brazilian Institute of Geography and Statistics (IBGE). The goal was to identify the possibilities associated with using small-scale upflow anaerobic sludge blanket (UASB) reactors for sewage treatment, mainly focusing on biogas production and its conversion into energy for cooking, water heating and sludge sanitization. Results showed that most rural agglomerations lacking the appropriate sewage treatment were predominant from 500 to 1500 inhabitants in both northern and southern Brazilian regions. The thermal energy available in the biogas would be enough to sanitize the whole amount of sludge produced in the sewage treatment plants (STPs), producing biosolids for agricultural purposes. Furthermore, the surplus of thermal energy (after sludge sanitization) could be routed for cooking (replacing LPG) and for water heating (replacing electricity) in the northern and southern regions, respectively. This would benefit more than 200,000 families throughout rural areas of the country. Besides the direct social gains derived from the practice of supplying biogas for domestic uses in the vicinity of the STPs, there would be tremendous indirect gains related to the avoidance of greenhouse gas (GHG) emissions. Therefore, an anaerobic-based sewage treatment may improve public health conditions, life quality and generate added value products in Brazilian rural areas.

Keywords: anaerobic treatment; bioenergy; energy assessment; rural sanitation; sludge; wastewater

1. Introduction

Sanitation is closely related to public health, environmental safety and life quality worldwide. Particularly in developing countries, sanitation has been debated in terms of human rights, highlighting situations of extreme violation [1]. In Brazilian rural areas, the Federal Sanitation Policy (Law n° 11,445/2007) determines social inclusion and the reduction of regional inequalities, seeking to provide adequate conditions of environmental health to rural populations and small isolated urban centres. In this context, sanitation plans, programs and projects in areas occupied by low-income populations should be given priority. Moreover, solutions attending to indigenous people and traditional populations should be compatible with their social and cultural characteristics. Finally, the referred legislation indicates the unity and articulation of different institutional agents, as well as the development of their organisation, technical aspects, management, and financial and human resources capacity, considering local specificities [2].

Nevertheless, despite the legislation, sparsely populated areas, characterized by low demographic densities, are commonly neglected due to the principle of economy of scale. The rural invisibility to public policies results in a precarious health situation and a wide regional inequality. In addition, rural agglomerations and/or communities in urban areas but distant from the urban sanitation infrastructure are examples of possibilities for developing collective small-scale sustainable sewage treatment plants (STPs).

In this context, an anaerobic sewage treatment has been investigated and applied in developing countries, such as Brazil. In fact, this country has the largest number of installed upflow anaerobic sludge blanket (UASB) reactors treating sewage in the world. In a recent publication, anaerobic-based STPs using UASB reactors accounted for 667 systems among the 1667 systems acknowledged (i.e., 40%), comprising systems serving from 5000 to 1 million inhabitants [3]. Moreover, an investigation of 2734 STPs in six Latin American and Caribbean countries showed that, besides Brazil, UASB reactors have been extensively used in Mexico, Colombia, Dominican Republic and Guatemala. Such anaerobic reactors represented up to 20% of the total number of STPs for all of the assessed countries [4]. It is worth mentioning that full-scale UASB reactors have also been successfully applied in India [5].

The advantages and potentialities of this technology are related to several aspects, noteworthy are the low sludge production and implementation and operation costs, compared with conventional aerobic (e.g., activated sludge) or physicochemical processes [6]. Moreover, an anaerobic sewage treatment generates biogas, which may be converted into electric or thermal energy for use in the STP itself or in the nearby community. The energy conversion process and its application depend on many factors, such as the STP size, energy policies and subsides, climatic conditions and socio-economic local characteristics. In general, biogas use in small-scale STPs in rural areas is designated to thermal energy conversion, such as water heating, cooking and sludge sanitization. For small- and medium-scale STPs (PE > 2000; PE < 100,000, where PE represents population equivalent), biogas use for the cogeneration of electricity and heat is generally not feasible, due to energy costs, lack of incentive programs for energy recovery from biogas and the poorly developed market for combined heat and power (CHP) engines [7].

The present study aimed at characterizing the potential for biogas generation in Brazilian rural agglomerations which are currently unattended by sanitation services. These agglomerations were organized in three categories using secondary data from the Brazilian Institute of Geography and Statistics (IBGE): (a) rural areas in urban area; (b) isolated rural areas with large settlements and; (c) isolated rural areas with small settlements. The goal was to identify the possibilities associated with using small-scale UASB reactors for sewage treatment, mainly focusing on biogas production and its conversion into energy for cooking, water heating and sludge sanitization, considering the different climate conditions amongst Brazilian regions. Additionally, carbon emissions were also assessed for the proposed technological flowsheets.

2. Material and Methods

2.1. Identification of Rural Agglomerations: Secondary Data

The most recent Brazilian demographic census was used to identify and analyze sewage sanitation infrastructure in the rural areas [8]. Secondary data collected aimed at classifying situations that lack an appropriate treatment, such as rudimentary pit, ditch or direct discharge into rivers, lakes or the sea. The available data of rural agglomerations were further reclassified by the National Program of Rural Sanitation (PNSR) [9], in order to better match the demographic densities of the different Brazilian localities/agglomerations. After the reclassification, four categories of rural households were adopted, identifying them with each specific sanitation demand and possible biogas uses, as shown in Table 1.

Table 1. Rural household classification and recommended sanitation solutions according to the National Program of Rural Sanitation.

Population Category	Description	Recommended Sanitation Solutions
A	Peripheric agglomerations in urban territory	The same as those practiced in cities (urban areas)
B	Isolated agglomerations with urban characteristics	Economy of scale justify the use of decentralised solutions and self-sufficient management model
C	Isolated agglomerations with rural characteristics	Individual and collective actions coexist; the management may require external support
D	Dispersed rural settlements	Individual actions prevail

Note: For this study, only "A", "B" and "C" were considered. Isolated stands for agglomerations far from the central core of the municipality (urban area).

As can be seen, Table 1 summarizes the description and recommended sanitation solutions considered for each population category. Category A comprises agglomerations located in peripheric regions of an urban area and, therefore, technological options for sewage treatment may be carried out as in cities. In those cases, UASB reactors may be developed as a decentralized option for recovering possible by-products for local use, such as biogas. Categories B, C and D are isolated or far from urban areas. Agglomerations classified as category B have a more urban-like lifestyle, while C and D have an agricultural economy and fewer services (e.g., transportation). For category B, an option for sanitation solutions may be decentralized systems comprising UASB reactors, while STP management may be conducted by a group of users that benefit from valued by-products. Agglomerations from category C may also use collective STP systems with UASB reactors when settlements are nearby, while a familiar approach would be applied for isolated houses. Individual solutions may require external technical support from public services, although collective solutions may possibly be managed through a local auto-organization approach. In fact, a recent study is being conducted in an *urban* occupation without wastewater treatment services, showing the possible scenarios of users and residents that manage the system and benefits from generated by-products, such as biofertilizer. Category D consists of isolated and dispersed occupation, leading to individual sanitation solutions. Therefore, this category was not considered for implementing UASB reactors and biogas recovery.

For this study, the surveyed agglomerations considered all populations from 500 to 3000 inhabitants associated with categories A, B and C, as they can be potentially served by decentralized small-scale STPs Moreover, data were analyzed according to the geographic region and, therefore, separated in two groups: northern region (North, Northeast and Centre-West) and southern region (South and South-East), as illustrated in Figure 1. These were chosen to discuss the different uses for the potential biogas produced, based on the different climate conditions amongst Brazilian regions. The average annual temperature in the southern region is around 20 °C, while in the northern regions it raises to around 28 °C [10]. Of course, the wide variety of local geographic conditions alongside the country is implied in the variations in such values. In any case, anaerobic digestion in UASB reactors has been successfully carried out throughout the country [3].

A previous study developed by our group [11] identified that the highest environmental, economic and also social gains of biogas recovery are associated with its primary use for sludge sanitization, as this allows the production of safe biosolids that can be used for agricultural purposes. This practice, besides contributing for closing nutrient (nitrogen and phosphorus) cycles, also plays a role for reducing the demand on chemical fertilizers, and for avoiding sludge transportation and disposal in landfills. Further potential uses of biogas produced in small anaerobic-based STPs are for cooking (an attractive alternative for all Brazilian regions, North and South) and for water heating, a choice especially appealing for the South Region, where cooler temperatures prevail. These recommended biogas uses according to each population category are summarized in Table 2.

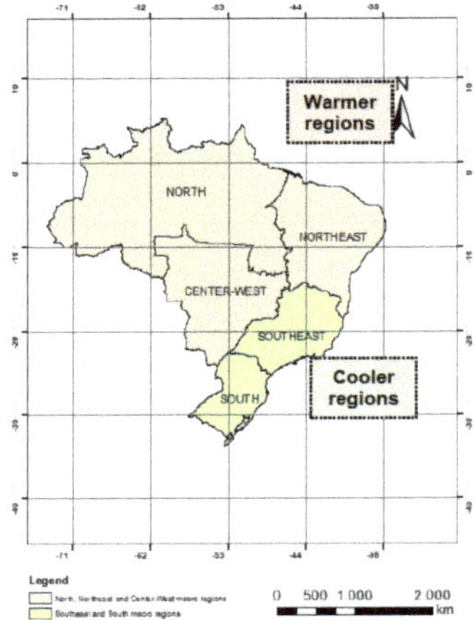

Figure 1. Groups of Brazilian regions considered for the distribution of the rural population under the categories A, B and C (detailed in Table 1).

Table 2. Recommended biogas end uses based on population categories and geographic regions.

Geographic Region	Population Category	Recommended/Potential Biogas End Uses
North	A, B and C	Sludge sanitization Cooking
South	A, B and C	Sludge sanitization Water heating

2.2. Energy Assessment and Carbon Emissions Evaluation

Biogas potential applications and end use depend on the amount of biogas generated, which is primarily a direct function of the STP size. Biogas production and energy recovery options for small-scale STPs using anaerobic reactors were estimated as follows. The equations used to perform the calculations were adapted from Soares et al. [12] (Equations (1)–(10)) and the parameters used are summarised in Table 3.

Table 3. Parameters used for determining the biogas production and energy recovery potential in small-scale sewage treatment plants (STPs) (adapted from [7,8]).

Parameters	Variable Name	Unit	Value	Reference
Daily per capita sewage generation	Q_{PC}	$L\ PE^{-1}\ d^{-1}$	160	[13]
Daily biogas consumption for cooking	$BC_{cooking}$	Nm^3 biogas $family^{-1}\ d^{-1}$	0.25	[14]
Unitary methane yield	Y_{CH4}	NL $CH_4\ m^{-3}$ sewage	64	[15]
Methane content in biogas	$\%_{CH4}$	%	75	[16]
Lower calorific value of methane	LCV_{CH4}	MJ $Nm^{-3}CH_4^{-1}$	35.8	[17]
Lower calorific value of LPG¹	LCV_{LPG}	MJ $Nm^{-3}CH_4^{-1}$	120.4	[17]
Daily per capita sludge (as DS²) generation in UASB reactors	DS_{PE}	gDS $PE^{-1}\ d^{-1}$	15	[18]
Water specific heat	H_w	kJ $kg^{-1}\ °K^{-1}$	4.18	[17]
Sludge specific heat	H_s	kJ $kg^{-1}\ °K^{-1}$	1.05	[12]
Sludge temperature	T_s	°C	20	[19]
Sanitized sludge temperature	T_{ss}	°C	70	[19]
Excess sludge concentration	C_{sludge}	%	4	[20]
Sludge specific mass	γ_s	kg m^{-3}	1020	[18]
Energy loss through the walls of the sanitizing tank	$EL_{sanit\text{-}tank}$	%	15	[12]
Difference between tap water and bath temperatures	Δw	°C	30	Assumed value
Thermal efficiency of boilers	$TE_{boilers}$	%	90	Standard engine reference
Emission factor for LPG burn	EF_{LPG}	$kgCO_2$ eq m^{-3} LPG	1507.1	[21]
Emission factor for the electricity generation in Brazil	EF_{elec}	gCO_2 eq $kW^{-1}\ h^{-1}$	125	[22]

Note: ¹ LPG: liquefied petroleum gas; ² DS: dry solids.

The energy potential of methane ($E_{CH4\text{-}potential}$) was calculated in terms of unitary methane yield (Y_{CH4}), daily per capita sewage generation (Qpc), person equivalent (PE) and the lower calorific value of methane (LCV_{CH4}), as in Equation (1).

$$E_{CH4\text{-}potential}\ (MJ\ d^{-1}) = Y_{CH4} \times Qpc/1000 \times PE \times LCV_{CH4} \quad (1)$$

The thermal energy potential ($E_{th\text{-}potential}$) was calculated in terms of the energy potential of methane and the thermal efficiency of boilers ($E_{boilers}$), as in Equation (2).

$$E_{th\text{-}potential}\ (MJ\ d^{-1}) = E_{potential\text{-}CH4} \times E_{boilers}/100 \quad (2)$$

Daily sludge production in UASB reactors ($P_{sludge\text{-}UASB}$) was calculated in terms of the daily dry sludge production per capita (DS_{PE}) and the person equivalent (PE), as in Equation (3).

$$P_{sludge\text{-}UASB}\ (kgDS\ d^{-1}) = DS_{PE}/1000 \times PE \quad (3)$$

The mass of water in sludge (M_{water}) was assumed using the daily sludge produced in reactors ($P_{sludge\text{-}UASB}$), the sludge specific mass (γ) and the excess sludge concentration (C), as in Equation (4).

$$M_{water}\ (kg) = P_{sludge\text{-}UASB}/(\gamma \times C/100) \quad (4)$$

The daily energy demand for sludge sanitization ($E_{th\text{-}sludge}$) was calculated in terms of the daily sludge produced in reactors ($P_{sludge\text{-}UASB}$), sludge specific heat (H_s), the difference between the sludge temperature and the sanitized sludge temperature (Δs), the mass of water in sludge (M_{water}), water specific heat (H_w) and, the energy loss through the walls of the sanitizing tank ($EL_{sanit\text{-}tank}$), as in Equation (5).

$$E_{th\text{-}sludge}\ (MJ\ d^{-1}) = [(P_{sludge\text{-}UASB} \times H_s \times \Delta s) + (M_{water} \times H_w \times \Delta s)] \times (1 + EL_{sanit\text{-}tank}/100)/1000 \quad (5)$$

The daily surplus of thermal energy ($E_{th\text{-}surplus}$) was calculated in terms of the daily thermal energy potential ($E_{th\text{-}potential}$) subtracted from the daily energy demand for sludge sanitization ($E_{th\text{-}sludge}$), as in Equation (6).

$$E_{th\text{-}surplus}\ (MJ\ d^{-1}) = E_{th\text{-}potential} - E_{th\text{-}sludge} \quad (6)$$

The daily water heating potential ($W_{potential}$) was calculated in terms of the daily surplus of thermal energy ($E_{th\text{-}surplus}$), the daily thermal efficiency of boilers ($E_{boilers}$), water specific heat (H_w) and the difference between the tap water temperature and the temperature of a bath (Δw), as in Equation (7).

$$W_{potential}\ (m^3\ _{water}\ d^{-1}) = [(E_{th\text{-}surplus} \times E_{boilers}/100)/(H_w \times \Delta w)]/1000 \quad (7)$$

The daily use of biogas for cooking ($BU_{cooking}$) was calculated in terms of the daily surplus of thermal energy ($E_{th\text{-}surplus}$), the lower calorific value of methane (LCV_{CH4}), the daily biogas consumption for cooking ($BC_{cooking}$) and the methane content in biogas ($\%_{CH4}$), as in Equation (8).

$$BU_{cooking}\ (MWh\ d^{-1}) = (E_{th\text{-}surplus} \times 1000/LCV_{CH4})/(BC_{cooking} \times \%_{CH4}) \quad (8)$$

The montly avoided CO_2 emission due to replacement of LPG for cooking (Avoided $CO_{2\ cooking}$) was calculated in terms of the surplus of thermal energy ($E_{th\text{-}surplus}$), the lower calorific value of methane (LCV_{CH4}), the lower calorific value of LPG (LCV_{LPG}) and the CO_2 equivalent emission factor for LPG burn (EF_{LPG}), as in Equation (9).

$$\text{Avoided } CO_{2\ cooking}\ (kgCO_2\ month^{-1}){:}\ E_{th\text{-}surplus} \cdot 30 \times LCV_{CH4}/LCV_{LPG} \times EF_{LPG} \quad (9)$$

The monthly avoided CO_2 emission due to electricity replacement for water heating (Avoided $CO_{2\ heating}$) was calculated in terms of the surplus of thermal energy ($E_{th\text{-}surplus}$) and the CO_2 equivalent emission factor for the electricity generation in Brazil (EF_{elec}), as in Equation (10).

$$\text{Avoided } CO_{2\ heating}\ (kgCO_2\ month^{-1}) = E_{th\text{-}surplus} \times 30 \times EF_{elec.} \quad (10)$$

For small volumes of produced biogas (1–5 $Nm^3\ d^{-1}$), directly burning them after the hydrogen sulphide (H_2S) removal is the most traditional biogas use for domestic applications, such as for heat production especially for cooking. In population agglomerations with agricultural practices, biogas could also be routed for sludge sanitization. These technological arrangements are illustrated in Figure 2. A simplified desorption column followed by a biofilter was considered in the flowsheet for the H_2S and CH_4 abatement due to the presence of these gases in the anaerobic effluent, therefore, no additional energy gains were achieved because methane is not recovered. The idea in this case is just to avoid greenhouse gas and odorous emissions.

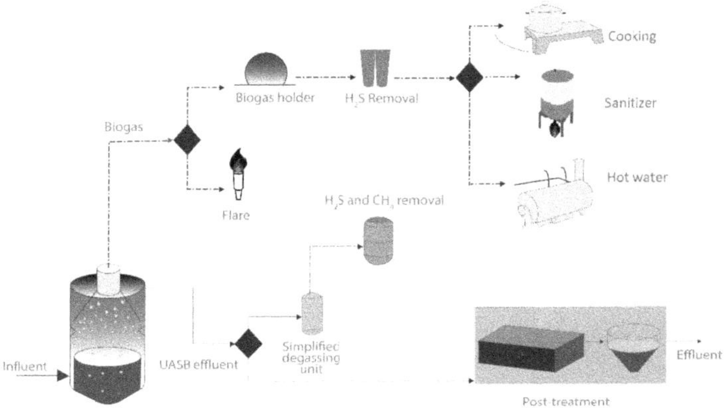

Figure 2. Flowsheet of the proposed biogas uses for small-scale STPs (adapted from [12]).

These gaseous management schemes were also evaluated in terms of carbon emissions using the tool "Sulphide and Carbon Emission Avoidance and Energy Recovery in STPs" [23]. The tool estimates the corresponding amount of methane supposed to be emitted into the atmosphere by anaerobic-based STPs, which is then converted to a CO_2 equivalent, allowing the assessment of the carbon footprint of the STP.

3. Results and Discussion

3.1. Rural Agglomerations in Brazil

The secondary data gathered from the demographic census [8] show that rural agglomerations lacking an appropriate sewage treatment were predominant from 500 to 1500 inhabitants, for all three categories analyzed (A, B and C) in both groups (northern and southern regions), as depicted in Figure 3. This population range embraces more than 5.0 and 2.5 million inhabitants in the northern and southern regions of Brazil, respectively, accounting for more than 92% of the Brazilian rural population gathered in categories A, B and C. This is a clear indication of the need for appropriate sewage treatment solutions for such small settlements.

A clearer picture of the population distribution in the whole assessed range (500 to 3000 inhabitants) considering the three categories is shown in Figure 4. Most of the population live in "rural areas of urban extension" (category A—Table 1), totalling almost 5.3 million inhabitants, followed by the population that live in "isolated rural areas where small settlements prevail" (category C—Table 1), accounting for around 2.4 million inhabitants. It is worth mentioning that approximately 94% of the population in category C is located in the northern region. A much lower population contingent (less than 600,000 inhabitants) lives in "isolated rural areas where large settlements prevail" (category B—Table 1). Bearing these numbers, one can realize that the population living in all three categories of such small settlements could potentially benefit from using the by-products generated in sustainable small-scale anaerobic-based STPs, especially biogas (for cooking, water heating and/or sludge sanitization), sanitized sludge and treated effluent (both for agricultural purposes), as further discussed in the following section.

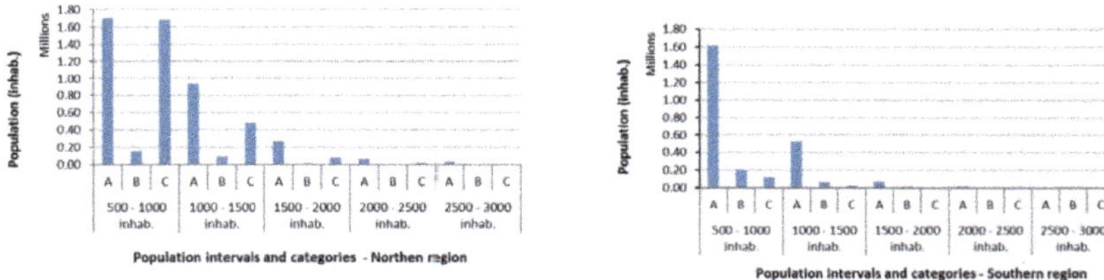

Figure 3. Distribution of the Brazilian rural population of the northern and southern regions according to intervals and categories. Categories A, B and C as shown in Table 1.

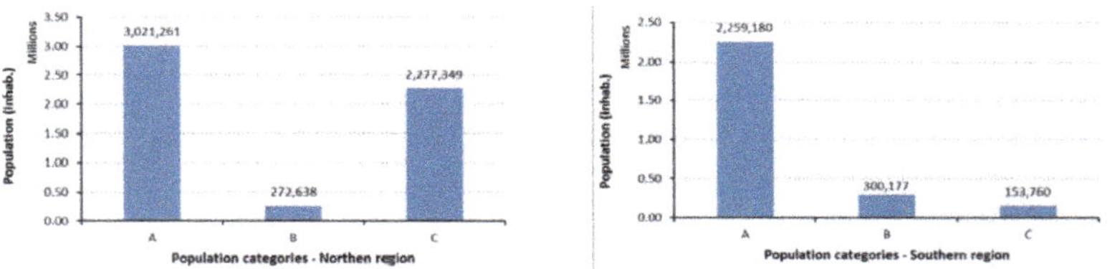

Figure 4. Distribution of the Brazilian rural population of the northern and southern regions according to categories. Categories A, B and C as shown in Table 1.

3.2. *Proposed Flowsheet for Sewage Treatment and by-Products Recovery/Use*

In order to better exemplify some of the environmental, economic and social gains associated with the use of by-products (biogas, sludge and water) generated in small-scale anaerobic-based STPs, we considered the treatment and by-product end uses flowsheet depicted in Figure 5. Although there are other possibilities of destination/uses of such by-products, depending on many factors (e.g., economic activities nearby the STP, applied process units, STP size, etc.), we carried out the study considering only the alternatives schematically represented in Figure 5 and further summarized in Table 4, specially focussing on potential biogas uses.

Table 4. By-products' (biogas, sludge and water) end uses considered in this study.

Geographic Region	Population Category	Population Equivalent (Inhabitants)	By-Products end Uses		
			Biogas	Sludge	Effluent
North	A B C Total	3,021,261 272,638 2,277,349 5,571,248	Sludge sanitization Cooking	Agriculture	Fertirrigation
South	A B C Total	2,259,180 300,177 153,760 2,713,117	Sludge sanitization Water heating	Agriculture	Fertirrigation

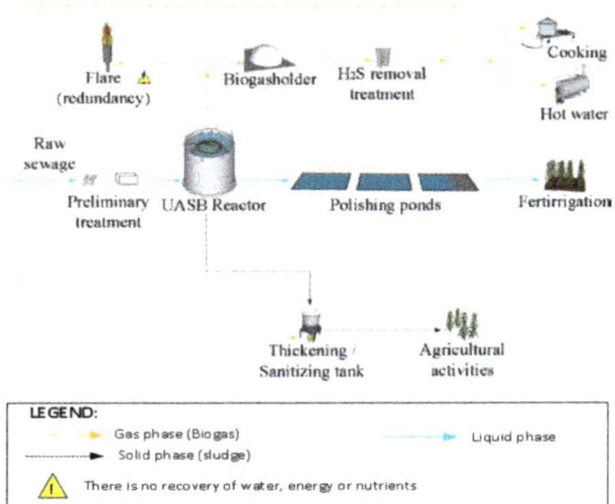

Figure 5. Schematic flowsheet representation of the small-scale anaerobic-based STP and by-products' end uses considered in the study. Source: Adapted from [24]

3.3. Potential Uses of the Biogas Produced in Small-Scale Anaerobic-Based STPs in the Northern and Southern Regions of Brazil

3.3.1. Use of Biogas for Sludge Sanitization

Thermal sludge sanitization can be achieved by means of a boiler fed on biogas and a simple heated concrete tank. This tank should be preferably fed once a day, in order to avoid the need of a big biogas holder. The sludge needs to be heated to 70 °C for 30 min (pasteurization) by means of a heat exchanger installed in the tank. After the sanitization process, the sludge can be routed to simple dehydration units (e.g., drying beds) or it can be directly spread on the agricultural land to be fertilized.

The results presented in Figure 6 show that the thermal energy generated by a boiler fed on biogas is much higher than the demand for sludge sanitization (less than 30% of the available energy), considering the typical parameters presented in Table 3. Therefore, the surplus of thermal energy (more than 70%) can be used for other purposes, such as for cooking (northern region) and/or for water heating (southern region), as discussed in the following section.

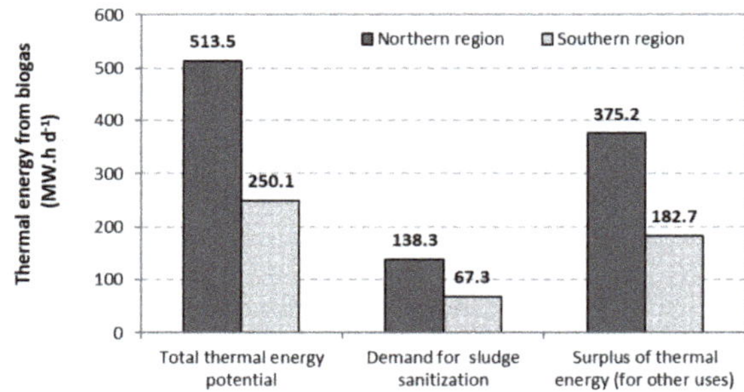

Figure 6. Potential thermal energy generation from biogas and supplied demands (sludge sanitization and other uses—cooking/water heating).

3.3.2. Use of Biogas for Cooking and Water Heating

The results presented in Figure 7 show that the direct use of biogas (after attending the demand for sludge sanitization) would allow more than 200,000 families to cook without the need of another external source of heat in the northern area of Brazil. This means that an equivalent population of approximately 800,000 inhabitants, or close to 15% of the total rural population that live in northern

Brazil (categories A, B and C altogether), could be supplied with the generated biogas. This possibility of biogas use is of great importance for this region, since the delivery (and costs) of liquefied petroleum gas (LPG) is a matter of concern. In this case, biogas can be supplied at much lower costs than LPG when considering the acquisition and transportation for delivering the fuel. In Brazilian rural areas, there are no gas pipelines and generally road infrastructure is inadequate. In addition, biogas is considered a clean and renewable source of energy, therefore, its use in the replacement of LPG would represent remarkable environmental gains due to the extremely high CO_2 emission factor of the latter (a petroleum-derived gas). Finally, this study considered agglomerations with no infrastructure in terms of sanitation. Therefore, the recovery of biogas for cooking is an added benefit of having a sewage treatment solution, which can also foster the implementation of new decentralized STPs. As a matter of fact, a recent study on a rural household in Costa Rica found that a family would save the equivalent of USD 26/month with the acquisition and transportation of LPG if biogas is used for cooking [14].

Likewise, the results depicted in Figure 7 indicate that the surplus of thermal energy (after attending the demand for sludge sanitization) would be enough to produce almost 5000 m^3 of hot water per day (50 °C). Considering a family with four persons and a consumption of 30 liters of hot water per bath/shower, the amount of produced hot water would be enough to supply almost 40,000 families per day in the southern region of Brazil (around 160,000 persons, or approximately 6% of the total rural population living in this region). This also has an associated positive social impact, as electricity usually represents a large share (~50%) in the monthly bill paid by families.

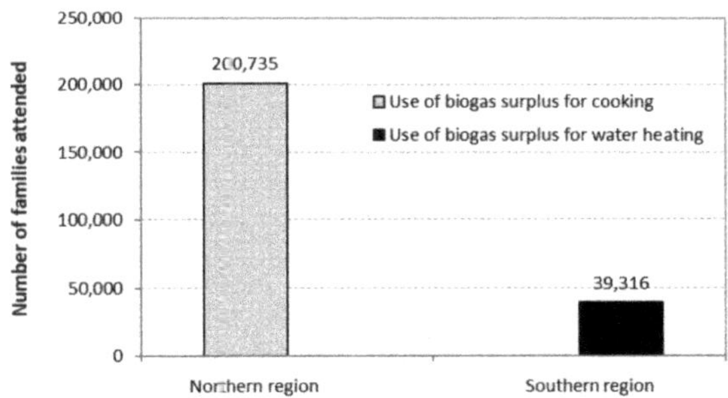

Figure 7. Thermal energy surplus (after sludge sanitization) and families attended with heat for cooking (northern region) and for water heating (southern region).

3.3.3. Avoided Emissions of GHG

According to the results presented in Figure 8, a remarkable negative carbon footprint would be achieved if the surplus biogas (after its main use for sludge sanitization) was used for cooking in the northern region of Brazil. In this case, approximately 6.1 Gt $CO_{2eq}{\cdot}y^{-1}$ (per capita of 1100 kg $CO_{2eq}{\cdot}PE^{-1}{\cdot}y^{-1}$) would be avoided to be emitted into the atmosphere. Such a CO_{2eq} reduction relates to the replacement of LPG by biogas. This may be compared with the per capita GHG emission in the Brazilian energy sector, which is approximately 60 kg $CO_{2eq}{\cdot}PE^{-1}{\cdot}y^{-1}$.

Although a much lower avoidance of GHG emissions would be achieved with the use of the surplus thermal energy from biogas for water heating, there would still be a contribution for neutralizing the overall STP carbon footprint, avoiding the emission of approximately 8 Mt $CO_{2eq}{\cdot}PE^{-1}{\cdot}y^{-1}$ (per capita of 3 kg $CO_{2eq}{\cdot}PE^{-1}{\cdot}y^{-1}$). In this case, the CO_{2eq} reduction is associated with the replacement of electricity by biogas, considering the emission factor of the Brazilian electric matrix and the energy consumption of an electric shower (Table 3).

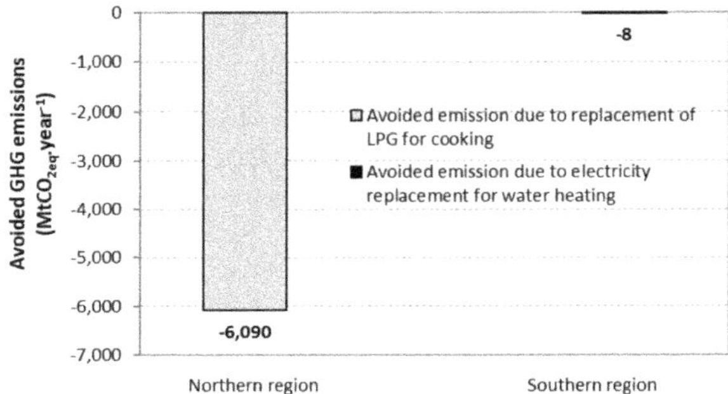

Figure 8. Avoided greenhouse gas (GHG) emissions due to the use of biogas instead of liquefied petroleum gas (LPG) for cooking (northern region) and for water heating (southern region).

3.4. Comparison with Biogas Use in Other Rural Contexts

Biogas has been used in rural areas in different regions worldwide and with different feedstock supplies to the anaerobic reactors. As may be observed, in most cases, biogas use in rural areas worldwide is applied to individual/family biodigesters treating agricultural and manure feedstock. For instance, China has the highest number of household biogas plants in the world, with 19% of the total population in rural areas (0.9 billion people) using biogas. However, an anaerobic digestion application relies on biodigesters fed with animal manure and agricultural residues [25]. Similarly, rural biodigesters developed in the Latin America region have also been used for treating animal manure, with some applications in agricultural residue and cooking grease. However, no data were found with biodigesters fed with sewage, not even co-digested with other substrates [26]. Nonetheless, both reviews identified similar bottlenecks, such as the low anaerobic biodegradability of lignocellulosic biomass, low temperatures in the winter season (~10 °C), low understanding of proper biogas use and limited management and technical support. Some of them such as the interest and involvement of the local population, as well as technical support and management, are also possible bottlenecks that rural STPs with UASB reactors may have.

4. Final Remarks

The overall balance of biogas production and thermal energy generation that could be achieved via the implementation of small-scale anaerobic-based STPs to attend to the Brazilian rural population grouped in categories A, B and C (around 8.3 million inhabitants) is extremely relevant and should not be neglected by designers and policy makers. The thermal energy available in the biogas would be enough to sanitize the whole amount of sludge produced in the STPs, making this material (biosolid) available to small farmers or even to encourage the practice of family farming nearby the plants. Besides contributing for closing the nutrient (N and P) cycles and lowering the production costs of agricultural products, there would still be a huge indirect benefit derived from the destination shift of this material, nowadays simply transported and disposed of in landfills.

Moreover, the surplus of thermal energy (after sludge sanitization) would be sufficient to attend to the demand of more than 200,000 families in the northern region with biogas for cooking (replacing LPG), and around 40,000 families in the southern region with biogas for water heating (replacing electricity). Again, besides the direct social gains derived from the practice of supplying biogas for domestic uses in the vicinity of the STPs, there would be tremendous indirect gains related to the avoidance of GHG emissions, especially when biogas is used to replace LPG. In this case, we estimated negative (avoided) GHG emissions equivalent to 6.1 Gt $CO_{2eq}y^{-1}{\cdot}y^{-1}$.

Likewise, an anaerobic treatment process may benefit small communities not only due to biogas and biosolids production but also with water reuse for agriculture. In this case, a simple post-treatment system (e.g., polishing pond) can meet the disinfection standards for restricted irrigation purposes [27].

Additionally, nitrogen that remains in the effluent can be considered a bonus for land irrigation. Therefore, an anaerobic sewage treatment can be faced as a low-cost technology that generates added value by-products and may improve public health conditions and life quality in Brazilian rural areas.

Author Contributions: All authors obtained and discussed the results and wrote the paper. S.R. specifically contributed to analyzing the secondary data from the Brazilian demographic census, while T.B.-R., F.P. and C.A.L.C. led the part on biogas recovery in UASB STPs. All authors have read and agreed to the published version of the manuscript.

Funding: This research received no external funding.

Acknowledgments: The authors acknowledge the support obtained from the following Brazilian institutions: Conselho Nacional de Desenvolvimento Científico e Tecnológico—CNPq; Fundação de Amparo à Pesquisa de Minas Gerais—FAPEMIG; Instituto Nacional de Ciência e Tecnologia em Estações Sustentáveis de Tratamento de Esgoto—INCT ETEs Sustentáveis and the National Health Foundation (Funasa), that promoted the National Program of Rural Sanitation (PNSR).

Conflicts of Interest: The authors declare no conflict of interest.

References

1. United Nations General Assembly (UNGA). *Human Right to Water and Sanitation*; UN Document A/RES/64/292, UNGA: Geneva, Switzerland, 2010. Available online: https://undocs.org/en/A/RES/64/292 (accessed on 17 November 2019).
2. Presidência da República, Casa Civil, Subchefia para Assuntos Jurídicos. LEI N° 11.445, DE 5 DE JANEIRO DE 2007. Available online: http://www.planalto.gov.br/ccivil_03/_ato2007-2010/2007/lei/l11445.htm (accessed on 21 November 2019).
3. Chernicharo, C.A.L.; Bressani-Ribeiro, T.; Garcia, G.B.; Lermontov, A.; Pereira, C.B.; Platzer, C.J.; Possetti, G.R.C.; Leites, M.A.L.; Rosseto, R. Panorama do tratamento de esgoto sanitário nas regiões Sul, Sudeste e Centro-Oeste do Brasil: Tecnologias mais empregadas (Overview of sewage treatment in the south, south-east and center-west regions of Brazil: Most employed technologies). *Rev. DAE* **2018**, *213*, 5–19.
4. Noyola, A.; Padilla-Rivera, A.; Morgan-Sagastume, J.M.; Güereca, L.P.; Hernández-Padilla, F. Typology of Municipal Wastewater Treatment Technologies in Latin America. *Clean Soil Air Water* **2012**, *40*, 926–932. [CrossRef]
5. Seghezzo, L.; Zeeman, Z.; Van Lier, J.B.; Hamelers, H.V.M.; Lettinga, L. A reiview: The anaerobic treatment of sewage in UASB and EGSB reactors. *Bioresour. Technol.* **1998**, *65*, 175–190. [CrossRef]
6. Chernicharo, C.A.L.; Bressani-Ribeiro, T.; von Sperling, M. Introduction to anaerobic sewage treatment. In *Anaerobic Reactors for Sewage Treatment: Design, Construction and Operation*; Chernicharo, C.A.L., Bressani-Ribeiro, T., Eds.; IWA Publishing: London, UK, 2019; pp. 1–7.
7. Possetti, G.R.C.; Rietow, J.; Cabral, C.B.G.; Moreira, H.C.; Platzer, C.; Bressani-Ribeiro, T.; Chernicharo, C.A.L. Energy recovery from biogas in UASB reactors treating sewage. In *Anaerobic Reactors for Sewage Treatment: Design, Construction and Operation*; Chernicharo, C.A.L., Bressani-Ribeiro, T., Eds.; IWA Publishing: London, UK, 2019; pp. 194–236.
8. IBGE. *2010 Population Census. Results of the Universe*; IBGE: Rio de Janeiro, Brazil, 2011.
9. National Rural Sanitation Program. Available online: http://www.funasa.gov.br/documents/20182/38564/MNL_PNSR_2019.pdf/08d94216-fb09-468e-ac98-afb4ed0483eb (accessed on 12 November 2019).
10. INMET (National Institute of Meteorology). Average Annual Temperature. Available online: http://www.inmet.gov.br/portal/index.php?r=clima/page&page=anomaliaTempMediaAnual (accessed on 17 March 2020). (In Portuguese)
11. Bressani-Ribeiro, T.; Lobato, L.C.S.; Chamhum-Silva, L.A.; Chernicharo, C.A.L. Sustainable STPs and public policies. In *Soluções Baseadas na Natureza e os Desafios das Águas: Acelerando a Transição Para Cidades Mais Sustentáveis*; Herzog, H., Freitas, T., Wiedman, G., Eds.; European Commission—Directorate-General for Research and Innovation: Brussels, Belgium, 2020.
12. Soares, A.; Mota, C.R.; Fdz-Polanco, F.; Huang, X.; Passos, F.; Bressani-Ribeiro, T.; Chernicharo, C.A.L. Closing cycles in anaerobic-based sewage treatment systems. In *Anaerobic Reactors for Sewage Treatment: Design, Construction and Operation*; Chernicharo, C.A.L., Bressani-Ribeiro, T., Eds.; IWA Publishing: London, UK, 2019; pp. 379–406.

13. Von Sperling, M.; Chernicharo, C.A.L. *Biological Wastewater Treatment in Warm Climate Regions (2 vols)*; IWA Publishing: London, UK, 2005.
14. Jiménez, F.M.; Zambrano, D. A Consumo de biogas en hogares rurales y sus implicaciones económicas y ambientales. *Caso El Porvenir, Limón. RedBioLAC* **2018**, *2*, 52–58.
15. Lobato, L.C.S.; Chernicharo, C.A.L.; Souza, C.L. Estimates of methane loss and energy recovery potential in anaerobic reactors treating domestic wastewater. *Water Sci. Technol.* **2012**, *66*, 2745–2753. [CrossRef] [PubMed]
16. Cabral, B.G.C.; Chernicharo, C.A.L.; Hoffmann, H.; Neves, P.N.P.; Platzer, C.; Bressani-Ribeiro, T.; Rosenfeldt, S. *Resultados do Projeto de Medições de Biogás em Reatores Anaeróbios/Probiogás (Results of the Biogas Measurement Project in Anaerobic Reactors/Probiogás)*; Deutsche Gesellschaf für Internationale Zusammenarbeit GmbH, Ed.; GIZ/Ministério das Cidades: Brasília, Brazil, 2016.
17. Moran, M.J.; Shapiro, H.N.; Boettner, D.D.; Bailey, M.B. *Fundamentals of Engineering Thermodynamics*, 8th ed.; Wiley: Hoboken, NJ, USA, 2014.
18. Andreoli, C.V.; von Sperling, M.; Fernandes, F. *Lodo de Esgoto: Tratamento e Disposição Final (Sewage Sludge: Treatment and Final Disposal)*; Department of Sanitary and Environmental Engineering of UFMG: Curitiba, Brazil, 2001; Volume 6, p. 484.
19. Borges, E.S.M.; Godinho, V.M.; Chernicharo, C.A.L. Thermal higienization of excess anaerobic sludge: A possible self-sustained application of biogas produced in UASB reactors. *Water Sci. Technol.* **2005**, *52*, 227–234. [CrossRef] [PubMed]
20. Chernicharo, C.A.L.; Bressani-Ribeiro, T. *Anaerobic Reactors for Sewage Treatment: Design, Construction and Operation*; IWA Publishing: London, UK, 2019; p. 420, ISBN 9781780409221.
21. U.S. EPA. 2014 Emission Factors for Greenhouse Gas Inventories. In *Report of the United States Environmental Protection Agency*; United States Environmental Protection Agency: Washington, DC, USA, 2014.
22. Ministério da Ciência, Tecnologia e Inovação. *Brasil 2017 Annual Estimates of Greenhouse Gas Emissions in Brazil (Estimativas Anuais de Emissões de Gases de Efeito Estufa no Brasil)*, 2nd ed.; Ministério da Ciência, Tecnologia e Inovação: Brasília, Brazil, 2017. (In Portuguese)
23. Chernicharo, C.A.L.; Brandt, E.M.F.; Bressani-Ribeiro, T.; Melo, V.R.; Bianchetti, F.J.; Mota Filho, C.; McAdam, E. Development of a tool for improving the management of gaseous emissions in UASB-based sewage treatment plants. *Water Pract. Technol.* **2017**, *12*, 917–926. [CrossRef]
24. Bressani-Ribeiro, T.; Mota Filho, C.R.; Melo, V.R.; Bianchetti, F.J.; Chernicharo, C.A.L. Planning for achieving low carbon and integrated resources recovery from sewage treatment plants in Minas Gerais, Brazil. *J. Environ. Manag.* **2018**, *242*, 465–473. [CrossRef] [PubMed]
25. Garfí, M.; Martí-Herrero, J.; Garwood, A.; Ferrer, I. Household anaerobic digesters for biogas production in Latin America: A review. *Renew. Sustain. Energy Rev.* **2016**, *60*, 599–614. [CrossRef]
26. Chen, Y.; Yang, G.; Sweeney, S.; Feng, Y. Household biogas use in rural China: A study of opportunities and constraints. *Renew. Sustain. Energy Rev.* **2010**, *14*, 545–549. [CrossRef]
27. Von Sperling, M.; Almeida, P.G.S.; Bressani-Ribeiro, T.; Chernicharo, C.A.L. Post-treatment of anaerobic effluents. In *Anaerobic Reactors for Sewage Treatment: Design, Construction and Operation*; Chernicharo, C.A.L., Bressani-Ribeiro, T., Eds.; IWA Publishing: London, UK, 2019; pp. 285–348.

Article

Effects of Increasing Nitrogen Content on Process Stability and Reactor Performance in Anaerobic Digestion

Ievgeniia Morozova *, Nadiia Nikulina, Hans Oechsner, Johannes Krümpel and Andreas Lemmer

State Institute of Agricultural Engineering and Bioenergy, University of Hohenheim, 70599 Stuttgart, Germany; nadiia.nikulina@uni-hohenheim.de (N.N.); Hans.Oechsner@uni-hohenheim.de (H.O.); j.kruempel@uni-hohenheim.de (J.K.); andreas.lemmer@uni-hohenheim.de (A.L.)

* Correspondence: ievgeniia.morozova@uni-hohenheim.de; Tel.: +49 711 459 23348; Fax: +49-(0)711-459-22111

Received: 1 December 2019; Accepted: 26 February 2020; Published: 3 March 2020 

Abstract: The aim of this study was to analyse the effect of different nitrogen increase rates in feedstock on the process stability and conversion efficiency in anaerobic digestion (AD). The research was conducted in continuously stirred tank reactors (CSTR), initially filled with two different inocula: inocula #1 with low and #2 with high nitrogen (N) concentrations. Three N feeding regimes were investigated: the "0-increase" feeding regime with a constant N amount in feeding and the regimes "0.25-increase" and "0.5-increase" where the N concentrations in feedstock were raised by 0.25 and 0.5 $g \cdot kg^{-1}$, respectively, related to fresh matter (FM) every second week. The N concentration inside the reactors increased according to the feeding regimes. The levels of inhibition (Inhibition) in specific methane yields (*SMY*), related to the conversion efficiency of the substrates, were quantified. At the N concentration in digestate of 10.82 ± 0.52 $g \cdot kg^{-1}$ FM measured in the reactors with inoculum #2 and "0.5-increase" feeding regime, the level of inhibition was equal to 38.99% ± 14.99%. The results show that high nitrogen increase rates in feeding regime are negatively related to the efficiency of the AD process, even if low volatile fatty acid (VFA) concentrations indicate a stable process.

Keywords: biogas; methane; ammonia; inhibition; acclimatization; trace elements

1. Introduction

Utilization of protein-rich substrates, such as kitchen waste, poultry manure, microalgae, green legumes, oilseeds, etc. may lead to high concentrations of nitrogen (N) in the reactor during anaerobic digestion (AD) [1–4]. High concentrations of N inside the reactor negatively affect process stability and efficiency due to ammonia formation. Total ammonia nitrogen (TAN), which is generally defined as the sum of free ammonia nitrogen (FAN, NH_3-N) and ammonium nitrogen (NH_4^+-N), is formed during the hydrolysis of proteins, urea and nucleic acids [5–8]. Ammonia freely passes through the cell membranes of methanogens and causes a proton imbalance [5,8,9]. Free ammonia changes the intracellular pH of methanogenic bacteria and inhibits specific enzymatic reactions [10]. Therefore, high concentrations of ammonia in anaerobic reactors lead to inhibition of methanogenesis and may cause complete failure of AD [6,11,12]. As reported by Chen et al. [13], temperature change has a direct impact on both microbial growth rates and free ammonia concentration: increased process temperature affects the metabolic rate of the microorganisms in a positive way; however, it also results in higher ammonia levels.

The chemical balance between NH_3 (free ammonia) and NH_4^+ (ammonium) is shown in Equation (1) [14,15].

$$NH_4^+ + OH^- \leftrightarrow NH_3 + H_2O \tag{1}$$

The shift of this equilibrium depends mainly on the process conditions, i.e., temperature and pH [7,9,14]. The concentration of free ammonia is positively correlated with temperature and pH [5,16].

Under high ammonia concentrations in the reactor, the acetoclastic methanogens (e.g., *Methanosarcina*, *Methanosaeta* spp.) are unable to degrade acetate, which results in its accumulation, depletion of buffer capacity and a subsequent drop in pH [16–19].

According to the literature [9,10,12,13,20], inhibition of the AD process by ammonia is indicated by the decrease in the specific methane yields along with the increase in volatile fatty acid (VFA) concentrations and a pH drop due to inhibition of bacterial growth. However, the limiting concentrations of TAN and FAN for maintaining AD without inhibition are subject to discussion (Table 1). In addition, there is a controversy whether TAN or FAN mainly inhibits methanogenesis [20].

Most authors in previous studies tend to agree that TAN $\geq$ 3.00 $g{\cdot}L^{-1}$ and FAN $\geq$ 0.20 $g{\cdot}L^{-1}$ have an inhibitory effect on AD (see Table 1). According to Table 1, very few studies have measured the level of inhibition in methane production when treating N-rich substrates.

For maintaining stable and efficient biogas production under high and/or increasing TAN and FAN concentrations, acclimatization strategies can be applied. A frequently used approach is to feed the reactor with a specific N- or ammonia-increase rate. However, no information on the maximum increase rates is available [1,4,12,21–24].

High nitrogen concentrations in the digestate are generally the result of a narrow carbon-to-nitrogen (C/N) ratio of the feedstock [9,12,25,26]. To reduce the concentrations of TAN and FAN in the digestate and thus to maximise biogas and methane yields, Shanmugam and Horan [26] recommend keeping the C/N ratio of the feedstock in the range of 15 to 20, while according to Kayhanian [9], this ratio should be between 27 and 32.

Currently, many operators of biogas plants suffer from AD inhibition and methane losses when utilizing N-rich substrates. The application of the acclimatization strategy with an optimal N-increase rate could stabilize AD and prevent or minimize methane losses. In this study, natural N-sources and microbial communities from full-scale biogas digesters were utilized. The research was conducted in continuously operated reactors under conditions similar to those in full-scale biogas plants. The aim of this investigation is to determine the effect of different nitrogen increase rates on anaerobic digestion in order to achieve an optimal process performance. The nitrogen increase in the feedstock was carried out every two weeks at rates of 0.25 and 0.5 $g{\cdot}kg^{-1}$ related to fresh matter. The N content of the digestate rose continuously in response to the two different increase rates in the feedstock. At the same time, the C/N ratio of the feedstock consistently decreased throughout the experiment. Thus, the influence of high nitrogen content on the stability of the fermentation process with regards to the C/N ratio could be investigated. By comparing the values of specific methane yield (*SMY*) obtained from the continuous experiment with those obtained from the batch experiment, the effect of increasing N content in the feedstock on the conversion efficiency of the substrates was studied.

Table 1. Limiting total ammonia nitrogen (TAN)- and free ammonia nitrogen (FAN)-concentrations ($g \cdot L^{-1}$) for maintaining stable anaerobic digestion (AD) under different temperature conditions.

Mesophilic conditions											
	TAN	**Treated Substrate**	**Operating Temperature**	**Inhibition in CH_4 Production**	**Reference**		**FAN**	**Treated Substrate**	**Operating Temperature**	**Inhibition in CH_4 Production**	**Reference**
≤	1.00	Mashed biowaste, residual food waste	Not indicated	Not indicated	[8]	≤	0.03	Mashed biowaste, residual food waste, steers manure	Not indicated, 35 °C	Not indicated	[8,21]
≤	2.00	Food waste	37 °C	Not indicated	[17]	≤	0.49 (b)	Animal manure, food waste	37 ± 1 °C	Not indicated	[22]
≤	2.40	Chicken manure, spent poppy straw	36 ± 1 °C	Not indicated	[27]	≤	1.10	Thin stillage	38 °C	Not indicated	[28]
≤	3.00	Municipal wastewater biosolids	36 ± 1 °C	Not indicated	[29]	≤	1.20	Pig slurry, maize silage, other agricultural wastes	38.0 ± 0.5 °C	Not indicated	[30]
≤	3.20 (b)	Municipal wastewater	~22 °C	Not indicated	[23]						
≤	3.50	Municipal wastewater biosolids	37 ± 1 °C	Not indicated	[1]						
≤	4.56 (b)	Jatropha press cake	37 °C	Not indicated	[31]						
≤	5.00	Animal/poultry manure, organic waste, municipal wastewater	30–38 °C	50% inhibition at TAN of 3.0 $g \cdot L^{-1}$	[11,32–34]						
≤	6.00	Pig slurry, maize silage, other agricultural wastes	38.0 ± 0.5 °C	Not indicated	[30]						
≤	7.00	Chicken manure, maize silage	37–41 °C	10–20% at TAN ≥ 7.0 $g \cdot L^{-1}$, FAN ~ 600 $mg \cdot L^{-1}$; 50% at TAN ≥ 8.8 $g \cdot L^{-1}$	[35]						

Table 1. *Cont.*

	TAN	Treated Substrate	Operating Temperature	Inhibition in CH_4 Production	Reference		FAN	Treated Substrate	Operating Temperature	Inhibition in CH_4 Production	Reference
≤	10.00 [(b)]	Animal waste, food waste	37 ± 1 °C	Not indicated	[4]						
≤	11.80 [(b)]	Beet-sugar factory wastewater	30 ± 1 °C	Not indicated	[24]						
Thermophilic Conditions											
	TAN	**Treated Substrate**	**Operating Temperature**	**Inhibition in CH_4 Production**	**Reference**		**FAN**	**Treated Substrate**	**Operating Temperature**	**Inhibition in CH_4 Production**	**Reference**
≤	1.80–2.40	Dairy manure	55 °C	Not indicated	[36]	≤	0.20 [(a)]	Steer manure	55 °C	Not indicated	[21]
≤	4.32 [(b)]	Animal manure, food industrial organic waste	53 ± 1 °C	Not indicated	[22]	≤	0.39 [(b)]	Steer manure	55 °C	Not indicated	[21]
						≤	0.85	Municipal wastewater biosolids	55 °C	Not indicated	[1]
						≤	1.20	Cattle manure	53–55 °C	Not indicated	[37]
						≤	1.43 [(b)]	Animal manure, food industrial organic waste	53 ± 1 °C	Not indicated	[22]

if stated: [(a)] unacclimatised, [(b)] acclimatized.

2. Materials and Methods

2.1. Reactor Design

The experiment was conducted in 12 horizontal, stainless steel, continuously stirred tank reactors (CSTR) of 20 L total volume (working volume 17 L) each, as described in [38], in duplicate repetition according to the Guideline 4630 issued by the Association of German Engineers (VDI) [39]. Different N-increase rates in the CSTR were achieved by different feeding regimes described in Section 2.2. During the experimental period, the organic loading rate related to volatile solids (OLR_{VS}) was kept at 3 $kg{\cdot}m^{-3}{\cdot}d^{-1}$ with a hydraulic retention time (HRT) of 40 days. The temperature in each reactor was mesophilic at 37 ± 1 °C.

2.2. Inocula and N-Increase in Feeding Regimes

Each digester was filled with 17 L of inoculum at the beginning of the experiment. Inoculum #1 and inoculum #2 from two full-scale biogas plants were used in this trial. These inocula differed in total Kjeldahl nitrogen (TKN) concentrations, with inoculum #2 containing twice as much nitrogen as inoculum #1 (Table 2). Inoculum #1 was taken from a digester treating cattle manure (35–40%), maize silage (40%), grain whole plant silage (5%) and triticale (rest). Inoculum #2 was taken from a digester treating turkey manure (10%), cattle manure (8%), cereals (10%) and maize silage (62%).

Table 2. Characteristics of the substrates. Gas volumes are given under standard temperature and pressure conditions (0 °C, 101.325 kPa). Units are given in square brackets. Values are given as mean; the standard deviation is given in round brackets.

Parameter	Inoculum		Maize Silage	Soybean Meal
	#1	#2		
DM_{FM} [(a)] [$g{\cdot}kg^{-1}$]	59.80 (2.99)	103.75 (5.18)	377.61 (18.88)	887.99 (18.46)
VS_{DM} [(b)] [$g{\cdot}kg^{-1}$]	738.74 (36.93)	789.74 (39.48)	907.38 (45.37)	927.19 (4.06)
TKN_{FM} [(c)] [$g{\cdot}kg^{-1}$]	3.34 (0.70)	7.14 (0.36)	4.00 (0.20)	67.83 (3.39)
${NH_4}^+{}_{FM}$ [(d)] [$g{\cdot}kg^{-1}$]	1.35 (0.07)	5.00 (0.25)	0.60 (0.03)	9.50 (0.48)
pH	7.44 (0.37)	8.42 (0.42)	NA	NA
SMY_{VS} [e)] [$L{\cdot}kg^{-1}$]	25.78 (1.30)	88.43 (4.42)	330.66 (15.08)	423.16 (21.11)

[(a)] Dry matter (DM) related to fresh matter (FM), [(b)] volatile solids (VS) related to DM, [(c)] total Kjeldahl nitrogen (TKN) related to FM, [(d)] ammonium related to FM, [e)] specific methane yield (*SMY*) related to VS.

The substrates were fed into the reactor daily are described in [38]. The daily feedstock consisted of fresh inoculum, maize silage (low nitrogen content) and soybean meal (N-rich substrate). Tap water was added in order to keep the HRT and OLR_{VS} constant, thus resulting in 425 g of fresh matter daily feedstock. Characteristics of the substrates are described in Table 2. The values of the specific methane yield for the feeding substrates were determined by the Hohenheim biogas yield test [40,41].

For each inoculum, the different feeding regimes were separately analysed, as shown in Figure 1. These feeding regimes represent the rate of N increase in the feeding ratio. The investigated feeding regimes were "0-increase", "0.25-increase" and "0.5-increase". Under the "0-increase" feeding regime, the nitrogen content did not change over the whole course of the experiment. For the other two regimes, there was an increase in nitrogen content in the feedstock (see Figure 1a). The increase in nitrogen concentration was achieved by adding soybean meal and simultaneously decreasing the share of maize silage. In the feeding regimes "0.25-increase" and "0.5-increase", the share of soybean meal was increased stepwise, thus leading to N-increase rates of 0.25 and 0.5 $g{\cdot}kg^{-1}$ FM every two weeks, respectively. By contrast, the C/N ratio in feedstock was decreasing as shown in Figure 1b.

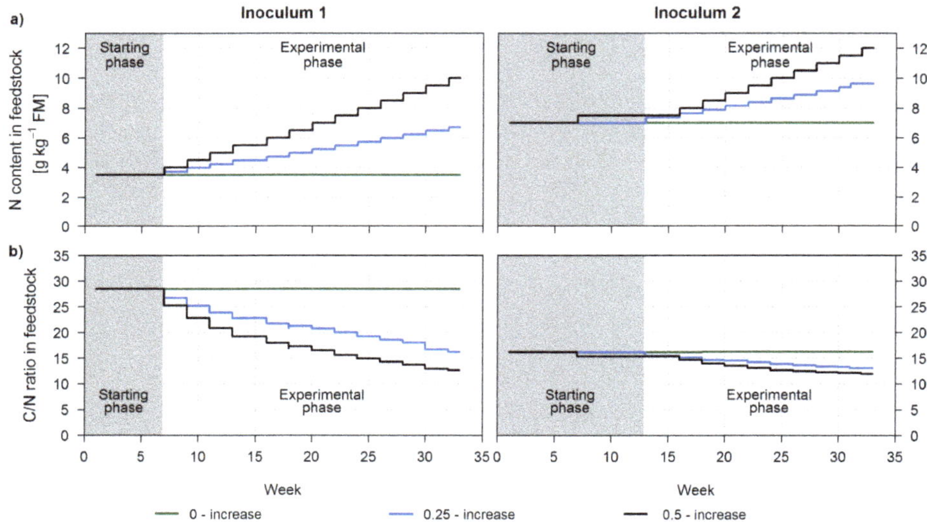

Figure 1. Experimental procedure: (**a**) N content in feedstock; (**b**) C/N ratio in feedstock. The results are given separately for inoculum #1 and inoculum #2. Different line colours in the graphs and in the legend correspond to the N-increase rates "0-increase", "0.25-increase" and "0.5-increase" in the feeding regimes. Grey and white backgrounds in the graphs are related to the starting phase and the experimental phase, respectively.

2.3. Trace Elements Supplementation

For AD process stability, the importance of micronutrients, i.e., iron, nickel, molybdenum, cobalt and selenium is described in literature [27,42–45]. After observing process instability for the reactors with inoculum #1 at the beginning of week 17, micronutrient levels were tested. In response to the identified deficiency in trace elements (TE) in the reactors with inoculum #1 and for keeping the TE in the range as recommended by Vintiloiu et al. [42], 1.23 g of BC.Pro Akut® was added to all the reactors (with #1 and #2) weekly, starting from the end of week 17 up to the end of the experimental trials. BC.Pro Akut® is a mixture of TE and other components comprising the following active substances in the ionic form: aluminium, boron, calcium, iron, cobalt, copper, magnesium, manganese, molybdenum, sodium, nickel, selenium, tungsten and zinc.

2.4. Analytical Methods

The produced biogas was collected in gas bags as described by Haag et al. [38]. A gas measuring unit automatically analysed the gas quantity (Hoentzsch FA MS40, Waiblingen, Germany), as well as the content of CH_4 and CO_2 (AGM 10, Sensors Europe, Erkrath, Germany). The measurements were carried out once per day before feeding.

Samples were taken from the reactors weekly. The dry matter content related to fresh matter (DM_{FM}) and volatile solids content related to dry matter (VS_{DM}) of the collected samples were determined by differential weighing before and after drying at 105 °C for 24 h and by subsequent ashing at 550 °C for 8 h, respectively. The pH was measured in each reactor three times per week with a WTW 323, using a SenTix 41 pH-electrode (WTW, Weilheim, Germany). Concentrations of VFA in the samples were determined by gas chromatography. The gas chromatograph Shimadzu GC-2010plus (Tokyo, Japan) was equipped with a FFAP 50 m × 0.32 mm column with a chemically bonded polyethylene glycol CP-Wax 58 FFAP CB 1.2 µm film, a flame ionization detector and helium as a carrier gas. Total ammonium concentrations in the digestate were determined by the automatic distillation system Gerhardt Vapodest 50s (Koenigswinter, Germany). Total Kjeldahl nitrogen (TKN) is expressed as total nitrogen or N if not stated otherwise. The total nitrogen in the samples was determined by Kjeldahl analysis. The potassium determination was done by means of flame atomic absorption spectroscopy (AAS, Eppendorf, ELEX 6361, Wesseling-Berzdorf, Germany), operated with an acetylene gas. For the determination of phosphorus, a cuvette test [46] and a spectrophotometer UV–VIS 1240 (Shimadzu, Tokyo, Japan) were used. All the analyses were carried out according to standard methods [46]. The analysis on trace element content in the samples was done by an external laboratory in accordance with standard methods [47–49].

2.5. Calculation of FAN and TAN

The NH_3 (free ammonia) concentration was calculated by using the equation described in [14].

$$NH_3 = K_{NH_4} \cdot \frac{NH_4^+}{H^+} \tag{2}$$

where NH_4^+ is the ammonium concentration in g·kg^{-1} related to FM; K_{NH_4} is the ionization constant of ammonium (for 37 °C, $K_{NH_4} = 1.14 \cdot 10^{-9}$ [21]); H^+ is the hydrogen ion concentration ($H^+ = 10^{-pH}$ [14]). NH_3 was recalculated to NH_3-N (FAN), and NH_4^+ was recalculated to NH_4^+-N (ammonium nitrogen) according to their molar masses. The concentration of TAN was calculated as the sum of FAN (NH_3-N) and NH_4^+-N.

2.6. Statistical Analysis

For data processing and visualization, Microsoft EXCEL 2016, SAS 9.4, R and RStudio (version 1.1.463) were used.

2.6.1. Inhibition in SMY

The inhibition in specific methane yields (Inhibition) for different N increase-rates in feeding regimes is defined by Equation (3):

$$\text{Inhibition} = \frac{1}{n} \sum_{i=1}^{n} \frac{SMY_t - SMY_m}{SMY_t} \cdot 100\% \tag{3}$$

where n is the number of observations over the experimental period taken for the analysis. The theoretical methane yields (SMY_t) were calculated based on the amounts of VS_{DM} added to the reactors and the SMY_{VS} of the substrates obtained by the Hohenheim biogas yield test (Table 2). The measured SMY (SMY_m) was based on the measured value of methane yield divided by the amount of VS added to the reactor. This inhibition can also be described as the conversion efficiency between the theoretical SMY_t values obtained from the batch experiment and the measured SMY_m values obtained from the continuous experiment.

The one-sided Tukey test was applied to identify whether the difference between the SMY_t and SMY_m was statistically significant. The analysis was done in Excel and Rstudio.

2.6.2. Analysis of the effect of TAN and FAN on inhibition

Based on the experimental data for the three investigated feeding regimes along with inocula #1 and #2, the effects of TAN and FAN concentrations in the reactor on the level of inhibition were analysed. For this purpose, mixed modelling for repeated measurements was applied [50]. This model was selected for serial correlation among observations on the same experimental unit. The experimental unit, in our case, was the reactor [50]. Analyses were based on the experimental data starting from week 17 of the trials after the TE supplementation was started. The applied data were checked by using the normality test on the studentized residuals. For meeting the requirements of the mixed model, the square-root transformation of the data on inhibition in *SMY* (sqrt_Inhibition) was used. Several types of models (independent, compound symmetry, autoregressive, unstructured) were checked; on the grounds of the normally distributed residual plots and the lowest Akaike information criterion (AIC) value, the compound symmetry type was selected as the best-fitting model.

The applied model is given in Equation (4):

$$y_{itk} = \mu + \alpha_i + r_t + b_{tk} + e_{itk} \tag{4}$$

where y_{itk} is the dependent variable; i is the i-th observation, t is the weekly measurement and k is related to the interaction between the fixed factor and the point in time (t); μ describes the general effect of the model; α_i is the i-th observation of the fixed factor; r_t is the replicate of a weekly measurement; b_{tk} is the random effect of a week and the interaction between week and the fixed factor; e_{itk} is the random deviation associated with y_{itk}.

The sqrt_Inhibition was used as the dependent variable; the TAN and FAN were separately analysed as the fixed factor. The influence of time and interaction between time ("WEEK", in our case) and a fixed factor was analysed on a random effect in the model. The "MIXED" procedure of SAS was used to fit the model.

3. Results and Discussion

The reactors were continuously monitored over the whole period of the trials. The measured values for N, TAN, FAN, acetic acid (HAc), pH, SMY_m and inhibition are shown in Figure 2. The trial period included a starting phase and an experimental phase.

3.1. The Starting Phase

During the starting phase, the OLR_{VS} was increased until the aimed values were achieved.

The starting phase was needed for the microorganisms to adapt to the operating conditions. During this phase, all reactors were fed with a constant N feeding ratio equivalent to the "0-increase" variant (Figure 2a) to establish stable conditions. The stable operation was determined by monitored VFA concentrations (Figure 2d) and specific methane production (Figure 2f). In week four to six, the TE concentrations were additionally tested, which showed sufficient nutrient levels according to Vintiloiu et al. [42] (see Table 3). For inoculum #1, the starting phase lasted for 48 days. For inoculum #2, the starting phase took 90 days.

The values provided in Figure 2 for the starting phase can be relevant for farmers and biogas operators when utilizing protein-rich substrates in biogas plants. However, these values are excluded from the statistical analysis described in Section 3.3.

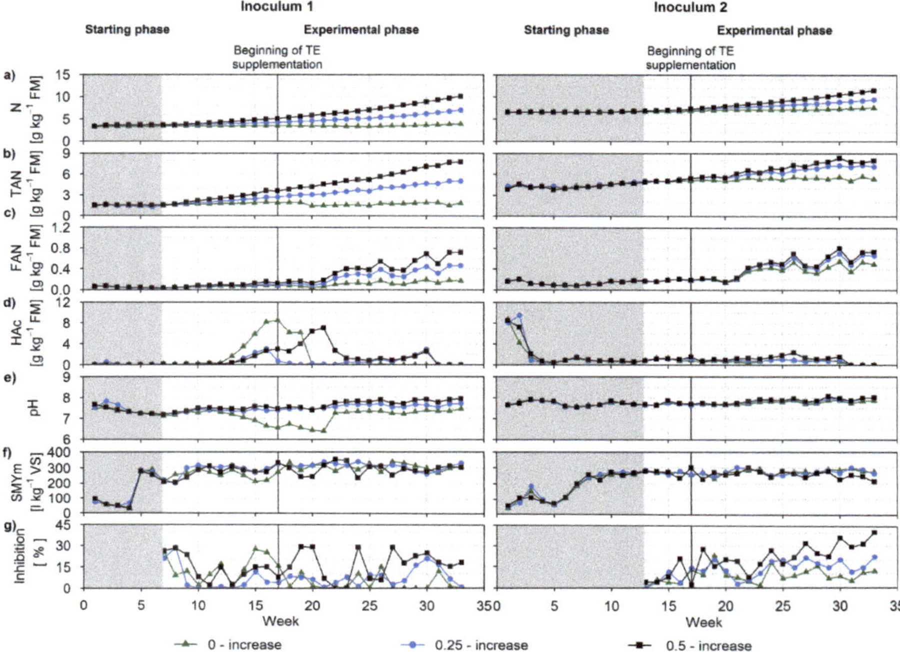

Figure 2. Measured values of the following parameters in the continuously stirred tank reactors (CSTR) under different N-increase rates in feeding regimes: (**a**) total nitrogen (N); (**b**) total ammonia nitrogen (TAN); (**c**) free ammonia nitrogen (FAN); (**d**) acetic acid (HAc); (**e**) pH; (**f**) the measured values of specific methane yield (SMY_m); (**g**) inhibition in specific methane yield (Inhibition). The results are given separately for inoculum #1 and inoculum #2. Different line colours along with different marks in the graphs and in the legend correspond to the N-increase rates "0-increase", "0.25-increase" and "0.5-increase" in the feeding regimes. Grey and white backgrounds in the graphs are related to the starting phase and the experimental phase, respectively. The vertical line in the graphs corresponds to the beginning of regular weekly trace elements (TE) supplementation.

Table 3. Concentrations of trace elements in the reactors over the trial period, in mg·kg^{-1} related to dry matter. Feeding regime expresses the N-increase rate in a feeding ratio. Values are given as mean; standard deviation is given in brackets.

Inoculum	Feeding	Week				
	Regime	4	5	6	17	24
			Fe			
1	0-increase	1827.94 (91.40)	1625.38 (81.27)	1463.40 (73.17)	936.00 (77.78)	2105 (106.07)
1	0.25-increase	1515.01 (75.75)	1512.09 (75.60)	1449.05 (72.45)	1120.00 (367.70)	1810.00 (84.85)
1	0.5-increase	1625.33 (81.27)	1368.86 (68.44)	1402.16 (70.11)	837.50 (74.25)	1835.00 (134.35)
2	0-increase	3055.97 (102.47)	2986.89 (246.08)	2779.68 (138.98)	2835.00 (473.76)	3795.00 (700.04)
2	0.25-increase	NA	NA	NA	2840.00 (339.41)	3540.00 (650.54)
2	0.5-increase	3116.76 (155.84)	2846.76 (142.34)	2705.63 (135.28)	2910.00 (14.14)	3380.00 (183.85)
			Ni			
1	0-increase	12.06 (0.60)	11.70 (0.59)	10.99 (0.55)	5.33 (2.26)	21.25 (0.21)
1	0.25-increase	6.81 (0.34)	6.92 (0.35)	6.57 (0.33)	15.79 (14.02)	21.15 (8.41)
1	0.5-increase	7.72 (0.39)	6.29 (0.31)	6.76 (0.34)	8.17 (3.16)	21.20 (1.27)
2	0-increase	8.14 (0.29)	8.62 (0.06)	8.45 (0.42)	9.22 (1.05)	21.10 (0.28)
2	0.25-increase	NA	NA	NA	9.97 (0.18)	20.80 (0.85)
2	0.5-increase	13.82 (0.69)	13.81 (0.69)	13.1 (0.66)	12.85 (0.49)	20.45 (3.18)
			Mo			
1	0-increase	4.91 (0.25)	4.83 (0.24)	4.83 (0.24)	2.85 (0.08)	5.02 (0.15)
1	0.25-increase	4.32 (0.22)	4.66 (0.23)	4.62 (0.23)	4.39 (0.92)	6.49 (0.59)
1	0.5-increase	4.93 (0.25)	4.39 (0.22)	4.84 (0.24)	4.06 (0.33)	7.145 (0.49)
2	0-increase	5.18 (0.39)	5.57 (0.34)	5.59 (0.28)	5.22 (0.66)	7.32 (1.01)
2	0.25-increase	NA	NA	NA	5.29 (0.25)	7.61 (0.95)
2	0.5-increase	5.70 (0.28)	5.77 (0.29)	5.78 (0.29)	5.67 (0.40)	7.56 (0.35)
			Co			
1	0-increase	1.14 (0.06)	1.17 (0.06)	1.12 (0.06)	0.51 (0.02)	1.90 (0.08)
1	0.25-increase	0.98 (0.05)	1.07 (0.05)	1.03 (0.05)	0.76 (0.31)	1.71 (0.15)
1	0.5-increase	1.05 (0.05)	0.95 (0.05)	1.04 (0.05)	0.65 (0.09)	1.83 (0.15)
2	0-increase	1.23 (0.10)	1.28 (0.01)	1.25 (0.06)	0.91 (0.11)	2.11 (0.25)
2	0.25-increase	NA	NA	NA	0.94 (0.02)	1.90 (0.28)
2	0.5-increase	1.36 (0.07)	1.38 (0.07)	1.35 (0.07)	0.99 (0.01)	1.79 (0.06)

Table 3. *Cont.*

Inoculum	Feeding	Week				
	Regime	4	5	6	17	24
			Se			
1	0-increase	0.51 (0.03)	0.49 (0.02)	0.73 (0.04)	0.34 (0.04)	1.60 (0.14)
1	0.25-increase	0.62 (0.03)	0.42 (0.02)	0.64 (0.03)	0.41 (0.05)	1.85 (0.07)
1	0.5-increase	0.61 (0.03)	0.50 (0.03)	0.68 (0.03)	0.40 (0.03)	2.00 (0.28)
2	0-increase	1.19 (0.18)	1.22 (0.07)	1.12 (0.06)	0.85 (0.13)	2.20 (0.42)
2	0.25-increase	NA	NA	NA	0.85 (0.05)	2.10 (0.28)
2	0.5-increase	1.25 (0.06)	1.12 (0.06)	1.18 (0.06)	0.88 (0.02)	1.90 (0.00)

3.2. The Experimental Period

After the starting phase, the reactors were continuously operated and monitored for 26 and 20 weeks for inoculum #1 and #2, respectively.

The lack of TE in the reactors with inoculum #1, which resulted in the accumulation of acetic acid up to 8.53 g·kg^{-1} FM, along with a drop in pH up to 6.40 (as described in [12,42]), was identified at the beginning of week 17 of the trials (see Figure 2d,e and Table 3). The weekly supplementation of the CSTR with TE was established thereafter in order to compensate for the deficiency in TE in the reactors with inoculum #1 and to ensure a sufficient TE supply for the remainder of the experiment. The vertical line shown at week 17 in Figure 2 marks the beginning of weekly TE supplementation. The positive effect of TE to AD process stability can be seen in Figure 2d,e in the stabilization of pH and HAc in the weeks following supplementation. The analysis of TE measured in week 24 showed that the amounts of these nutrients in the reactors were well-balanced (see Table 3).

Additionally, the total phosphorus and potassium concentrations inside the reactors were analysed. The availability of these nutrients may be of great interest when using digestate as a fertilizer. The concentrations of these macro elements within the research period were the following: for inoculum #1, P = 0.62 ± 0.13 g·kg^{-1} FM, K = 3.20 ± 0.35 g·kg^{-1} FM; for inoculum #2, P = 0.99 ± 0.09 g·kg^{-1} FM, K = 3.73 ± 1.11 g·kg^{-1} FM.

Over the experimental period, the concentration of N in the digestate was accumulating, as shown in Figure 2a. The accumulation of N in the reactors was related to the analysed N-increase rates. The average N-increase rate in the daily feedstock under the "0.5-increase" feeding regime was 35.7 mg·kg^{-1}·d^{-1} related to the fresh matter of the input substrates. At the end of the experiment, the highest values of total nitrogen in the digestate were 10.09 ± 0.08 g·kg^{-1} FM and 11.49 ± 0.01 g·kg^{-1} FM for the reactors with inoculum #1 and #2, respectively. Accordingly, a maximum "nitrogen loading rate" (NLR) can be given; the NLR was equal to 0.25 g·L^{-1}·d^{-1} for the reactors with inoculum #1 and 0.30 g·L^{-1}·d^{-1} for those with inoculum #2.

Concurrently, the TAN and FAN concentrations in the digestate increased, as shown in Figure 2b,c. At the end of the experiment, the highest values of TAN were 7.72 ± 0.33 g·kg^{-1} FM (for the reactors with inoculum #1) and 7.95 ± 1.08 g·kg^{-1} FM (for the reactors with inoculum #2). The highest FAN concentration in the final samples was 0.72 ± 0.03 g·kg^{-1} FM and 0.74 ± 0.12 g·kg^{-1} FM for the reactors with inoculum #1 and #2, respectively.

The concentration of HAc in the reactors over the period of the trials is shown in Figure 2d. The average concentrations of acetic and propionic acids in the CSTR during the experimental period were 0.88 ± 0.46 g·kg^{-1} FM and 0.17 ± 0.32 g·kg^{-1} FM, respectively, independent of the inoculum. In the reactors with #1, acetate accumulation caused by TE deficiency decreased to a minimum after the start of TE supplementation, with no acetate found in weeks 31–33. In the reactors with inoculum #2, acetate remained at a stable low concentration over the entire experimental phase with zero-values at the end of the trials. The concentrations of other VFA, i.e., iso-butyric, n-butyric, iso-valeric, n-valeric and caproic acids were low over the research period; the concentration of these acids was 0.04 ± 0.16 g·kg^{-1} FM for both inocula.

The pH-values during the experimental phase were slightly higher than those in the starting phase. Over the entire experimental period, the pH levels in the CSTR were stable, except for the reactors with inoculum #1 under the TE deficiency with a drop in pH up to 6.40 (see Figure 2e). The average pH was 7.45 ± 0.21 for the experiments based on inoculum #1 and 7.77 ± 0.11 for those based on inoculum #2.

The values of SMY_m during the experimental phase are given in Figure 2f. The mean SMY_m was 289.93 ± 35.13 L·kg^{-1} VS and 267.20 ± 19.86 L·kg^{-1} VS for the reactors with #1 and #2, respectively.

The values of inhibition during the experimental phase are given in Figure 2g. At the end of the experiment, the values of inhibition for inoculum #1 were 0.57% ± 1.22% (in weeks 32–33), 18.02% ± 22.64% (in weeks 32–33) and 26.96% ± 22.88% (in weeks 29–33) for the "0-increase", "0.25-increase" and "0.5-increase" variants, respectively. At the final phase of the experiment (in weeks 29–33) the values of inhibition for inoculum #2 were 10.91% ± 4.58%, 19.38% ± 8.93%, 38.99% ± 14.99% for the "0-increase", "0.25-increase" and "0.5-increase" variants, respectively. The 38.99% ± 14.99% inhibition determined in the reactors with #2 and "0.5-increase" feeding regime was related to N, TAN and FAN concentrations of 10.82 ± 0.52 g·kg^{-1} FM, 7.92 ± 0.27 g·kg^{-1} FM and 0.69 ± 0.10 g·kg^{-1} FM, respectively. As seen in Figure 2g, inhibition levels in the reactors with both inocula appear to have reached higher levels at the higher N increase rate.

3.3. *Results of Statistical Analysis*

3.3.1. Results of analysis on inhibition in *SMY*

The results of the analysis on inhibition in *SMY* over the experimental phase are given in Table 4 and are shown in Figure 3. According to the results of the Tukey test, for all the analysed feeding regimes the difference between the SMY_t and SMY_m was statistically significant. The large variation in the SMY_m for the reactors with inoculum #1 and the "0-increase" variant can be explained by the instability of the AD process under the TE deficiency. The highest inhibition was determined in the reactors with inoculum #2 and the "0.5-increase" variant.

Table 4. The results of analysis on inhibition in specific methane yield (*SMY*). Feeding regime expresses the N-increase rate in a feeding ratio. Degrees of freedom (DF). SMY_t and SMY_m are the theoretical and measured values of specific methane yield, respectively. Gas volumes are given under standard temperature and pressure conditions (0 °C, 101.325 kPa). Units are given in square brackets. Values of SMY_t and SMY_m are given as mean; the standard deviation is given in round brackets.

Inoculum	Feeding Regime	DF	SMY_t (L·kg$_{VS}^{-1}$)	SMY_m (L·kg$_{VS}^{-1}$)	t-Value	*p*-Value
1	0-increase	148	304.65 (11.80)	298.68 (4.44)	1.70	0.05
1	0.25-increase	182	323.71 (18.17)	302.41 (51.70)	5.42	0.00 *
1	0.5-increase	182	333.36 (21.78)	289.81 (55.09)	9.53	0.00 *
2	0-increase	141	296.65 (10.57)	264.50 (55.87)	6.55	0.00 *
2	0.25-increase	141	313.72 (16.48)	269.99 (46.99)	10.29	0.00 *
2	0.5-increase	141	325.72 (22.40)	257.87 (60.36)	11.62	0.00 *

* Significant at *p*-value = 0.0001.

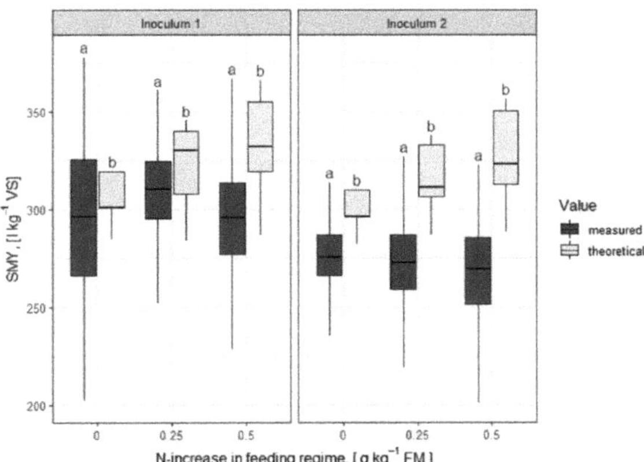

Figure 3. The results of analysis on inhibition in specific methane yield (*SMY*). The results are given separately for inoculum #1 and inoculum #2. Tick marks "0", "0.25" and "0.5" on the x-axis correspond to the N-increase variants of "0-increase", "0.25-increase" and "0.5-increase" in feeding regimes. The "measured" value is the measured *SMY* (SMY_m); the "theoretical" value is the theoretical *SMY* (SMY_t). Letters "a" and "b" denote the significant differences between the SMY_m and SMY_t for the same variant of N-increase according to the results of the one-sided Tukey test.

According to the results of the analysis, the N-increase rate in feeding regime had a negative effect on the AD process efficiency.

3.3.2. Results of analysis of the effect of TAN and FAN on inhibition

The results of the fitted model were the following: The increase in TAN levels resulted in an increase of inhibition in *SMY*, *p*-value = 0.0001 (Table 5 and Figure 4). The increase in FAN concentration in the AD reactor resulted in an increase of the inhibition level, *p*-value = 0.0012 (Table 5 and Figure 5). The observed noise in Figures 4 and 5 can be associated with the fact that the inhibition does not derive only from TAN or FAN concentrations inside the reactors; this inhibition can be also affected by other parameters.

Table 5. The effect of total ammonia nitrogen (TAN) and free ammonia nitrogen (FAN) on the inhibition in specific methane yield: the results of the fitted model. Degrees of freedom (DF). The square-root transformed values of inhibition in specific methane yield (sqrt_Inhibition); the transformation was done for meeting the requirements of the model.

Dependent Variable	Effect	Numerator DF	Denominator DF	F-Value	R^2	*p*-Value
sqrt_Inhibition	TAN	1	30.7	19.08	0.20	0.0001
sqrt_Inhibition	FAN	1	16.5	15.11	0.15	0.0012

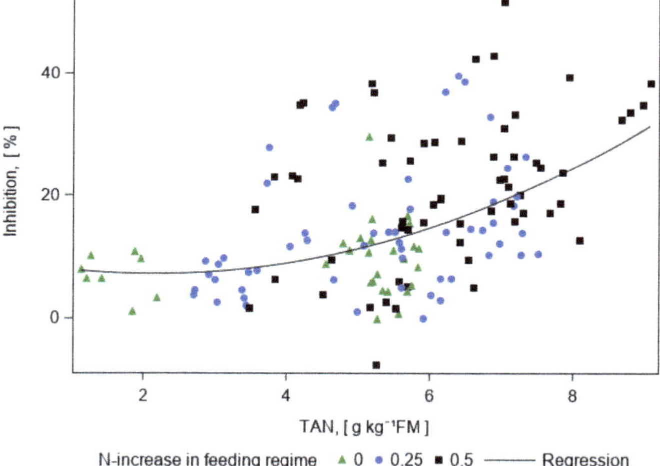

Figure 4. The correlation between the total ammonia nitrogen (TAN) and the inhibition in specific methane yield (Inhibition). The "0", "0.25" and "0.5" marks in the legend correspond to the N-increase variants of "0-increase", "0.25-increase" and "0.5-increase" in the feeding regimes. The regression line was built based on the results obtained from the model.

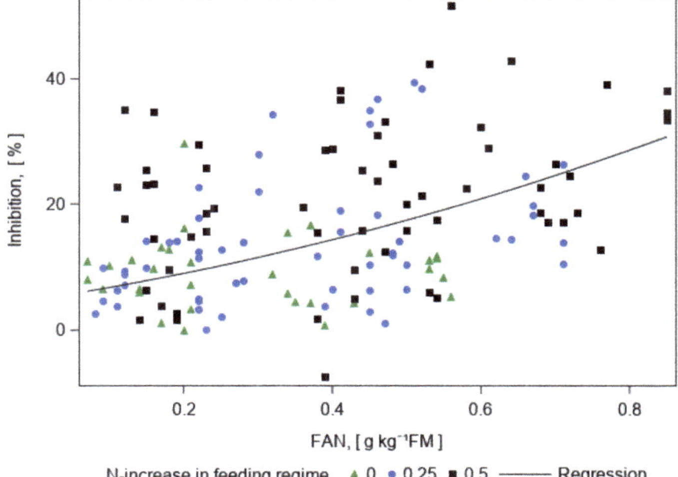

Figure 5. The correlation between the free ammonia nitrogen (FAN) and the inhibition in specific methane yield (Inhibition). The "0", "0.25" and "0.5" marks in the legend correspond to the N-increase variants of "0-increase", "0.25-increase" and "0.5-increase" in the feeding regimes. The regression line was built based on the results obtained from the model.

The results of the data analysis show that the analysed N-increase rates can be recommended for a stable AD process. However, the level of inhibition in *SMY* depends on the concentration of TAN and FAN inside the reactors and the N-increase rate in the feeding regimes (see Figures 4 and 5).

3.4. Discussion

The inhibitory effect of urea, NH_4Cl, TAN, FAN and high N concentration in feeding, as well as the effect of elevated ammonium (NH_4^+), elevated ammonium nitrogen (NH_4^+-N) and elevated TAN on biogas and methane yields, have been previously studied [1,4,15,17,22,32,51–53]. However, no results on the effects of N concentration in inoculum and N-increase rate in feedstock on the level of inhibition in specific methane yield were found.

Contrary to the research results reported by Siegrist et al. [18], Chen et al. [17], Meng et al. [19] and Theuerl et al. [16], the stable AD-process was found in this study as indicated by stable pH values and a minimal accumulation of acetate (except for the reactors with #1 under the TE deficiency) (see Figure 2e,d). During the experimental phase, the specific methane yields were kept stable in all the reactors, and their values were in a normal range (see Figure 2f). Based on the results obtained it can be assumed that the analysed feeding regimes enabled the microorganisms to adapt to changing N-conditions, which is indicated by a stable AD process. The regular supplementation of reactors with TE positively contributed to the process stability. The proposed increase rates did not have any negative effect on the process stability. Hereby the N-increase variants of "0.25-increase", "0.5-increase" and the NLR up to 0.30 $g{\cdot}L^{-1}{\cdot}d^{-1}$ can be recommended for maintaining a biogas plant in a stable way.

In contrast, the efficiency of AD, which in this study corresponded to the inhibition in specific methane yield (Inhibition), was affected by N-increase rate and the level of TAN and FAN inside the reactor. The conversion process in the reactors, which in this study is described as inhibition, became more and more inefficient due to the closer C/N ratio in feedstock (see Figure 1b). Chen et al. [17] has stated that the methane production was intensely inhibited when TAN increased to 5 $g{\cdot}L^{-1}$ and they recommended to maintain the ammonium concentration below 2 $g{\cdot}L^{-1}$ in the reactors for preventing the ammonium shock to the AD process. According to the review made by Chen et al. [15], in different studies there is controversial information on the level of inhibition in methane production depending on the TAN and FAN concentrations in the AD reactor: 50% of methane inhibition was observed at TAN of 1.44 $g{\cdot}L^{-1}$, 2.48 $g{\cdot}L^{-1}$ and 5.60 $g{\cdot}L^{-1}$ and FAN of 0.03 $g{\cdot}L^{-1}$ and 0.64 $g{\cdot}L^{-1}$; 100% of methane inhibition was identified at TAN values above 5.20 $g{\cdot}L^{-1}$ and FAN of 0.20 $g{\cdot}L^{-1}$ and 0.62 $g{\cdot}L^{-1}$. Fotidis et al. [32,53] specify that at the $NH_4{}^+$-N in the range of 3–5 $g{\cdot}L^{-1}$, an ammonia induced inhibited-steady state in the AD reactors was observed with inhibition in methane production of 30–40%, and the authors recommend a bioaugmentation strategy for overcoming an ammonia inhibiting effect. However, in our research, under $NH_4{}^+$-N of 5.03 ± 0.06 $g{\cdot}kg^{-1}$ FM, our TAN and FAN concentrations in the reactors were 5.34 ± 0.15 $g{\cdot}kg^{-1}$ FM and 0.31 ± 0.14 $g{\cdot}kg^{-1}$ FM, respectively, and the value of inhibition was equal to 9.46% ± 5.60%. According to the results of the data analysis, both TAN and FAN had a significant effect on the level of inhibition. As FAN levels are mostly affected by temperature and pH fluctuations [7,9,14,21], the effect of FAN was less significant than TAN in our research, since the reactors were operated under mesophilic conditions at stable temperature and pH (except for the pH values in the reactors with inoculum #1 under the TE deficiency). As the OLR_{VS}, HRT, temperature and pH in the reactors were kept stable, the results show that the N-increase rate in the feeding regime was negatively related to the efficiency of the AD process even if low VFA concentrations indicated a stable process. In further studies, the influence of the increasing N concentrations in the digestate on the microbial population should be investigated.

The results of this study can be applied by biogas operators running their systems at high nitrogen concentrations up to 11.5 $g{\cdot}kg^{-1}$ FM or utilizing substrates with varying nitrogen contents.

4. Conclusions

In this study, we analysed the effect of different inocula and different N-increase rates in feeding regimes on AD process stability and efficiency. The stepwise acclimatisation strategy used for microorganisms to adapt to a new nitrogen concentration according to the feeding regime prevented failure of the AD process under high and elevated ammonia levels. The research approach applied in this study enabled us to run the CSTR in a stable way under the elevated nitrogen loading rates up to 0.30 $g{\cdot}L^{-1}{\cdot}d^{-1}$. The highest N, TAN and FAN in the digestate at the end of the experiment were equal to 11.50 $g{\cdot}kg^{-1}$ FM, 9.07 $g{\cdot}kg^{-1}$ FM and 0.85 $g{\cdot}kg^{-1}$ FM. However, the study indicates that the N-increase rate was negatively related to the AD process efficiency. The level of inhibition in specific methane yield was positively correlated to the TAN and FAN concentrations in the digestate.

Author Contributions: Data curation: I.M., N.N. and A.L.; formal analysis: I.M.; funding acquisition: H.O. and A.L.; project administration: A.L. and I.M.; resources: H.O. and A.L.; supervision: A.L.; writing—original draft: I.M; writing—review and editing: I.M., A.L. and J.K. All authors have read and agreed to the published version of the manuscript.

Funding: This research was supported by the Agency for Renewable Resources (Fachagentur Nachwachsende Rohstoffe e. V. or FNR) under the framework of the project "Optimized substrate management and influence of digestate composition on the soil nitrogen and soil humus balance" (grant no. 22402412). All experiments were conducted at the State Institute of Agricultural Engineering and Bioenergy, University of Hohenheim. The APC was funded by the State Institute of Agricultural Engineering and Bioenergy, University of Hohenheim.

Acknowledgments: We want to thank Prof. Dr. Hans-Peter Piepho for his valuable advice and help during the statistical analysis. Ievgeniia Morozova would like to thank the German Academic Exchange Service (Deutscher Akademischer Austauschdienst, DAAD) for the financial support.

Conflicts of Interest: The authors declare no conflicts of interest.

References

1. Yirong, C.; Zhang, W.; Heaven, S.; Banks, C.J. Influence of ammonia in the anaerobic digestion of food waste. *J. Environ. Chem. Eng.* **2017**, *5*, 5131–5142. [CrossRef]
2. Zhang, C.; Wang, F.; Pei, M.; Qiu, L.; Qiang, H.; Yao, Y. Performance of Anaerobic Digestion of Chicken Manure Under Gradually Elevated Organic Loading Rates. *Int. J. Environ. Res. Public Health* **2019**, *16*, 2239. [CrossRef]

3. Zhang, Q.; Hu, J.; Lee, D.-J. Biogas from anaerobic digestion processes: Research updates. *Renew. Energy* **2016**, *98*, 108–119. [CrossRef]
4. Tian, H.; Fotidis, I.A.; Mancini, E.; Treu, L.; Mahdy, A.; Ballesteros, M.; González-Fernández, C.; Angelidaki, I. Acclimation to extremely high ammonia levels in continuous biomethanation process and the associated microbial community dynamics. *Bioresour. Technol.* **2018**, *247*, 616–623. [CrossRef] [PubMed]
5. Korres, N. *Bioenergy Production by Anaerobic Digestion*; Routledge: Abingdon, UK, 2013; ISBN 9780203137697.
6. Carlos, A.; De Lemos, C. *Anaerobic Reactors. Biological Wastewater Treatment Series*; IWA Publishing: London, UK, 2007; Volume 4, ISBN 9781843391647.
7. Anthonisen, A.C.; Loehr, R.C.; Prakasam, T.B.; Srinath, E.G. Inhibition of Nitrification by Ammonia and Nitrous Acid. *J. Water Pollut. Control Fed.* **1976**, *48*, 835–852.
8. Poirier, S.; Madigou, C.; Bouchez, T.; Chapleur, O. Improving anaerobic digestion with support media: Mitigation of ammonia inhibition and effect on microbial communities. *Bioresour. Technol.* **2017**, *235*, 229–239. [CrossRef]
9. Kayhanian, M. Ammonia Inhibition in High-Solids Biogasification: An Overview and Practical Solutions. *Environ. Technol.* **1999**, *20*, 355–365. [CrossRef]
10. Krakat, N.; Demirel, B.; Anjum, R.; Dietz, D. Methods of ammonia removal in anaerobic digestion: A review. *Water Sci. Technol.* **2017**, *76*, 1925–1938. [CrossRef]
11. Fotidis, I.A.; Karakashev, D.; Kotsopoulos, T.A.; Martzopoulos, G.G.; Angelidaki, I. Effect of ammonium and acetate on methanogenic pathway and methanogenic community composition. *Fems Microbiol. Ecol.* **2013**, *83*, 38–48. [CrossRef]
12. Rajagopal, R.; Massé, D.I.; Singh, G. A critical review on inhibition of anaerobic digestion process by excess ammonia. *Bioresour. Technol.* **2013**, *143*, 632–641. [CrossRef]
13. Chen, Y.; Cheng, J.J.; Creamer, K.S. Inhibition of anaerobic digestion process: A review. *Bioresour. Technol.* **2008**, *99*, 4044–4064. [CrossRef] [PubMed]
14. Perry, L.; McCarty, P.L.; McKenney, R.E. Salt Toxicity in Anaerobic Digestion. *J. Artic.* **1961**, *33*, 399–415.
15. Chen, J.L.; Ortiz, R.; Steele, T.W.J.; Stuckey, D.C. Toxicants inhibiting anaerobic digestion: A review. *Biotechnol. Adv.* **2014**, *32*, 1523–1534. [CrossRef]
16. Theuerl, S.; Klang, J.; Prochnow, A. Process Disturbances in Agricultural Biogas Production—Causes, Mechanisms and Effects on the Biogas Microbiome: A Review. *Energies* **2019**, *12*, 365. [CrossRef]
17. Chen, H.; Wang, W.; Xue, L.; Chen, C.; Liu, G.; Zhang, R. Effects of Ammonia on Anaerobic Digestion of Food Waste: Process Performance and Microbial Community. *Energy Fuels* **2016**, *30*, 5749–5757. [CrossRef]
18. Siegrist, H.; Vogt, D.; Garcia-Heras, J.L.; Gujer, W. Mathematical Model for Meso- and Thermophilic Anaerobic Sewage Sludge Digestion. *Environ. Sci. Technol.* **2002**, *36*, 1113–1123. [CrossRef] [PubMed]
19. Meng, X.; Zhang, Y.; Sui, Q.; Zhang, J.; Wang, R.; Yu, D.; Wang, Y.; Wei, Y. Biochemical Conversion and Microbial Community in Response to Ternary pH Buffer System during Anaerobic Digestion of Swine Manure. *Energies* **2018**, *11*, 2991. [CrossRef]
20. Yenigün, O.; Demirel, B. Ammonia inhibition in anaerobic digestion: A review. *Process Biochem.* **2013**, *48*, 901–911. [CrossRef]
21. Hashimoto, A.G. Ammonia inhibition of methanogenesis from cattle wastes. *Agric. Wastes* **1986**, *17*, 241–261. [CrossRef]
22. Tian, H.; Fotidis, I.A.; Mancini, E.; Angelidaki, I. Different cultivation methods to acclimatise ammonia-tolerant methanogenic consortia. *Bioresour. Technol.* **2017**, *232*, 1–9. [CrossRef]
23. Angenent, L.T.; Sung, S.; Raskin, L. Methanogenic population dynamics during startup of a full-scale anaerobic sequencing batch reactor treating swine waste. *Water Res.* **2002**, *36*, 4648–4654. [CrossRef]
24. Koster, I.W.; Lettinga, G. Anaerobic digestion at extreme ammonia concentrations. *Biol. Wastes* **1988**, *25*, 51–59. [CrossRef]
25. Kayhanian, M. Performance of a high-solids anaerobic digestion process under various ammonia concentrations. *J. Chem. Technol. Biotechnol.* **1994**, *59*, 349–352. [CrossRef]
26. Shanmugam, P.; Horan, N.J. Optimising the biogas production from leather fleshing waste by co-digestion with MSW. *Bioresour. Technol.* **2009**, *100*, 4117–4120. [CrossRef] [PubMed]
27. Molaey, R.; Bayrakdar, A.; Sürmeli, R.Ö.; Çalli, B. Anaerobic digestion of chicken manure: Mitigating process inhibition at high ammonia concentrations by selenium supplementation. *Biomass Bioenergy* **2018**, *108*, 439–446. [CrossRef]

28. Moestedt, J.; Müller, B.; Westerholm, M.; Schnürer, A. Ammonia threshold for inhibition of anaerobic digestion of thin stillage and the importance of organic loading rate. *Microb. Biotechnol.* **2016**, *9*, 180–194. [CrossRef]
29. Li, Y.; Zhang, Y.; Sun, Y.; Wu, S.; Kong, X.; Yuan, Z.; Dong, R. The performance efficiency of bioaugmentation to prevent anaerobic digestion failure from ammonia and propionate inhibition. *Bioresour. Technol.* **2017**, *231*, 94–100. [CrossRef] [PubMed]
30. Lauterböck, B.; Ortner, M.; Haider, R.; Fuchs, W. Counteracting ammonia inhibition in anaerobic digestion by removal with a hollow fiber membrane contactor. *Water Res.* **2012**, *46*, 4861–4869. [CrossRef] [PubMed]
31. Grimsby, L.K.; Fjørtoft, K.; Aune, J.B. Nitrogen mineralization and energy from anaerobic digestion of jatropha press cake. *Energy Sustain. Dev.* **2013**, *17*, 35–39. [CrossRef]
32. Fotidis, I.A.; Wang, H.; Fiedel, N.R.; Luo, G.; Karakashev, D.B.; Angelidaki, I. Bioaugmentation as a solution to increase methane production from an ammonia-rich substrate. *Environ. Sci. Technol.* **2014**, *48*, 7669–7676. [CrossRef]
33. Niu, Q.; Qiao, W.; Qiang, H.; Hojo, T.; Li, Y.-Y. Mesophilic methane fermentation of chicken manure at a wide range of ammonia concentration: Stability, inhibition and recovery. *Bioresour. Technol.* **2013**, *137*, 358–367. [CrossRef] [PubMed]
34. van Velsen, A. Adaptation of methanogenic sludge to high ammonia-nitrogen concentrations. *Water Res.* **1979**, *13*, 995–999. [CrossRef]
35. Sun, C.; Cao, W.; Banks, C.J.; Heaven, S.; Liu, R. Biogas production from undiluted chicken manure and maize silage: A study of ammonia inhibition in high solids anaerobic digestion. *Bioresour. Technol.* **2016**, *218*, 1215–1223. [CrossRef]
36. Yao, Y.; Yu, L.; Ghogare, R.; Dunsmoor, A.; Davaritouchaee, M.; Chen, S. Simultaneous ammonia stripping and anaerobic digestion for efficient thermophilic conversion of dairy manure at high solids concentration. *Energy* **2017**, *141*, 179–188. [CrossRef]
37. Nielsen, H.B.; Angelidaki, I. Strategies for optimizing recovery of the biogas process following ammonia inhibition. *Bioresour. Technol.* **2008**, *99*, 7995–8001. [CrossRef] [PubMed]
38. Haag, N.L.; Nägele, H.-J.; Reiss, K.; Biertümpfel, A.; Oechsner, H. Methane formation potential of cup plant (*Silphium perfoliatum*). *Biomass Bioenergy* **2015**, *75*, 126–133. [CrossRef]
39. VDI-Society Energy and Environment. *VDI 4630 Fermentation of Organic Materials. Characterization of the Substrate, Sampling, Collection of Material Data, Fermentation Tests*; VDI: Düsseldorf, Germany, 2006.
40. Helffrich, D.; Oechsner, H. The Hohenheim Biogas Yield Test. *Landtechnik* **2003**, *58*, 148–149.
41. Mittweg, G.; Oechsner, H.; Hahn, V.; Lemmer, A.; Reinhardt-Hanisch, A. Repeatability of a laboratory batch method to determine the specific biogas and methane yields. *Eng. Life Sci.* **2012**, *12*, 270–278. [CrossRef]
42. Vintiloiu, A.; Lemmer, A.; Oechsner, H.; Jungbluth, T. Mineral substances and macronutrients in the anaerobic conversion of biomass: An impact evaluation. *Eng. Life Sci.* **2012**, *12*, 287–294. [CrossRef]
43. Voelklein, M.A.; O' Shea, R.; Jacob, A.; Murphy, J.D. Role of trace elements in single and two-stage digestion of food waste at high organic loading rates. *Energy* **2017**, *121*, 185–192. [CrossRef]
44. Wintsche, B.; Glaser, K.; Sträuber, H.; Centler, F.; Liebetrau, J.; Harms, H.; Kleinsteuber, S. Trace Elements Induce Predominance among Methanogenic Activity in Anaerobic Digestion. *Front. Microbiol.* **2016**, *7*, 2034. [CrossRef] [PubMed]
45. Choong, Y.Y.; Norli, I.; Abdullah, A.Z.; Yhaya, M.F. Impacts of trace element supplementation on the performance of anaerobic digestion process: A critical review. *Bioresour. Technol.* **2016**, *209*, 369–379. [CrossRef] [PubMed]
46. American Public Health Association; American Water Works Association; Water Environment Federation. *Standard Methods for the Examination of Water and Wastewater*, 20th ed.; American Public Health Association: Washington, DC, USA, 1998.
47. DIN EN 13650. *Soil Improvers and Growing Media—Extraction of Aqua Regia Soluble Elements*, German version EN 136502001; British Standards Institution: London, UK, 2001.
48. DIN EN ISO 17294-2. *Water Quality—Application of Inductively Coupled Plasma Mass Spectrometry. Determination of Selected Elements Including Uranium Isotopes*, German version EN ISO 172942016; International Organization for Standardization: Geneva, Switzerland, 2017.

49. DIN EN ISO 11885. *Water Quality—Determination of Selected Elements by Inductively Coupled Plasma Optical Emission Spectrometry*, German version EN ISO 118852009; International Organization for Standardization: Geneva, Switzerland, 2017.
50. Piepho, H.P.; Buchse, A.; Richter, C. A Mixed Modelling Approach for Randomized Experiments with Repeated Measures. *J. Agron. Crop. Sci.* **2004**, *190*, 230–247. [CrossRef]
51. Tian, H.; Fotidis, I.A.; Kissas, K.; Angelidaki, I. Effect of different ammonia sources on aceticlastic and hydrogenotrophic methanogens. *Bioresour. Technol.* **2018**, *250*, 390–397. [CrossRef]
52. Dai, X.; Yan, H.; Li, N.; He, J.; Ding, Y.; Dai, L.; Dong, B. Metabolic adaptation of microbial communities to ammonium stress in a high solid anaerobic digester with dewatered sludge. *Sci. Rep.* **2016**, *6*, 28193. [CrossRef]
53. Fotidis, I.A.; Treu, L.; Angelidaki, I. Enriched ammonia-tolerant methanogenic cultures as bioaugmentation inocula in continuous biomethanation processes. *J. Clean. Prod.* **2017**, *166*, 1305–1313. [CrossRef]

Article

Enhanced Biogas Production of Cassava Wastewater Using Zeolite and Biochar Additives and Manure Co-Digestion

Chibueze G. Achi [1,2], Amro Hassanein [1] and Stephanie Lansing [1,*]

[1] Department of Environmental Science & Technology, University of Maryland, College Park, MD 20742, USA; achicgjr@gmail.com (C.G.A.); ahassane@umd.edu (A.H.)

[2] Department of Civil Engineering, University of Ibadan, Ibadan 200284, Nigeria

* Correspondence: slansing@umd.edu

Received: 30 November 2019; Accepted: 14 January 2020; Published: 19 January 2020

Abstract: Currently, there are challenges with proper disposal of cassava processing wastewater, and a need for sustainable energy in the cassava industry. This study investigated the impact of co-digestion of cassava wastewater (CW) with livestock manure (poultry litter (PL) and dairy manure (DM)), and porous adsorbents (biochar (B-Char) and zeolite (ZEO)) on energy production and treatment efficiency. Batch anaerobic digestion experiments were conducted, with 16 treatments of CW combined with manure and/or porous adsorbents using triplicate reactors for 48 days. The results showed that CW combined with ZEO (3 g/g total solids (TS)) produced the highest cumulative CH_4 (653 mL CH_4/g VS), while CW:PL (1:1) produced the most CH_4 on a mass basis (17.9 mL CH_4/g substrate). The largest reduction in lag phase was observed in the mixture containing CW (1:1), PL (1:1), and B-Char (3 g/g TS), yielding 400 mL CH_4/g volatile solids (VS) after 15 days of digestion, which was 84.8% of the total cumulative CH_4 from the 48-day trial. Co-digesting CW with ZEO, B-Char, or PL provided the necessary buffer needed for digestion of CW, which improved the process stability and resulted in a significant reduction in chemical oxygen demand (COD). Co-digestion could provide a sustainable strategy for treating and valorizing CW. Scale-up calculations showed that a CW input of 1000–2000 L/d co-digested with PL (1:1) could produce 9403 m^3 CH_4/yr using a 50 m^3 digester, equivalent to 373,327 MJ/yr or 24.9 tons of firewood/year. This system would have a profit of $5642/yr and a $47,805 net present value.

Keywords: methane; fermentation; dairy; poultry; absorbent

1. Introduction

Eutrophication and organic pollution resulting from poor management of wastes from food processing industries, such as cassava processing industries, is a major problem in many developing countries [1,2]. Cassava (*Manihot esculenta*) is a starch-containing root crop of global importance that can be processed into food, feed, and other non-food products [3]. The cassava processing industry is a key industry in many developing countries, especially in Africa, but also in parts of Latin America and Asia. In 2017, the African region contributed 55% of the global production of cassava, equivalent to 121 million tons, with 25–37% of the crop discarded as waste in the form of peels and pulp [4,5], and approximately 60,000 L of effluent generated from each ton of cassava tubers processed [3,6]. Meanwhile, more than 70% of cassava production in sub-tropical and tropical regions of the world is conducted by small and medium-scale farmers [7,8]. These small and medium-scale cassava industries lack the capacity to treat the large waste streams resulting from daily cassava roots processing, which can lead to environmental degradation and pollution of nearby water bodies [9]. Additionally, farmers often depend solely on firewood as a source of energy to process cassava, which has associated negative environmental impacts.

Bioenergy from organic wastes materials through anaerobic digestion (AD) can be used to produce renewable energy from this organic-rich wastewater, while reducing the concentrations of organic pollutants [10]. Cassava wastewater (CW) has a high organic loading, with high concentrations of chemical oxygen demand (COD), biochemical oxygen demand (BOD), and total solids (TS), as well as a low pH [3,11]. In parts of Thailand, Brazil, Vietnam, and India, CW has been managed using stabilization ponds, aerobic systems, and AD [12]. Reported concerns associated with digestion of CW are the low nitrogen concentration and rapid acidification (low pH) of CW [13,14]. Co-digestion with a nitrogen-rich substrate, such as manure, could decrease the carbon-to-nitrogen (C:N) ratio and provide buffering capacity for stabilizing the pH in order to increase methane (CH_4) production. Previous studies have investigated cassava peels and pulps co-digested with livestock wastes [2,15], digestion of cassava starch effluent with separation of the acidogenic and methanogenic phases [16], re-circulation of methanogenic sludge [17], dolomitic limestone addition to increase alkalinity [13], and use of up flow anaerobic sludge blanket (UASB) digestion processing [18,19].

Porous adsorbents, such as biochar, zeolite, and activated carbon, have been used to enhance CH_4 production and general AD processes [20]. Biochar is a carbonaceous material obtained from agricultural biomass through pyrolysis and gasification. Mumme et al. reported that biochar is relatively cheaper to manufacture than other adsorbents, which has increased interest in land application of biochar, and more recently, inclusion in AD processing [21]. The use of biochar as an additive in AD has not been fully investigated, and no work has been done with biochar and AD of CW. There is potential for biochar to enhance the operational stability of the AD process and increase the quality of the digestate produced.

Zeolite has also been reported to possess favorable characteristics for microorganism adhesion [22], with the capacity to induce ion exchange during AD due to the presence of Na^+, Ca^{2+}, and Mg^{2+} cations in its crystalline structure. These properties could be useful for improving AD of wastewaters with high concentrations of nitrogen, such as poultry manure, as it prevents process inhibition. Application of natural zeolites as support media in digesters treating wastewaters has been reported to increase microbial population density and provide greater opportunity for microbial growth and attachment, cross feeding, co-metabolism, and interspecies hydrogen and proton transfer [22].

A prior study by Montalvo et al. reported that the addition of natural zeolite at doses between 2 and 4 g/L increased CH_4 production, with increasing inhibition at doses >6 g/L [23]. The use of porous materials, such as natural zeolites, to create surface area for microbial communities and increase retention of high biomass concentrations in the digestion of wastewater has been documented [22]. Zeolite was employed to enhance energy recovery, in the form of hydrogen (H_2), from cassava-ethanol wastewater during the dark fermentation process. [24]. To our knowledge, zeolite additions have not been used during the digestion of CW to enhance CH_4 production.

General implementation and adoption of large-scale biogas technology in most African countries have been limited due to the high costs associated with investments and operations of AD systems [25], especially when there is an additional cost of transporting wastes to offsite AD reactors. The availability of agricultural biomass, which is abundant in the rural cassava industry, along with the large volume of CW generated daily in this industry, would provide larger quantities of organic material to be treated onsite using AD. This study focused on investigating the impact of the co-digestion of CW with selected livestock manures and porous adsorbents on biogas production for potential implementation and adoption in cassava industries. The objectives of the study were to: (1) characterize cassava wastewater as a substrates for AD and identify appropriate substrates for co-digestion, (2) investigate the CH_4 potential of cassava digestion, with and without co-digestion with manure, biochar, and zeolite, in terms of cumulative production and retention time, (3) characterize the wastewater transformations during digestion in terms of organic and nutrient transformations, and (4) analyze the economic viability and environmental impact of employing digestion for the rural cassava industry.

2. Materials and Methods

2.1. Substrate and Inoculum Collection and Preparation

2.1.1. Cassava Wastewater Substrate

The cassava tubers were obtained from a farmer's market in Adelphi, Maryland, USA. The cassava tubers were manually peeled and soaked for 5 days in the laboratory using 1 L/kg of deionized water to replicate the rural cassava processing steps for 'fufu' production, a popular African dish derived from fermented cassava paste. Fermented tubers were manually squeezed, and the cassava wastewater (CW) was collected and used for the experiments. The substrate characteristics and the experiment design are given in Tables 1 and 2. The CW had a COD range of 29.8–33.4 g/L, volatile solids (VS) of 17.3 g/kg, total solids (TS) of 17.8 g/kg, and a pH of 5.5.

Table 1. Experimental design showing grams (g) of substrate addition into each triplicate 250 mL reactor.

Treatment	Cassava Wastewater (CW) (g)	Zeolite (ZEO) (g)	Biochar (B-Char) (g)	Poultry Litter (PL) (g)	Dairy Manure (DM) (g)	Inoculum (g)	Water (g)
CW + PL + ZEO (HC)	28.9	3.4	0.0	0.8	0.0	92.1	28.2
CW + PL + B-Char (HC)	28.9	0.0	3.4	0.8	0.0	92.1	28.2
CW:PL (1:1)	28.9	0.0	0.0	0.8	0.0	92.1	28.2
CW:PL (2:1)	38.6	0.0	0.0	0.5	0.0	92.1	18.9
CW + DM + ZEO (HC)	28.9	3.4	0.0	0.0	4.4	92.1	24.7
CW:DM (1:1)	28.9	0.0	0.0	0.0	4.4	92.1	24.7
CW:DM (2:1)	38.6	0.0	0.0	0.0	2.9	92.1	16.5
CW + ZEO (HC)	57.8	0.3	0.0	0.0	0.0	92.1	0.1
CW + ZEO (HC)	57.8	1.5	0.0	0.0	0.0	92.1	0.1
CW + B-Char (LC)	57.8	0.0	0.3	0.0	0.0	92.1	0.1
CW + B-Char (HC)	57.8	0.0	1.5	0.0	0.0	92.1	0.1
CW-only	57.8	0.0	0.0	0.0	0.0	92.1	0.1
Inoculum-only	0.0	0.0	0.0	0.0	0.0	92.1	0.0

Table 2. Characterization of the substrates, cassava wastewater (CW), poultry litter (PL), and dairy manure (DM) and the inoculum source, including total solids (TS), volatile solids (VS), chemical oxygen demand (COD), pH, total Kjeldahl nitrogen (TKN), total phosphorus (TP), and the carbon to nitrogen ratio (C:N).

	TS (g/kg)	VS (% TS)	COD (g/L)	pH	TKN (mg N/L)	TP (mg P/L)	C:N Ratio
CW	17.8 ± 0.7	97.2 ± 0.7	33.7 ± 0.8	5.53	375	222	27.8
PL	776 ± 1	80.0 ± 0.2	NA	8.25	3675	1245	13.0
DM	131 ± 2	87.3 ± 0.6	NA	7.33	3450	603	15.2
Inoculum	29.5 ± 0.1	73.6 ± 7.0	25.1 ± 0.3	7.55	3050	1225	3.91

2.1.2. Dairy and Poultry Manure Substrates and Inoculum Source

The dairy manure (DM) used as a co-substrate was obtained from the 100-cow dairy at the US Department of Agriculture (USDA) Beltsville Agricultural Research Service (ARS) in Beltsville, MD. Poultry litter (PL) was obtained from a poultry (broiler) farm at the University of Maryland Extension—Talbot county, Easton Maryland. The poultry litter consisted of poultry droppings and beddings from wood shavings. Both manure substrates were collected onsite and stored at 4 °C before use. The inoculum used for the experiment was digestate of a complete mixed wastewater sludge digester (Alexandria, VA, USA) and was stored at 4 °C prior to use.

2.1.3. Biochar and Zeolite Additives

Two porous materials, biochar (B-Char) and clinoptitolite zeolite (ZEO), were added to the CW as co-treatments. The biochar (B-Char) substrate was derived from corn stover prepared through pyrolysis under an O_2-free atmosphere at 500 °C, with a holding time of 10 min (ArtiCHAR, Prairie

City, Iowa, USA). The biochar particle size varied from 841 mm to <74 mm, with a VS and TS of 690 and 980 g/kg, respectively. The zeolite was a high purity 97% clinoptilolite zeolite produced at Amargosa Valley (Nye county, NV, USA). The zeolite used was in the form of granules with an angular shape and gray color. The pore diameter was between 4.0–7.0 angstroms.

2.2. Experimental Design

A batch digestion experiment was conducted based on the biochemical methane potential (BMP) test following methods by Moody et al. [26]. The digestion tests were conducted at the University of Maryland's Department of Environmental Science and Technology (ENST) Water Quality Laboratory (College Park, MD USA). Prior to starting BMP tests, the TS and VS for CW, PL, DM, and inoculum were determined and used to combine the co-substrate ratios based on VS. The experiment was designed for 16 treatments, with three replicates for each treatment (48 total digestion reactors). Each digestion reactor consisted of a 250 mL serum bottle, with the substrates and inoculum loaded at a 2:1 inoculum to substrate ratio (ISR) based on VS and operated in mesophilic conditions (35 °C). For all treatments, an equal volume of inoculum (92.1 g) was added to each triplicate reactor.

The biochar (B-Char) and zeolite (ZEO) treatments were prepared using a low concentration (LC) (0.5 g adsorbent/g TS of substrate) and a high concentration (HC) (3 g adsorbent/g TS of substrate) added to 57.82 g CW. For the manure co-digestion treatments, 3.42 g of PL and 3.26 g of DM were digested alone and co-digested with 28.91 g CW with and without the HC of ZEO. Additionally, PL was co-digested with 28.91 g CW and the HC of B-Char. Inoculum-only reactors were also incubated, and the CH_4 production from the inoculum was subtracted from each treatment to account for residual CH_4 production from organics in the inoculum.

Prior to incubation, the headspace in each vessel was purged with N_2 for three minutes to ensure anaerobic conditions and immediately capped with a rubber septum, and the bottles were placed on a shaker (120 rpm) in a controlled environmental chamber at 35 °C for 48 days. The daily biogas volume was measured by volumetric displacement using a graduated, gas-tight, wet-tipped 50 mL glass syringe inserted through the top of the rubber septum. Biogas production was quantified volumetrically at normal temperature and pressure conditions using a glass gas-tight syringe, equilibrated to atmospheric pressure [26]. All CH_4 production values are reported in normal temperature and pressure conditions (1 atm and 20 °C).

2.3. Analytical Methods

The pH of substrates and inoculum were determined with an Accumet AB 15 pH meter (Fisher Scientific, Hampton, NH). For all samples, TS (Method 2540B) and VS (Method 2540E) concentrations were determined using standard methods for the examination of water and wastewater [27]. Total Kjeldahl nitrogen (TKN) and total phosphorus (TP) samples were analyzed on a Lachat autoanalyzer (Quikchem 8500, Hach Company, Loveland, CO, USA) using QuikChem methods 13-107-06-2-D for TKN and 13-115-01-1-B for TP. The COD concentration was measured using a Hach DR 5000 spectrophotometer (Hach Company, Loveland, CO, USA).

The carbon content of the CW, PL, DM, and inoculum were calculated using the equation from Adams et al. [28], where % Carbon = % VS/1.8.

Biogas was analyzed for CH_4 and CO_2 content by injecting 0.10 mL of gas sample using a luer-lock, gas-tight syringe into an Agilent HP 7890 A gas chromatograph (Agilent Technologies, Santa Clara, CA, USA) equipped with a thermal conductivity detector (TCD) and single HP porous layer open tubular (PLOT) Q column with an injection temperature of 250 °C, a detector temperature of 250 °C, an oven temperature of 60 °C, and conveyed using He gas at a flow rate of 8.6 mL He/min [10].

2.4. Statistical Analysis

Cumulative CH_4 production was analyzed using analysis of variance (ANOVA) to determine significantly differences, with p-values < 0.05 considered significant. Tukey's honestly significant difference (HSD) post-hoc tests were performed for multiple comparisons between variables based on different digestion periods during the 48 days of the experiment. All results presented in the tables and charts are average values with standard error (SE).

3. Results and Discussion

3.1. Characterization of Substrate and Inoculum

The pH of the CW substrate was between 5.5 and 6.5 (Table 2). Some studies have reported lower pH values for CW, ranging from 3.9–4.5 [29]. The pH of mixed substrates before and after AD was within the ideal pH range (6.5–8) for CH_4 production [30]. The TS and VS of the cassava wastewater (17.8 and 17.3 g/L, respectively) was 75.8% and 60.4% lower than PL respectively, and 11.3 and 9.7% lower than DM, respectively. As the CW was a liquid wastewater, it was a more dilute waste stream than the manure substrates and had comparatively less TS and VS.

The TKN and TP of the CW were 375 and 222 mg/L respectively (Table 2), whereas, the PL had higher TKN and TP values (3675 and 1245 mg/L, respectively), which were similar to DM (3450 and 603 mg/L, respectively). The low nitrogen content of CW observed in this study was consistent with findings from others CW studies [1,2,8]. The C:N ratio of the CW substrate was 27.8, which was similar to the value 29.1 reported by Lin et al. [31]. The carbon to nitrogen ratio is a key factor affecting anaerobic digestion [32], with C:N ratios between 25 and 30 reported as the most suitable for CH_4 production [2,31]. The C:N ratio for PL and DM in this experiment was 13.0 and 15.2 respectively, which were lower than the optimal conditions, while the CW was higher. When the substrates were combined, the C:N ratio of the mixtures were 20.4 and 21.1, for PL and DM, respectively.

3.2. Effect of Livestock Manure Co-Digestion with Cassava Wastewater on Biogas Production

3.2.1. Cumulative CH_4 Production Based on VS Addition into the Digestion Reactor

After 48 days of digestion, the cumulative CH_4 production (on a per g VS-basis) from CW-only (620 mL CH_4/g VS) was 15.8% higher than co-digestion of CW:DM at a 2:1 ratio (522 mL CH_4/g VS; p-value < 0.001; Figure 1; Table 3). The CW-only digestion had 5.8% higher CH_4 production than CW:PL at a 2:1 ratio (590 mL CH_4/g VS), but this difference was not statistically significant (p-value = 0.864). Similarly, CW-only was 14.4% and 25.6% higher respectively, than CW co-digested at a 1:1 ratio with DM and PL (461 and 531 mL CH_4/g VS; p-values < 0.001 and 0.001, respectively).

Table 3. Reductions in volatile solids (VS) and chemical oxygen (COD) during digestion for the substrate and inoculum in each reactor. The cumulative methane (CH_4) production is given using two normalizations: per g VS added and per g of total substrate added. Superscript letters (a through g) indicate significant differences within each column at p-value < 0.05.

Substrate	VS Reduction (%)	Influent COD (g/L)	COD Reduction (%)	Cumulative CH_4 (mL CH_4/g VS)	Cumulative CH_4 (mL CH_4/g Substrate)
CW-only	65.5 ± 0.1 ^{a}	29.6 ± 0.4 ^{ab}	40.6 ± 2.9 ^{a}	620 ± 6.0 ^{abc}	10.7 ± 0.1 ^{a}
CW + B-Char (HC)	37.5 ± 0.1 ^{b}	41.9 ± 0.7 ^{c}	23.6 ± 7.5 ^{b}	611 ± 27 ^{bc}	10.6 ± 0.5 ^{a}
CW + B-Char (LC)	62.7 ± 0.3 ^{a}	34.9 ± 2.4 ^{d}	48.8 ± 5.3 ^{a}	611 ± 16 ^{a}	10.6 ± 0.3 ^{a}
CW + ZEO (HC)	66.0 ± 1.9 ^{a}	33.7 ± 0.8 ^{de}	49.2 ± 1.6 ^{a}	653 ± 4 ^{a}	11.3 ± 0.1 ^{ab}
CW + ZEO (LC)	66.2 ± 2.6 ^{a}	32.8 ± 1.0 ^{de}	46.2 ± 1.0 ^{a}	634 ± 6 ^{ab}	11.0 ± 0.1 ^{ab}
CW + PL + B-Char (HC)	6.88 ± 1.4 ^{c}	41.6 ± 0.6 ^{c}	−31.1 ± 1.2 ^{*c}	471 ± 16 ^{d}	15.9 ± 0.5 ^{dce}
CW + PL + ZEO (HC)	61.4 ± 0.6 ^{a}	31.6 ± 2.1 ^{ae}	44.0 ± 3.4 ^{a}	518 ± 8 ^{e}	17.4 ± 0.3 ^{dce}
CW + DM + ZEO (HC)	57.3 ± 0.2 ^{a}	25.8 ± 0.3 ^{f}	21.9 ± 6.1 ^{b}	473 ± 5 ^{d}	14.2 ± 0.2 ^{dbe}
CW:PL (1:1)	63.5 ± 0.6 ^{a}	28.8 ± 2.0 ^{b}	42.7 ± 3.2 ^{a}	531 ± 10 ^{e}	17.9 ± 0.3 ^{c}
CW:PL (2:1)	64.8 ± 0.5 ^{a}	28.9 ± 0.3 ^{b}	39.6 ± 4.9 ^{a}	590 ± 6 ^{c}	15.1 ± 0.2 ^{dce}
CW:DM (1:1)	59.9 ± 1.0 ^{a}	29.5 ± 0.9 ^{ab}	37.4 ± 0.5 ^{a}	461 ± 17 ^{d}	13.8 ± 0.5 ^{abe}
CW:DM (2:1)	63.4 ± 0.3 ^{a}	32.5 ± 1.9 ^{e}	47.8 ± 2.7 ^{a}	522 ± 14 ^{e}	12.6 ± 0.3 ^{abe}
DM-only	48.4 ± 5.3 ^{d}	28.0 ± 0.7 ^{bf}	20.9 ± 2.9 ^{b}	100 ± 5 ^{f}	22.9 ± 1.1 ^{f}
PL-only	63.0 ± 8.4 ^{a}	28.2 ± 1.3 ^{b}	20.7 ± 0.1 ^{b}	156 ± 3 ^{g}	193 ± 4 ^{g}

* a negative value indicates a percent increase due to addition of biochar not included in pre-COD.

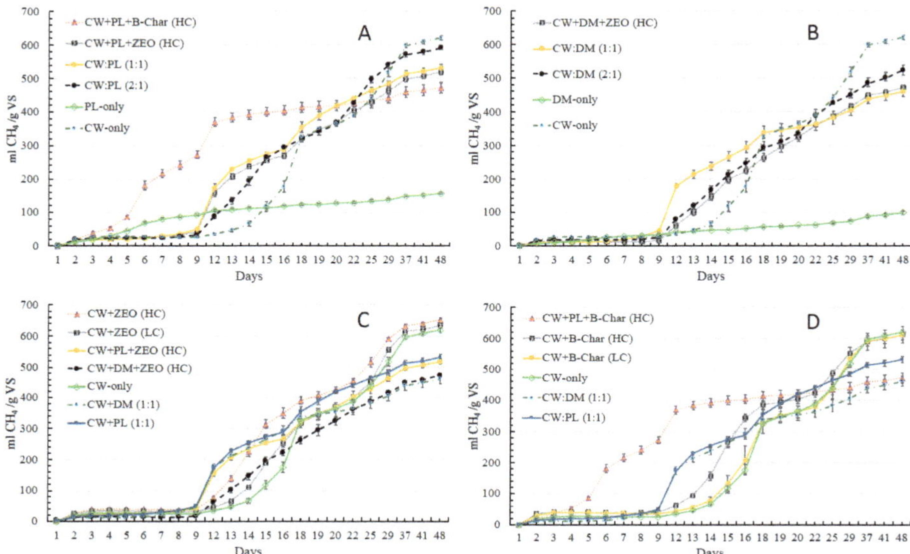

Figure 1. Cumulative CH_4 production based on volatile solids (VS) added to each reactor for cassava wastewater (CW) digested alone and co-digested with poultry litter (PL) shown in (**A**), co-digestion with dairy manure (DM) shown in (**B**), and co-digestion with zeolite (ZEO) and biochar (B-Char) at low and high concentrations (LC and HC) shown in (**C**) and (**D**), respectively.

When CH_4 production is normalized on a VS basis, the efficiency of the organic material to CH_4 conversion process is shown. These results show that CW can be co-digested or digested alone, and co-digestion of CW with manure resulted in similar or slightly lower CH_4 production efficiency values (5.8%–25.6% decrease with co-digestion). It should be noted that the inoculum included in the digestion reactor helped to lower the C:N ratio from 27.8 in the CW substrate to 21.7 in the digestion reactor with the inoculum and CW mixture, with an increase in the pH value from 5.53 to 7.75 due to inoculum inclusion. The significance of pH as a key determining factor for AD process, especially in full-scale continuous reactors were highlighted in Calabrò et al. [33,34]. In field conditions, it would be important to have a viable inoculum source for initiating digestion and to consider a co-digestion material that can help to neutralize the low pH and is high in nitrogen to ensure that the microbes are not nitrogen-limited.

3.2.2. Cumulative CH_4 Production Based on the Mass of Substrate Added to the Digestion Reactor

Due to the high VS concentration of the manure substrates, the PL-only reactors had an order a magnitude higher CH_4 production on a mass basis (193 mL CH_4/g substrate) than DM-only (22.9 mL CH_4/g substrate) and all CW reactors (Table 3). The DM and PL manure substrates had 84.8% to 97.2% higher VS concentrations respectively, than the CW substrate (Table 2). Digestion of CW yielded 83.9% and 74.8% more cumulative CH_4 (on a VS-basis) than DM-only and PL-only digestion (100 and 156 mL CH_4/g VS, respectively; Table 2), as the organic matter in the CW substrate was converted into CH_4 more efficiently than the manure substrates, likely due to the more recalcitrant nature of the VS in the complex manure substrates compared to the cassava wastewater. The CW:PL (1:1) had the highest CH_4 production (17.9 mL CH_4/g substrate) of the CW co-digestion treatments (on a mass basis). The PL co-substrate had higher CH_4 production efficiencies than CW co-digested with DM at both the 1:1 and 2:1 ratios (p-values = 0.017 and 0.025, respectively), indicating that a continuously-fed CW digestion system would benefit from co-digesting with PL due to the high organic loading of the PL substrate.

3.2.3. Cumulative CH_4 Production Based on Digestion Period

While the overall CH_4 production efficiency from the CW-only was higher than the manure substrates, the CW-only treatment produced 118 mL CH_4/g VS (19% of the cumulative CH_4) in the first 15 days of the 48-day digestion period (Table 4; Figure 2). During this first third of the digestion period (Days 1–15), the CH_4 production from the CW-only treatment was significantly lower (38.6%) than CW:DM at 1:1 (265 mL CH_4/g VS) and 32.6% lower CW:PL at 1:1 (274 mL CH_4/g VS; p-values < 0.001). When the ratio of CW to manure was doubled (2:1), the CH_4 production from Days 1–15 in the CW:PL (2:1) and CW:DM (2:1) reactors was 263 and 211 mL CH_4/g VS respectively, which was 25.7 and 21.5% higher than CW-only (p-value < 0.001 and 0.002, respectively).

Table 4. Cumulative methane (CH_4) during the 48-day digestion period for designated time periods, with the percent of the total cumulative CH_4 production in parenthesis. Superscript letters (a through g) significant differences within each column at p-value < 0.05.

Treatment	Cumulative CH_4 in mL CH_4/g VS and (% of Total CH_4 Production)				
	9 Days	15 Days	20 Days	37 Days	48 Days
CW-only	26.8 (4.3%) ^{abc}	118 (19%) ^{a}	364 (58.7%) ^{ab}	598 (96.4%) ^{abc}	620 (100%) ^{abc}
CW + B-Char (HC)	39.9 (6.5%) ^{adef}	265 (43.4%) ^{b}	403 (66%) ^{bc}	589 (96.4%) ^{bc}	611 (100%) ^{bc}
CW + B-Char (LC)	37.8 (6.2%) ^{acdef}	133 (21.7%) ^{a}	363 (59.4%) ^{a}	591 (96.8%) ^{bc}	611 (100%) ^{bc}
CW + ZEO (HC)	39.3 (6%) ^{acdef}	314 (48.1%) ^{c}	425 (65.2%) ^{c}	634 (97.1%) ^{a}	653 (100%) ^{a}
CW + ZEO (LC)	31.5 (5%) ^{ace}	189 (29.8%) ^{d}	370 (58.4%) ^{ab}	614 (96.8%) ^{ab}	634 (100%) ^{ab}
CW + PL + B-Char (HC)	273 (57.9%) ^{g}	400 (84.8%) ^{e}	419 (88.9%) ^{c}	459 (97.4%) ^{de}	471 (100%) ^{d}
CW + PL + ZEO (HC)	43.2 (8.3%) ^{def}	255 (49.2%) ^{b}	370 (71.4%) ^{ab}	497 (95.9%) ^{f}	518 (100%) ^{e}
CW + DM + ZEO (HC)	17 (3.6%) ^{b}	197 (41.7%) ^{d}	323 (68.5%) ^{d}	447 (94.7%) ^{de}	473 (100%) ^{d}
CW:PL (1:1)	49.8 (9.4%) ^{c}	274 (51.6%) ^{bc}	419 (78.8%) ^{c}	513 (96.5%) ^{f}	531 (100%) ^{e}
CW:PL (2:1)	32.6 (5.5%) ^{acde}	263 (44.7%) ^{b}	366 (62%) ^{ba}	571 (96.8%) ^{c}	590 (100%) ^{c}
CW:DM (1:1)	46.5 (10.1%) ^{df}	265 (57.6%) ^{b}	353 (76.6%) ^{da}	439 (95.3%) ^{d}	461 (100%) ^{d}
CW:DM (2:1)	25.4 (4.9%) ^{bc}	211 (40.5%) ^{d}	336 (64.3%) ^{da}	484 (92.8%) ^{ef}	522 (100%) ^{e}
DM-only	34.8 (34.6%) ^{acde}	48.4 (48.2%) ^{f}	62.6 (62.3%) ^{e}	89.1 (88.8%) ^{g}	100 (100%) ^{f}
PL-only	92.3 (59%) ^{h}	114 (72.9%) ^{a}	128 (82%) ^{f}	148 (95%) ^{h}	156 (100%) ^{g}

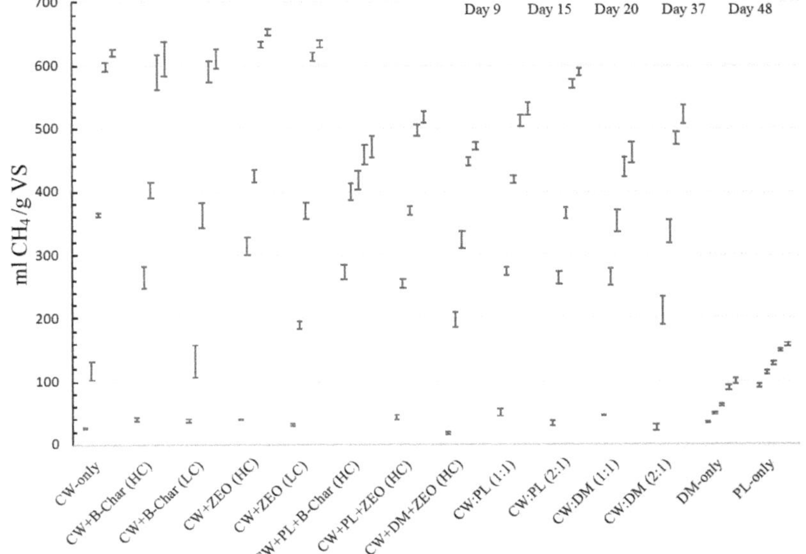

Figure 2. Cumulative methane (CH_4) production based on volatile solids (VS) added to each reactor for cassava wastewater (CW) digested alone and co-digested with poultry litter (PL), dairy manure (DM), zeolite (ZEO), and biochar (B-Char) at low and high concentrations (LC and HC) at five time points (Days 9, 15, 20, 37, and 48) in the 48-day digestion period.

There was no significant difference in cumulative CH_4 production between CW:PL at the 1:1 and 2:1 ratios (p-value = 0.087). The CW:PL (2:1) had 4.1% less CH_4 in the first 15 days of digestion, showing a slight decrease in lag phase during digestion without a significant effect on the overall CH_4 production potential (Figure 2). A similar trend was observed when comparing the CW:DM at 1:1 and 2:1 ratios, with 265 and 211 mL CH_4/g VS respectively, in the first 15 days of digestion, which were not significantly different (p-value of 0.061).

Generally, co-digestion of CW with PL or DM reduced the lag phase for CH_4 production. The highest rate of CH_4 production occurred within the first two weeks, which is consistent with the results from Witarsa and Lansing [30], where a large percentage of CH_4 production from digestion of separated and unseparated dairy manure (DM) occurred in the first 16 days of a 216 day digestion period (40% and 36%, respectively). Cassava wastewater (CW-only) generated 19% of the total cumulative CH_4 in the first 15 days of our 48-day digestion period, while DM-only generated 48.2% of the total cumulative CH_4 and CW:DM (1:1) generated 57.6% of the cumulative CH_4 production in the first 15 days.

The observed increase in CH_4 production with co-digestion of CW within the first two weeks was consistent with other findings, which emphasized the advantages of co-digestion over single digestion of substrates [2,32,35–37]. Panichnumsin et al. [2] examined the potential of co-digestion of cassava pulp and swine manure in a semi-continually fed stirred tank reactor in mesophilic conditions (37 °C) at a constant organic loading rate of 3.5 kg VS/m^3d for 15 days and reported a 41% increase in CH_4 yield compared with digestion of swine manure alone. Similar to our study, a batch experiment conducted by Riano et al. [37] at 35 °C for 55 days reported that co-digestion of winery wastewater (10–40%) and swine manure increased CH_4 production by 45–75% and improved digestion stability compared to digestion of swine manure alone.

Abouelenien et al. [35] co-digested, poultry manure (PM) with mixed agricultural wastes comprised of coconut wastes, cassava wastes, and coffee grounds. The cassava waste used in their study was root residue and wet cake from cassava, while our study utilized cassava wastewater. Similar to our study, co-digestion was conducted under mesophilic conditions (35 °C) and saw an increase in CH_4 yield of up to 50% (506 mL CH_4/g VS) compared to PM-only after 40 days of digestion. Their results were comparable to our study, with cumulative CH_4 production of 531 mL CH_4/g VS for CW:PL (1:1) after 48 days, which was significantly (p-value = 0.001) higher than PL-only, which yielded only 156 mL CH_4/g VS. Contrary to our findings, Abouelenien et al. [37] reported an elongation of the lag phase due to co-digestion, which was attributed to the complex organic matter in the mixed agricultural wastes compared to the PM substrate. Whereas in our study, a reduction in lag phase was recorded due to the liquid state of the CW substrate, which was more readily accessible for the rate-limiting hydrolysis phase of digestion.

3.3. *Impact of Porous Adsorbent on AD of Cassava Wastewater*

3.3.1. Zeolite Addition with Cassava Wastewater Digestion

Digesting CW and a high concentration (HC) of zeolite (CW + ZEO-HC) produced the highest cumulative CH_4 (653 mL CH_4/g VS) for all treatments after 48 days, followed by the treatment with a lower concentration (LC) of zeolite (CW + ZEO-LC), which produced 634 mL CH_4/g VS, with no significant different between the two zeolite concentrations (p-value = 1.00; Figure 1; Table 2). The two porous adsorbents used in this study at the HC were also not significantly different (p-value = 0.50), with the cumulative CH_4 produced from CW + ZEO-HC (653 mL CH_4/g VS) only slightly higher than CW + B-Char-HC (611 mL CH_4/g VS).

After 9 days of digestion, a significantly higher percentage of the total CH_4 production (37.8%) was observed in CW + ZEO-HC compared to CW-only (Table 3; Figure 2). This observation is consistent with Milan et al., where doses of zeolite between 2 and 4 g/L increased CH_4 production of swine manure and of zeolite, while doses above 6 g/L inhibited the process [38]. In our study, the effect of zeolite on digestion of CW, singly and co-digested with manure, the 3 g/L of zeolite (CW + ZEO-HC) had 90.7% and 15.5% more CH_4 production on Days 15 and 20 than CW-only (p-value < 0.001 and 0.045, respectively). The effect of zeolite addition was significant during the early stages of digestion process, up to the first three weeks. At Days 37 and 48, there was no significant difference between CW-only and CW + ZEO-HC, with the cumulative CH_4 from CW + ZEO-HC only 5.1% higher than CW-only (p-values = 0.68 and 0.82, respectively).

3.3.2. Biochar Addition with Cassava Wastewater Digestion

Similarly, an increase in CH_4 production with a shortened lag phase was observed due to biochar addition (Figure 2). The effect of biochar addition in reducing the lag phase in AD has been previously reported [39,40]. Jang et al. showed a 24.9% increase in cumulative CH_4 (467 mL CH_4/g VS) with 10 g/L of biochar compared to 1 g/L of biochar (395 mL CH_4/g VS) with mesophilic conditions and 40 days of digestion [40]. Our findings showed that on Day 15, CW + B-Char-HC had 76.8% more CH_4 production than the CW-only treatment, which Jang et al. suggested was due to the high alkalinity of biochar enhancing CH_4 production and shortening the lag phase [40].

Comparing the concentrations of porous adsorbent added, there was a significant difference in cumulative CH_4 production between the low and high concentrations of biochar (p-value < 0.001) on Day 15 of digestion (Figure 2). In the first two weeks of digestion, CW + B-Char-HC yielded 265 mL CH_4/g VS, while CW + B-Char-LC yielded 133 mL CH_4/g VS, illustrating the decrease in lag phase with an increase in the quantity of biochar added.

Comparing the LC and HC of zeolite showed no significant difference (p-value = 1.000) after 48 days of digestion. The ZEO-LC and ZEO-HC at Day 15 produced 8.1% and 4.7% more CH_4 than B-Char-LC and HC, respectively. Yet by 48 days, there were no significant differences between LC and HC of B-Char and ZEO (p-values = 0.974 and 1.000, respectively). The observed lag in digestion in the mixtures containing CW alone or low concentrations of zeolite or biochar could be as a result of the rapid acidification of CW and inadequate buffer to provide the necessary buffer for microbial community and methanogens, and thus, a longer lag phase for microbial recovery.

The combined effects of manure and biochar showed that CW + PL + B-Char produced significantly more CH_4 at Days 15 and 20 (273 and 400 mL CH_4/g VS, respectively; p-values < 0.001) than CW + B-Char (39.9 and 265 mL CH_4/g VS, respectively). After Day 20, the daily CH_4 production of CW + PL + B-Char decreased, while CW + B-Char increased and resulted in higher cumulative CH_4 over 44 days. While the addition of PL or DM to CW increased CH_4 production in the first 15 days of digestion, the CW + PL + B-Char-HC treatment yielded 400 mL CH_4/g VS (84.8% of total cumulative CH_4) in the first 15 days, with this reduction is lag phase likely attributed to the combined presence of biochar and poultry litter.

The ability of biochar to catalyze digestion by providing surface area for the colonization of the microbial cell was previously reported in a review by Mumme et al. [21]. The CW substrate used in our experiment contained a low pH and when co-digested with biochar and manure showed an improved AD process due to the buffer provided by manure and biochar [20], as observed in the first two weeks of our experiment.

3.4. *Volatile Solids and COD Reduction during Digestion*

Chemical oxygen demand (COD) and VS reduction is associated with CH_4 production. The substrate mixture containing CW + ZEO-HC showed the highest VS and COD reductions during digestion (66% and 49%), which corresponded with the highest cumulative CH_4 production (Table 2). Similar trends were reported in previous work [18,29]. Jiraprasertwong et al. used cassava wastewater in a three-stage up flow anaerobic sludge blanket (UASB) reactor and showed a steady reduction in COD removal with increasing COD loading and an increasing biogas production up to 15 kg COD/m^{3}d (one reactor) and 10 kg/m^{3}d (two reactors) [18]. For comparison, our batch study had a one-time COD loading for each substrate tested that ranged from 25 to 43 kg COD/m^3, respectively.

3.5. Scale-Up Model

A medium size cassava factory in Nigeria processes 3000–6000 kg of cassava tubers per day, yielding 1000–2000 L CW/d. The size of the digester needed to co-digest CW with PL was calculated to be 50 m^3 (40 m^3 liquid and 10 m^3 biogas headspace), as shown in Figure 3. The quantity of PL added to digester would vary from 56 kg/d during high cassava production (March to October) to 28 kg/d during low cassava production (November to February) to maintain a 1:1 ratio (by VS). The hydraulic retention time (HRT) would vary from 20 days during high cassava production to 40 days during low cassava production, which should result in 78.8% to 96.5% of the cumulative CH_4 production from the 48-day BMP test (Table 4). Using the results from CW:PL (1:1), the daily CH_4 production in the 50 m^3 digester would be 28.9 m^3 CH_4/d (20-day HRT) to 19.2 m^3 CH_4/d (40-day HRT), with an annual CH_4 production of 9403 m^3 CH_4/yr. Using Gibbs free energy (ΔG_{CH4} = 890.4 kJ/mol) [41], the annual CH_4 production would be equivalent to 373,327 MJ/yr and 24.9 tons of firewood/yr saved, based on 15 MJ/kg of firewood used for heating [42].

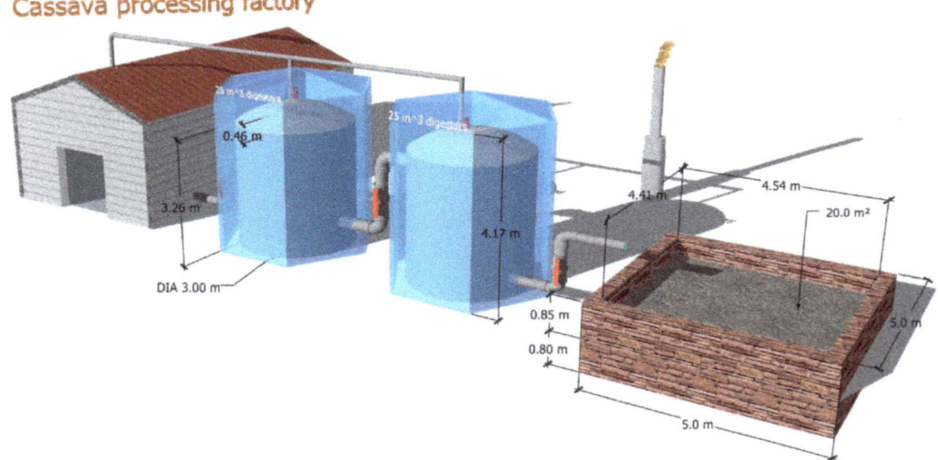

Figure 3. Scale-up model for a cassava processing factory, with two 25 m^3 digesters plumbed in series to treat cassava wastewater and poultry manure with the utilization of the digestate for fertilizer.

The estimated cost of the digester system, including two flexible PVC bag digesters, greenhouse enclosures, plumbing, and gas piping, was estimated to be \$7500 (\$150/m^3 digester) (Shenzhen Puxin Technology Co. Ltd., Shenzhen, China). This cost does not take into consideration the land value or the cost of a heating system. A heating system would likely not be necessary in Nigeria, with average maximum and minimum ambient temperatures of 33 and 25 °C respectively [43], which are within the mesophilic temperature range. While biogas production decreases with temperature, this decrease is more pronounced below the mesophilic range (<25 °C) [30]. The digester design also includes a greenhouse covering the digester, which our previous research has shown can significantly increase (6.8–24.5 °C) the digester temperature and help maintain a more consistent digestion temperature throughout the day [44]. Assuming a 10-year lifetime and 8% discount rate, the discounted capital investment would be \$8617. Based on the expected annual CH_4 production, the cost would be \$0.09/m^3 CH_4, which does not take into account the value of the produced fertilizer from the digester effluent. Assuming a natural gas price of \$0.6/m^3 (\$1.2/kg natural gas) [45], the system would have a yearly profit of \$5642/yr. The net present value (NPV) is calculated as \$47,805, which does not take into consideration the value of the digestate, which can provide valuable nutrients to produce cassava and/or other crops. The price of adding zeolite or biochar was not included, since these additives may not be available and did not increase the overall biogas production, only decreased the lag phase associated with biogas production. If a higher throughput is desired or if the pH is not stabilized, the addition zeolite or biochar could be added to the full-scale system, if available.

4. Conclusions

Anaerobic digestion of cassava wastewater was shown to be viable, with CH_4 production enhanced by additions of zeolite and biochar. Co-digestion with dairy and poultry manure decreased the lag phase of digestion but did not increase overall CH_4 production, likely due to the more recalcitrant materials within the manure feedstock compared to CW. While CH_4 production was more efficient with CW-only (higher CH_4/g VS included), more gross energy production (CH_4/g substrate added) occurred with the manure substrates due to the higher VS content of these substrates compared to the relatively dilute CW. Poultry litter generally contributed to higher CH_4 production when digested with CW compared to the use of DM, likely due to the higher N content of the PL raising the low C:N value of CW. All combinations of DM and PL showed that adding CW increased their overall CH_4 potential compared to the mono-digestion of PL or DM-only.

Co-digesting CW with PL with or without biochar in a typical rural cassava processing industry can significantly enhance the valorization of CW by yielding more CH_4 in less time and an estimated profit of >$5000/yr, with the creation of valuable fertilizer. The reduction in COD achieved through the digestion of CW could contribute significantly to reducing pollution of surface water due to indiscriminate disposal of untreated CW, as currently practiced in many rural settings. The potential application of digestate for land treatment should be further explored as a means of adding value to the overall cassava processing and value chain.

Author Contributions: Conceptualization, C.G.A., A.H., and S.L.; Data curation, C.G.A.; Formal analysis, C.G.A. and A.H.; Funding acquisition, C.G.A. and S.L.; Investigation, C.G.A.; Methodology, C.G.A., A.H., and S.L.; Project administration, S.L.; Resources, S.L.; Software, C.G.A. and A.H.; Supervision, S.L. and A.H.; Validation, C.G.A. and A.H.; Visualization, C.G.A., A.H., and S.L.; Writing—original draft, C.G.A.; Writing—review and editing, C.G.A., A.H., and S.L. All authors have read and agreed to the published version of the manuscript.

Funding: This research was funded by the US Department of State Fulbright Foreign Student Program, administered by International Institute for Education (IIE) and conducted at the Department of Environmental Science and Technology (ENST) at the University of Maryland, College Park, USA.

Acknowledgments: Special thanks to all the graduate students, especially Abhinav Choudhury, and to the undergraduate students who helped during sample preparation.

Conflicts of Interest: The authors declare no conflict of interest.

References

1. Glanpracha, N.; Annachhatre, A.P. Anaerobic co-digestion of cyanide containing cassava pulp with pig manure. *Bioresour. Technol.* **2016**, *214*, 112–121. [CrossRef] [PubMed]
2. Panichnumsin, P.; Nopharatana, A.; Ahring, B.; Chaiprasert, P. Production of methane by co-digestion of cassava pulp with various concentrations of pig manure. *Biomass Bioenergy* **2010**, *34*, 1117–1124. [CrossRef]
3. Sun, L.; Wan, S.; Yu, Z.; Wang, Y.; Wang, S. Anaerobic biological treatment of high strength cassava starch wastewater in a new type up-flow multistage anaerobic reactor. *Bioresour. Technol.* **2012**, *104*, 280–288. [CrossRef] [PubMed]
4. FAOStat 2017. Available online: http://www.fao.org/faostat/en/#data/QC/visualize (accessed on 4 November 2019).
5. Achi, C.G.; Coker, A.O.; Sridhar, M.K.C. *Utilization and Management of Bioresources. Proceedings of 6th IconSWM*; Kumar Ghosh, S., Ed.; Springer: Singapore, 2018; pp. 77–90.
6. Araujo, I.R.C.; Gomes, S.D.; Tonello, T.U.; Lucas, S.D.; Mari, A.G.; Vargas, R.J. Methane Production from Cassava Starch Wastewater in Packed-Bed Reactor and Continuous Flow. *Eng. Agric.* **2018**, *38*, 270–276. [CrossRef]

7. FAO; IFAD. A review of cassava in Africa with country case studies on Nigeria, Chana, the United Republic of Tanzania, Uganda and Benin. In *Processing of the Validation Forum on the Global Cassava Development Strategy*; The Food and Agricultural Organization of the United Nations: Rome, Italy; International Fund for Agricultural Development: Rome, Italy, 2005; Volume 2, Available online: http://www.fao.org/3/a-a0154e.pdf (accessed on 4 November 2019).
8. Zhang, M.; Xie, L.; Yin, Z.; Khanal, S.K.; Zhou, Q. Biorefinery approach for cassava-based industrial wastes: Current status and opportunities. *Bioresour. Technol.* **2016**, *215*, 50–62. [CrossRef] [PubMed]
9. Kaewkannetra, P.; Imai, T.; Garcia-Garcia, F.J.; Chiu, T.Y. Cyanide removal from cassava mill wastewater using Azotobactor vinelandii TISTR 1094 with mixed microorganisms in activated sludge treatment system. *J. Hazard. Mater.* **2009**, *172*, 224–228. [CrossRef] [PubMed]
10. Hassanein, A.; Lansing, S.; Tikekar, R. Impact of metal nanoparticles on biogas production from poultry litter. *Bioresour. Technol.* **2019**, *275*, 200–206. [CrossRef]
11. Paulo, P.L.; Colman-Novaes, T.A.; Obregao, L.D.S.; Boncz, M.A. Anaerobic Digestion of Cassava Wastewater Pre-treated by Fungi. *Appl. Biochem. Biotechnol.* **2013**, *169*, 2457–2466. [CrossRef] [PubMed]
12. Ubalua, A.O. Cassava wastes: Treatment options and value addition alternatives. *Afr. J. Biotechnol.* **2007**, *6*, 2065–2073.
13. Palma, D.; Fuess, L.T.; de Lima-Model, A.N.; da Conceicao, K.Z.; Cereda, M.P.; Tavares, M.H.F.; Gomes, S.D. Using dolomitic limestone to replace conventional alkalinization in the biodigestion of rapid acidification cassava processing wastewater. *J. Clean. Prod.* **2018**, *172*, 2942–2953. [CrossRef]
14. Amorim, M.C.C.; de S. e Silva, P.T.; Gavazza, S.; Sobrinho, M.A. Viability of rapid startup and operation of UASB reactors for the treatment of cassava wastewater in the semi-arid region of northeastern Brazil. *Can. J. Chem. Eng.* **2018**, *96*, 1036–1044. [CrossRef]
15. Glanpracha, N.; Basnayake, B.M.N.; Rene, E.R.; Lens, P.N.L.; Annachhatre, A.P. Cyanide degradation kinetics during anaerobic co-digestion of cassava pulp with pig manure. *Water Sci. Technol.* **2018**, *77*, 650–660. [CrossRef] [PubMed]
16. Carbone, S.R.; da Silva, F.M.; Tavares, C.R.; Dias Filho, B.P. Bacterial population of a two-phase anaerobic digestion process treating effluent of cassava starch factory. *Environ. Technol.* **2002**, *23*, 591–597. [CrossRef] [PubMed]
17. Jiang, H.; Qin, Y.; Gadow, S.I.; Ohnishi, A.; Fujimoto, N.; Li, Y.Y. Bio-hythane production from cassava residue by two-stage fermentative process with recirculation. *Bioresour. Technol.* **2018**, *247*, 769–775. [CrossRef] [PubMed]
18. Jiraprasertwong, A.; Maitriwon, K.; Chavadej, S. Production of biogas from cassava wastewater using a three-stage upflow anaerobic sludge blanket (UASB) reactor. *Renew. Energy* **2019**, *130*, 191–205. [CrossRef]
19. Rajagopal, R.; Saady, N.M.C.; Torrijos, M.; Thanikal, J.V.; Hung, Y. Sustainable Agro-Food Industrial Wastewater Treatment Using High Rate Anaerobic Process. *Water* **2013**, *5*, 292–311. [CrossRef]
20. Fagbohungbe, M.O.; Herbert, B.M.J.; Hurst, L.; Ibeto, C.N.; Li, H.; Umani, S.Q.; Semple, K.T. The challenges of anaerobic digestion and the role of biochar in optimizing anaerobic digestion. *Waste Manag.* **2017**, *61*, 236–249. [CrossRef]
21. Mumme, J.; Srocke, F.; Heeg, K.; Werner, M. Use of biochars in anaerobic digestion. *Bioresour. Technol.* **2014**, *164*, 189–197. [CrossRef]
22. Montalvo, S.; Guerrero, L.; Borja, R.; Sanchez, E.; Milan, Z.; Cortes, I.; Rubia, M.A. Application of natural zeolites in anaerobic digestion processes: A review. *Appl. Clay Sci.* **2012**, *58*, 125–133. [CrossRef]
23. Montalvo, S.; Diaz, F.; Guerrero, L.; Sanchez, E.; Borja, R. Effect of particle size and doses of zeolite addition on anaerobic digestion processes of synthetic and piggery wastes. *Process Biochem.* **2005**, *40*, 1475–1481. [CrossRef]
24. Lin, R.; Cheng, J.; Yang, Z.; Ding, L.; Zhang, J.; Zhou, J.; Cen, K. Enhanced energy recovery from cassava ethanol wastewater through sequential dark hydrogen, photo hydrogen and methane fermentation combined with ammonium removal. *Bioresour. Technol.* **2016**, *214*, 686–691. [CrossRef] [PubMed]
25. Roopnarain, A.; Adeleke, R. Current status, hurdles and future prospects of biogas digestion technology in Africa. *Renew. Sustain. Energy Rev.* **2017**, *67*, 1162–1179. [CrossRef]
26. Moody, L.B.; Burns, R.T.; Bishop, G.; Sell, S.T.; Spajic, R. Using biochemical methane potential assays to aid in co-substrate selection for co-digestion. *Appl. Eng. Agric.* **2011**, *27*, 433–439. [CrossRef]

27. American Public Health Association (APHA); American Water Works Association (AWWA); Water Environment Federation (WEF). *Standard Methods for the Examination of Water and Wastewater*, 21st ed.; APHA-AWWA-WEF: Washington, DC, USA, 2005.
28. Adams, R.C.; MacLean, F.S.; Dixon, J.K.; Bennett, F.M.; Martin, G.I.; Lough, R.C. The Utilization of Organic Wastes in N.Z.: Second Interim Report of the Inter-Departmental Committee. 1951. Available online: http://compost.css.cornell.edu/calc/carbon.html (accessed on 28 November 2019).
29. Peres, S.; Monteiro, M.R.; Ferreira, M.L.; do Nascimento, A.F., Jr.; Palha, A.P. Anaerobic Digestion Process for the Production of Biogas from Cassava and Sewage Treatment Plant Sludge in Brazil. *BioEnergy Res.* **2018**, *12*, 150–157. [CrossRef]
30. Witarsa, F.; Lansing, S. Quantifying methane production from psychrophilic anaerobic digestion of separated and unseparated dairy manure. *Ecol. Eng.* **2015**, *78*, 95–100. [CrossRef]
31. Lin, R.; Cheng, J.; Murphy, J.D. Unexpectedly low biohydrogen yields in co-fermentation of acid pretreated cassava residue and swine manure. *Energy Convers. Manag.* **2017**, *151*, 553–561. [CrossRef]
32. Hagos, K.; Zong, J.; Li, D.; Liu, C.; Lu, X. Anaerobic co-digestion process for biogas production: Progress, challenges and perspectives. *Renew. Sustain. Energy Rev.* **2017**, *76*, 1485–1496. [CrossRef]
33. Calabrò, P.S.; Fazzino, F.; Folino, A.; Paone, E.; Komilis, D. Semi-Continuous Anaerobic Digestion of Orange Peel Waste: Effect of Activated Carbon Addition and Alkaline Pretreatment on the Process. *Sustainability* **2019**, *11*, 3386. [CrossRef]
34. Calabrò, P.S.; Fazzino, F.; Folino, A.; Scibetta, S.; Sidari, R. Improvement of semi-continuous anaerobic digestion of pre-treated orange peel waste by the combined use of zero valent iron and granular activated carbon. *Biomass Bioenergy* **2019**, *129*. [CrossRef]
35. Abouelenien, F.; Namba, Y.; Kosseva, M.R.; Nishio, N.; Nakashimeda, Y. Enhancement of methane production from co-digestion of chicken manure with agricultural wastes. *Bioresour. Technol.* **2014**, *159*, 80–87. [CrossRef]
36. Abouelenien, F.; Namba, Y.; Nishio, N.; Nakashimeda, Y. Dry Co-Digestion of Poultry Manure with Agriculture Wastes. *Appl. Biochem. Biotechnol.* **2016**, *178*, 932–946. [CrossRef] [PubMed]
37. Riaño, B.; Molinuevo, B.; García-González, M.C. Potential for methane production from anaerobic co-digestion of swine manure with winery wastewater. *Bioresour. Technol.* **2011**, *102*, 4131–4136. [CrossRef] [PubMed]
38. Milan, Z.; Sanchez, E.; Weiland, P.; Borja, R.; Martõn, A.; Ilangovan, K. Infuence of different natural zeolite concentrations on the anaerobic digestion of piggery waste. *Bioresour. Technol.* **2001**, *80*, 37–43. [CrossRef]
39. Fagbohungbe, M.O.; Herbert, B.M.; Hurst, L.; Li, H.; Usmani, S.Q.; Semple, K.T. Impact of biochar on the anaerobic digestion of citrus peel waste. *Bioresour. Technol.* **2016**, *216*, 142–149. [CrossRef] [PubMed]
40. Jang, H.M.; Choi, Y.K.; Kan, E. Effects of dairy manure-derived biochar on psychrophilic, mesophilic and thermophilic anaerobic digestions of dairy manure. *Bioresour. Technol.* **2018**, *250*, 927–931. [CrossRef] [PubMed]
41. Hassanein, A.; Witarsa, F.; Guo, X.; Yong, L.; Lansing, S. Next generation digestion: Complementing anaerobic digestion (AD) with a novel microbial electrolysis cell (MEC) design. *Int. J. Hydrogen Energy* **2017**, *42*, 28681–28689. [CrossRef]
42. FAO 1998, Woodfuel Flow Study of Phnom Penh, Cambodia. 1998. Available online: http://www.fao.org/3/x5667e/x5667e04.htm (accessed on 28 November 2019).
43. World Weather online. Ibadan Monthly Climate Averages. 2019 November 2019. Available online: https://www.worldweatheronline.com/ibadan-weather-averages/oyo/ng.aspx (accessed on 28 November 2019).
44. Hassanein, A.; Qiu, L.; Junting, P.; Yihong, G.; Witarsa, F.; Hassanain, A. Simulation and validation of a model for heating underground biogas digesters by solar energy. *Ecol. Eng.* **2015**, *82*, 336–344. [CrossRef]
45. Asikhia, O.; Orugbo, D. Marketing Cost Efficiency of Natural Gas in Nigeria. *Pet.-Gas Univ. Ploiesti Bull.* **2011**, *LXIII*, 1–13.

Article

Evaluation of Hydrogen Sulfide Scrubbing Systems for Anaerobic Digesters on Two U.S. Dairy Farms

Abhinav Choudhury [1], Timothy Shelford [2], Gary Felton [1], Curt Gooch [2] and Stephanie Lansing [1,*]

[1] Department of Environmental Science and Technology, University of Maryland, College Park, MD 20742, USA; abhinavc@terpmail.umd.edu (A.C.); gfelton@umd.edu (G.F.)

[2] Department of Biological and Environmental Engineering, Cornell University, Ithaca, NY 14853, USA; tjs47@cornell.edu (T.S.); cag26@cornell.edu (C.G.)

* Correspondence: slansing@umd.edu

Received: 5 November 2019; Accepted: 2 December 2019; Published: 4 December 2019

Abstract: Hydrogen sulfide (H_2S) is a corrosive trace gas present in biogas produced from anaerobic digestion systems that should be removed to reduce engine-generator set maintenance costs. This study was conducted to provide a more complete understanding of two H_2S scrubbers in terms of efficiency, operational and maintenance parameters, capital and operational costs, and the effect of scrubber management on sustained H_2S reduction potential. For this work, biogas H_2S, CO_2, O_2, and CH_4 concentrations were quantified for two existing H_2S scrubbing systems (iron-oxide scrubber, and biological oxidation using air injection) located on two rural dairy farms. In the micro-aerated digester, the variability in biogas H_2S concentration (average: 1938 ± 65 ppm) correlated with the O_2 concentration (average: 0.030 ± 0.004%). For the iron-oxide scrubber, there was no significant difference in the H_2S concentrations in the pre-scrubbed (450 ± 42 ppm) and post-scrubbed (430 ± 41 ppm) biogas due to the use of scrap iron and steel wool instead of proprietary iron oxide-based adsorbents often used for biogas desulfurization. Even though the capital and operating costs for the two scrubbing systems were low (<$1500/year), the lack of dedicated operators led to inefficient performance for the two scrubbing systems.

Keywords: micro-aeration; biogas; iron; bioenergy; H_2S scrubber

---

1. Introduction:

Hydrogen sulfide (H_2S) is a corrosive gas that can corrode and damage, even in trace quantities, engine-generator sets (EGS) utilizing biogas from anaerobic digestion (AD) for electricity production. The produced H_2S can react with water vapor present in the biogas producing hydrosulfuric acid that can be further oxidized to sulfuric acid, which can cause corrosion. Hydrogen sulfide is also toxic to living organisms under certain concentrations and can result in range of adverse health effects. The US Occupational Safety and Health Administration (OSHA) lists the acceptable ceiling concentration for human exposure to H_2S to be 20 ppm for an 8-h duration [1]. In some industrial sectors, the total weighted average exposure limit is 10 ppm over 8 h. The acceptable peak concentration above the ceiling concentration is 50 ppm, but for a maximum time limit of 10 min. Concentrations exceeding 500 ppm in a closed environment can lead to death within 30–60 min, while concentrations exceeding 1000 ppm is instantly fatal [2]. Combustion of H_2S also leads to SO_x emissions, which has harmful environmental effects. Anaerobic digesters, used in conjunction with H_2S scrubbers, are effective at controlling odor problems, which is often perceived as an environmental issue by residents living close to dairy farms [3]. For digestion systems with EGS to operate effectively, it is important to remove H_2S from biogas before utilization.

Corrosion from H_2S has led to interrupted operation of farm-based EGS, resulting in increased maintenance costs and decreased revenues [4]. Biogas is a saturated (4% to 5% moisture content) mixture of 50% to 70% methane (CH_4) and 30% to 50% carbon dioxide (CO_2), with traces of H_2S (100–10,000 ppm; 0.01% to 1%). The variability of H_2S in biogas production and different efficiencies of scrubbers in reducing H_2S in the biogas over time can also affect EGS downtimes and overall lifetime [5,6]. The recommended upper limits of H_2S concentration for energy conversion technologies that use biogas are outlined in Table 1.

Table 1. Recommended hydrogen sulfide (H_2S) concentration limits for biogas utilization technologies [7,8].

Technology	H_2S Limit (ppmv)
Gas Heating Boilers	<1000
Combined Heat and Power (CHP)	<1000
Fuel Cells	<1
Natural Gas Upgrade	<4 (variations among countries)

The two H_2S scrubbing techniques discussed in this study include: (1) biological desulfurization (BDS) of H_2S using sulfur-oxidizing bacteria (SOB) to oxidize H_2S to elemental sulfur and sulfates, which can occur in a separate bio-trickling filter (BTF) or with air injection into the digester headspace, and (2) physical-chemical adsorption and oxidation using iron oxides.

Biological conversion of H_2S results from microbial oxidation in an oxygenated environment. Small concentrations of air (or oxygen) are injected into a biological scrubbing system, such as a BTF, or into the digester headspace [9]. The oxygen is used by SOB, which use H_2S, sulfur, and thiosulfate as their primary energy sources. Schieder et al. (2003) showed 90% reduction in H_2S concentrations (up to 5000 ppm) using BTF-based biogas scrubbers (BIO-Sulfex® biofilter modules (Promis Company, Warsaw, Poland), with inlet biogas flow rates ranging from 10 to 350 m^3/h [10]. A simpler method of BDS of biogas is the controlled addition of oxygen or air directly into the digester headspace, which creates a micro-aerobic environment for H_2S oxidation. However, air injection needs to be carefully controlled in order to prevent accidental formation of explosive gas mixtures of CH_4 and O_2 [3]. With differences based on the temperature, residence time, and the percentage of injected air, there have been full-scale digesters with micro-aeration that have observed reductions as high as 80% to 99%, reducing H_2S in the biogas from approximately 500 ppm to 20–100 ppm [2].

Iron oxide pellets or wood chips impregnated with iron oxide (also known as 'iron sponge') can also be used for biogas desulfurization [11]. The iron oxide in the media reacts with the H_2S and is converted into iron sulfide. Iron sponge is the most recognized iron oxide adsorbent in the industry with H_2S reductions >99.9% (3600 ppm to 1 ppm after scrubbing) reported in the literature [2]. The iron sponge adsorbent can also operate in conjunction with a small air flow into the system, along with the biogas input, to promote continuous regeneration. Sulfide removal rates up to 2.5 kg H_2S/kg Fe_2O_3 have been observed in continuously regenerated systems with <1% oxygen input [12]. Studies have shown that proprietary iron oxide-based scrubbing systems, such as SOXSIA® (Gastreatment Services, Bergambacht, Netherlands), can remove up to 2000 ppm of H_2S at 40 °C, with biogas flow rates of 1000 Nm^3/h in full-scale anaerobic digestion (AD) systems, resulting in 2 Nm^3 of H_2S removed per hour (2.9 kg H_2S/h) [8].

A previous study investigated the performance and economic benefits of two BTF systems on NY farms and found that the total annual cost to own and operate the scrubbers may not justify the capital and maintenance costs of the scrubber systems compared to increasing the frequency of oil changes [4]. It was suggested that longer monitoring periods may be necessary to understand the benefits of H_2S scrubbing on major generator overhauls. The study also highlighted the importance of a dedicated operator for keeping the systems functioning at peak efficiency. A report published on biomethane production in California estimated the cost of an H_2S scrubbing system to be around 10% of the total capital costs [3]. It was also suggested that the use of H_2S scrubbers was dependent on the end-use of the biogas, as more frequent oil changes (every 300 h instead of 600 h) could be sufficient for maintaining EGS health. Even though several H_2S scrubbing technologies exist, there is only limited field-scale data on long-term H_2S removal efficiency, and the costs associated with operating and maintaining a scrubbing system, especially on rural dairy farms in the United States [2].

The objective of this study was to quantify the efficacy and costs associated with H_2S scrubber systems using units on dairy farms with AD systems. Two different H_2S scrubber systems on rural US dairy farms were evaluated through quantification of scrubbing efficiency, capital costs, maintenance costs, and maintenance practices to determine how scrubber management affected the performance of these systems. The results can be used to understand the costs, maintenance requirements, and variations over time for these two H_2S scrubbing systems.

2. Methods

2.1. Farm and H_2S Scrubber Information

The iron oxide scrubber (IOS) on Farm 1 (S_{IOS}) treated biogas from an ambient temperature anaerobic digester. The 2574 m^3 AD system received a combination of food waste and the liquid fraction of dairy manure after solid-liquid separator. The unheated digester was exposed to ambient temperatures, which resulted in lower biogas production during winter months. In addition, there was no mixing of the substrate inside the digester. The farm (750 cows) operated a 110-kW EGS for electricity production, with the produced energy used on-farm.

The vessel for the H_2S scrubber was a 208 L plastic drum. PVC piping was used for the connection from the digester to the scrubber and then to the EGS. The iron oxide scrubber was filled with rusted scrap iron and steel scrapings (approximately 50% volume of the scrubber system). Additional rusted scrap iron (approximately 25% of the scrubber volume) was added by the farmer after 45 days of monitoring (without cleaning out used media in the vessel) to increase the efficiency of the scrubbing unit. After 105 days, the old media was removed and changed to fresh grade 000 steel wool (252 pads, 4.4 kg) (Homax, Bellingham, WA, USA) to determine if the increased surface area of this material would affect scrubber performance. The scrubber media covered three-quarter of the entire volume (156 L) of the scrubbing unit in order to enhance the contact time between the untreated biogas and the steel wool.

Biogas flow rate from the digester was measured before the biogas passed through the scrubber. There were no condensation traps before the scrubber to collect condensed water from water vapor present in the produced biogas. The biogas exited the digester and entered the bottom of the scrubber, flowing through the barrel over the scrubbing media before exiting from the top of the scrubber vessel. A regenerative blower (Gast Regenair Model—R5325R-50, Benton Harbor, MI, USA) installed at the outlet of the scrubber was used to pull the biogas through the scrubber and directed the biogas to the generator. The generator was operated only during the farm operational hours, which averaged 12 h per day.

The air injection pump for BDS (S_{BDS}) inside the digester headspace on Farm 2 was connected to a commercially designed, mixed anaerobic digester. Raw unseparated dairy manure (650 cows) was mixed with solid food waste (discarded produce) and fed into 1817 m^3 capacity digester. Electricity was generated using a 140-kW generator. The digester was heated to 35 °C using the waste heat from the EGS, with electricity sold to the grid. The generator was operated continuously, with breaks in operation for maintenance and repairs only.

The H_2S scrubber system consisted of an air pump that pumped air into the headspace of the digester. The pump (SST10 Aquatic Ecosystems Inc, Pentair, Apopka, FL, USA) was rated at 223 W, 51 Nm^3/h, and single phase (115/230 V). The air pump was set to inject air at a consistent rate of 2.86 m^3/h. A rotameter attached to the air pump, installed by the farmer, was used to measure the flowrate. The installed air pump did not have an automatic air flow regulator to change the airflow according to the amount of H_2S in the biogas. The pipe from the air pump to the digester headspace required regular maintenance to prevent clogs.

2.2. Performance Monitoring and Cost Information

The CH_4, and H_2S concentrations were logged for 179 and 73 days for S_{IOS} and S_{BDS}, respectively. The scrubber system capital costs were confirmed, and the scrubber maintenance costs were collected for at least one year from each farm. Untreated and treated biogas were analyzed to detect daily and seasonal differences using two portable continuous biogas testing and monitoring systems (Siemens Model #7MB2337-3CR13-5DR1, Siemens AG, Munich, Germany) for CH_4 (0% to 100%), CO_2 (0% to 100%), O_2 (0% to 100%), and H_2S (0–5000 ppm), with a Campbell Scientific CR1000 data logger and acquisition system, and gas meters (Model #9500, Thermal Instrument Co, Trevose, PA, USA; Model #FT2, Fox Thermal, Marina, CA, USA) and assembled as described in Shelford et al. 2019 [4]. The monitoring system were moved and installed at each farm for the study period (73 and 179 days). The Ultramat 23 was capable of an auto-calibration with air every eight hours, with regular monitoring and calibration of the units were conducted according to manufacturer's standards to maintain the accuracy of the H_2S sensors. The monitoring systems collected data for 15 min for each biogas stream (pre- and post-H_2S scrubbing). Operation and maintenance records of the AD and scrubbing systems was undertaken by the farmers, with records on the time and costs spent on their AD and scrubber system, including oil change costs, generator repair costs, and electrical energy generated over 12 months, if available.

At the end of December 2016, the gas analyzer system installed for project purposes on Farm 2 (S_{BDS}) started malfunctioning and the system had to be removed for repairs, likely due to H_2S corrosion. The on-farm biogas was then field tested using a Landtec handheld gas meter (Biogas 5000, Landtec, Dexter, MI, USA) during farm visits.

2.3. H_2S Removal Calculations

Hydrogen sulfide percent removal (η) was calculated using the formula:

$$\eta = \frac{(C_{in} - C_{out})}{C_{in}} \times 100\% \tag{1}$$

where C_{in} and C_{out} (ppm) are the scrubber inlet and outlet H_2S concentrations. The daily mass (grams/d) of sulfur removed (w) was calculated using the formula:

$$w = \frac{(C_{in} - C_{out}) \times 1.43 \times F}{1000} \tag{2}$$

where C_{in} and C_{out} (ppm) are the scrubber inlet and outlet H_2S concentrations, 1.43 kg/m^3 is the gas density at NTP (20 °C, 1 atm), and F is the biogas flow rate (m^3/d).

2.4. Statistical Analysis

Significant differences in pre- and post-scrubbed CH_4 and H_2S concentrations over time within each farm was determined using t-tests using SAS® statistical analysis software (version 9.4, SAS Institute Inc., Cary, NC, USA), with an alpha value set at 0.05. All values are presented as mean ± standard error.

3. Results and Discussion

3.1. Iron Oxide Scrubber (S_{IOS})

The mean H_2S concentrations in the pre-scrubbed and post-scrubbed biogas of S_{IOS} were 450 ± 42 ppm and 430 ± 41 ppm (based on 179 data points: n = 179), respectively, when averaged over the entire study period (August 2016–January 2017) (Figure 1). Prior to the media change from scrap iron to steel wool (n = 85), the H_2S concentrations in the pre-scrubbed biogas was 740 ± 53 ppm and post-scrubbed biogas was 719 ± 52 ppm. After the media change, the pre-scrubbed H_2S concentration (52 ± 9 ppm) was significantly higher (p-value < 0.0001) than the post-scrubbed H_2S concentration (33 ± 6 ppm). This rapid decrease (Days 102–120) in H_2S concentration is likely due to the temperature drop in the unheated digester at that time. The temperature of the digester effluent dropped from 28.1 °C in August to 10.5 °C in December, which corresponded with the ambient temperatures, which averaged 26.1 °C and 3.5 °C, respectively [13]. Sulfate reducing bacteria (SRB), the primary producers of H_2S in anaerobic digesters, have lowered activities at temperatures below 20 °C [14].

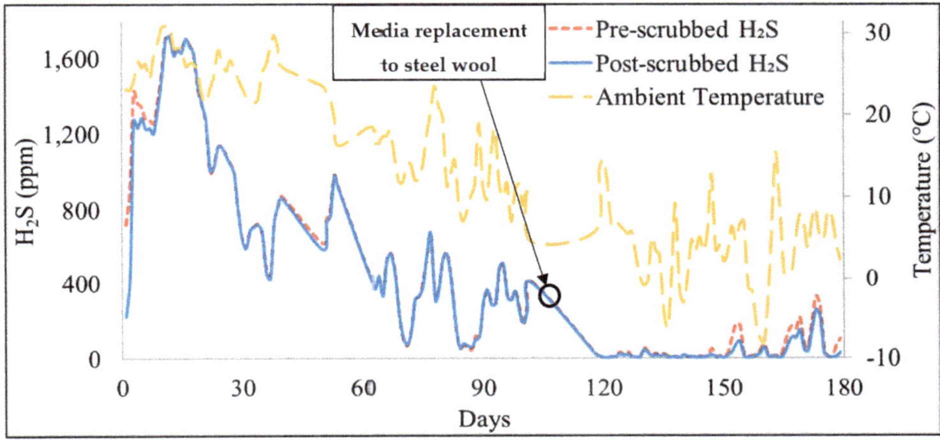

Figure 1. Hydrogen sulfide (H_2S) concentrations in the biogas from the iron oxide scrubber (S_{IOS}), with scrubber media replacement to steel wool after 105 days (mid-November).

The use of scrap iron and unoxidized steel wool as scrubbing media, instead of iron sponge or proprietary iron-oxide based adsorbents resulted in poor H_2S removal efficiencies for S_{IOS}. Dry iron-oxide based adsorbents are the most commonly used and effective scrubbing technique but can generate a hazardous waste stream [2]. Commercially available iron sponge media can be up to 100% effective, but the use of scrap iron and steel wool as the adsorption media resulted in low H_2S reduction efficiency (3%) for S_{IOS} [12]. Kohl and Nielsen (1997) also reported that wetted iron-oxide based adsorbents are not as effective as chemically hydrated oxides [15]. The steel wool media and the scrap iron media were not allowed to oxidize before being used for H_2S scrubbing, which could have contributed to the low scrubbing efficiency.

The media replacement to steel wool and the increased residence time due to the lowered biogas flow rates in the winter season resulted in a decrease in the biogas H_2S content even though the pre-scrubbed H_2S concentration was below 100 ppm. The biogas production varied from 1202 m^3/d in the summer (June to September, with an average temperature of 28 °C) to 51 m^3/d in the winter (January to February, with an average temperature of 10.9 °C) (Figure 2). The average biogas flow rate before the media change was 980 m^3/d (n = 4), which was reduced to 51 m^3/d (n = 4) due to the temperature drop that coincided with the media change. The residence time of the biogas in the scrubber increased from 0.25 min to 6 min, as the lower winter temperatures led to a sharp decline in the biogas production from the unheated digester. Commercially available iron oxide media usually require 1–15 min residence time and could have been more efficient at removing H_2S for S_{IOS}, especially during the summer months [12]. Zicari (2003) reported that a farm digester (capacity—554 m^3) with an average biogas production of 669 m^3/d could reduce H_2S concentrations from 3600 to <1 ppm, with a 4200 L iron oxide scrubber with a bed height of 240 cm [2]. The S_{IOS} volume was 208 L with an empty bed height of 88 cm (66 cm media height), with 4.2 kg of steel wool. The low adsorption efficiency seen in this study was affected by the high volume of biogas passing through the scrubber compared to the scrubber size. The total volume of biogas passing through the scrubber from August to November 2016 was 119,000 m^3, with 3.8 kg of H_2S removed from the biogas through the scrubber. After the media replacement with steel wool, a total of 1800 m^3 of biogas flowed through the scrubber in 36 days, with 68 g of H_2S removed. The low sulfur removal was likely due to the low concentrations of H_2S present in the biogas coupled with the comparatively low effectiveness of the fresh steel wool. Iron oxide-based adsorbents have been shown to remove 0.56 kg H_2S/kg adsorbent in a batch system, with a recommended bed height of 120–300 cm [15]. Based on the results from the study, the steel wool had an adsorption capacity of 0.016 kg H_2S/kg steel wool, which is an order of magnitude lower than the adsorption capacities of commercially available dry iron oxide-based sorbents.

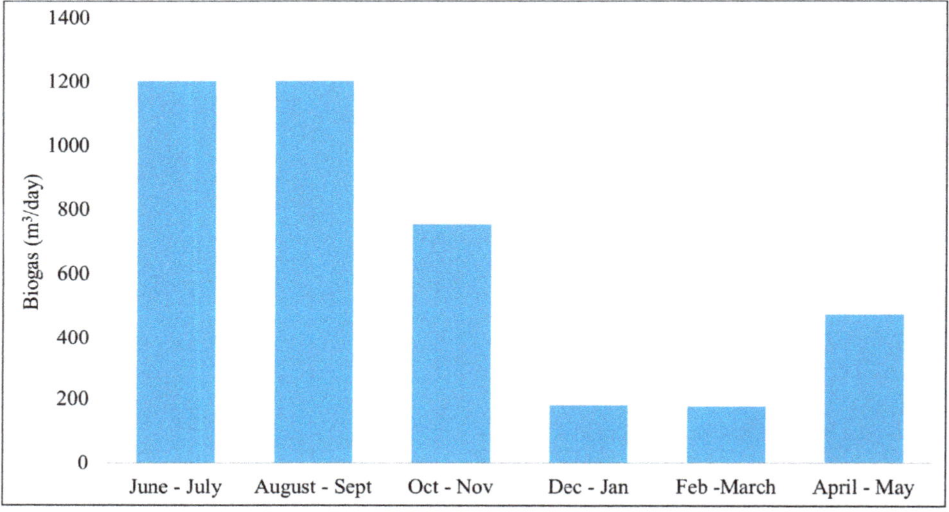

Figure 2. Average daily biogas production over two month period from June 2016 to May 2017 in the AD system with the iron oxide scrubber (S_{IOS}).

During the study period, the average CH_4 content in the pre-scrubbed biogas was 64.1 ± 0.2%, with 64.9 ± 0.2% CH_4 in the post-scrubbed biogas (Figure 3). The average daily CH_4 production rate calculated using the biogas production data over one year (June 2016 to May 2017) was 432 m^3/d or 0.58 m^3/cow.day. The daily CH_4 production rate from a mesophilic dairy manure AD system can vary from 1.5 m^3/cow·day to 3.9 m^3/cow·day [16]. As the AD system in this study was not heated, the average CH_4 yield was below this average range.

The generator produced a total of 47,158 kWh of electrical energy from the produced biogas from August to December 2016 (131 days), resulting in a daily average rate of 380 kWh/d. The EGS stopped functioning in December 2016, but the exact reason for generator failure was not determined. During daily operation, the generator did not run continuously, which could affect the EGS lifetime. The EGS had an average run-time of 12 h/d, corresponding with day-time farm operations, but variations in the EGS run-time were verified in the farmer's reports. From June to December 2016, the biogas flow rate was continuous during the EGS operational hours, with the regenerative blower suppling the biogas to the generator. The average daily CH_4 production during the monitoring period of generator activity was 542 m^3/d. The electricity generated from the biogas was 0.70 kWh/m^3 CH_4, but the flare was not metered, so the actual value may be lower than estimated.

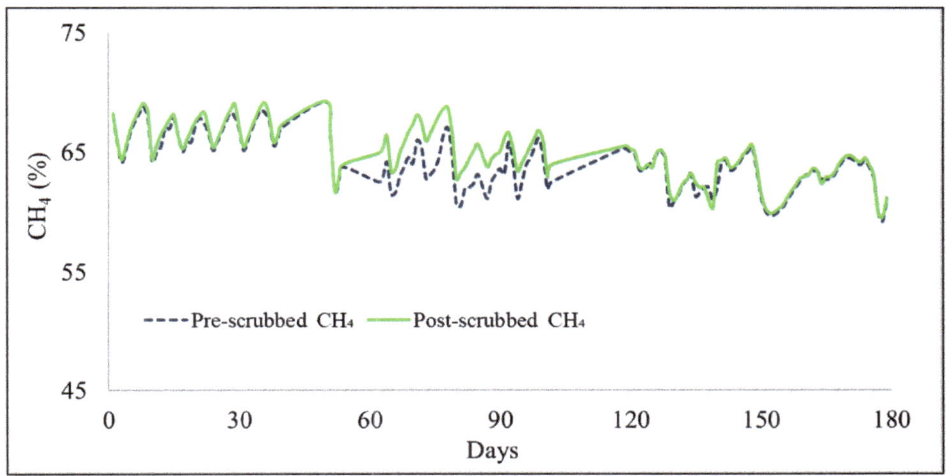

Figure 3. Daily average pre-scrubbed and post-scrubbed CH_4 concentration in biogas produced from the AD system with the iron oxide scrubber (S_{IOS}).

3.2. In-Vessel Biological Desulfurization System Using Air Injection (S_{BDS})

Overall, biogas H_2S concentrations (average: 1938 ± 65 ppm; n = 73) varied considerably during the study period from 171 to 3327 ppm, but the CH_4 (56.2 ± 0.1%) and O_2 concentrations (0.030 ± 0.004%) were consistent (October to December 2016). Correlations between the H_2S, CH_4, and O_2 were also observed, as expected (Figures 4 and 5). In mid-October (Day 7), the H_2S concentration decreased to 171 ppm, while the O_2 concentration rose to 0.51%, and the CH_4 concentration dropped to 50%, likely due to nitrogen (N_2) introduced into the biogas stream with air injection. It is likely that once the oxygen was depleted, further oxidation did not take place, and the H_2S concentration increased (after Day 9). Schieder et al. (2003) reported that micro-aeration by itself may not be sufficient to achieve complete desulfurization [10]. They collected data from biogas plants in the state of Baden-Württemberg in Germany and found that 54% of the micro-aerated AD systems had outlet H_2S concentrations >500 ppm. They suggested the use of an external biological scrubber to achieve outlet H_2S concentrations of <100 ppm and increase the life of combined heat and power (CHP) units and decrease the frequency of oil changes. In practice, digester manufacturing companies in the US have recommended limits of 500 ppm H_2S in the biogas [4]. The variable H_2S concentrations during the study period indicated variable treatment efficiency. The O_2 concentration was not always sufficient for adequate H_2S removal (<500 ppm) throughout the period after the initial rise to 0.51% O_2. The O_2 concentrations increased to 0.07% in mid-December for a short duration, which correlated with a decrease in the H_2S concentration from 2596 to 1645 ppm.

Ramos et al. (2013) showed that an outlet H_2S concentration of <200 ppm can be obtained with low O_2 (0.2% to 0.3%) concentrations in the output biogas [17]. The O_2 utilization efficiency for H_2S oxidation by the SOB increased with a decrease in the $O_{2input}/H_2S_{initial}$ ratio. Mulbry et al. (2017) also showed that an outlet H_2S concentration of <100 ppm can be obtained with 0.5% O_2 in the output biogas [18]. In S_{BDS}, the average outlet O_2 concentration was much lower (0.03%), as the air input was set at 2.86 m^3/h (2.75% of the average biogas flow rate), resulting in an average O_2 input of 0.58%. An increase in the air injection rate could have decreased H_2S concentrations further but at the cost of lowering CH_4 concentration due to N_2 dilution. The AD operator did not increase the air injection rate due to the low CH_4 concentration (50% to 55%) in the produced biogas. The EGS efficiency can be negatively affected when operated with a CH_4 concentration of <50% [15,16]. In such cases, a pure O_2 input may be desirable over air injection, but a pure O_2 input entails a higher operational cost.

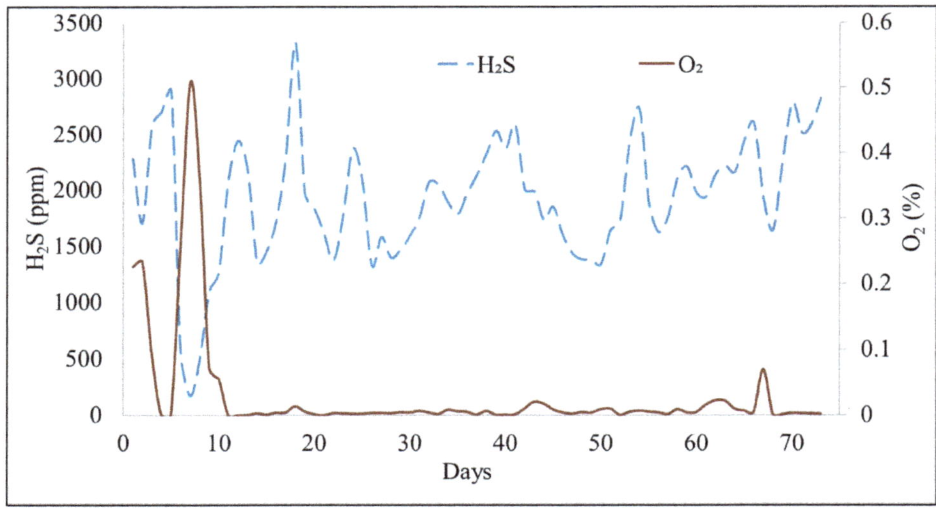

Figure 4. Hourly H_2S and O_2 concentrations in the biogas from the AD system with in-vessel biological desulfurization (S_{BDS}).

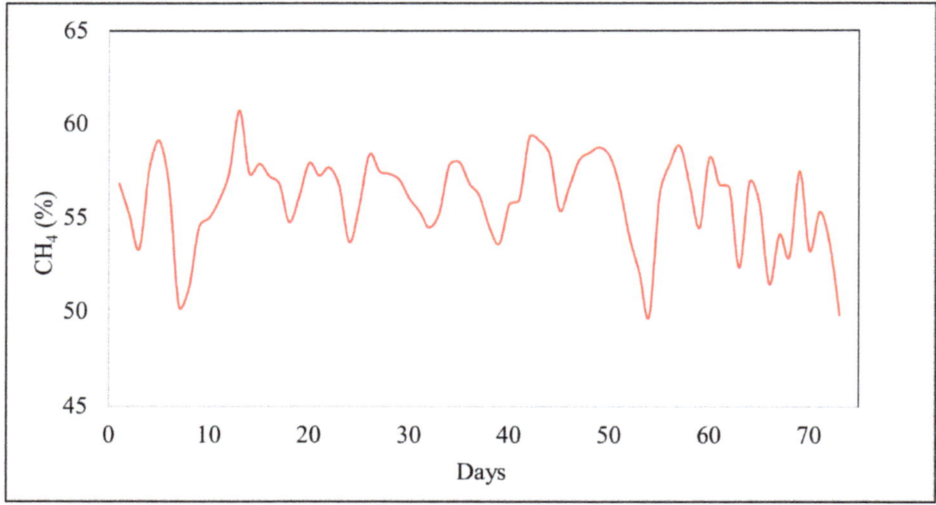

Figure 5. Hourly biogas CH_4 concentrations from the AD system with in-vessel biological desulfurization (S_{BDS}).

A constant air flow rate could have reduced the desulfurization efficiency in the digester headspace. A variable air flow rate based on the H_2S production can ensure sufficient desulfurization to meet recommended limits for heating or electricity production while minimizing N_2 dilution [19]. Ramos and Fdz-Polanco (2014) used a PID (proportional-integral-derivative) controller to vary the O_2 flow rate to meet the set output H_2S concentrations. The O_2 input was controlled using two methods: H_2S content in the biogas, and biogas production rate, and in both cases >99% removal of H_2S was obtained [20]. The ORP (oxidation–reduction potential) of the liquid wastewater was used by Khanal and Huang (2006) as a parameter to control the injection rate to prevent under-dosing/overdosing of O_2 [21].

However, instead of adding O_2 directly into the headspace, the authors injected it into the outlet of the reactor that contained a mixture of both biogas and the digester effluent. The resulting mixture was then sent to a separate sulfur oxidizing unit to separate the biogas, the effluent, and the elemental sulfur produced by the SOB. The method was able to reduce >99% of the total dissolved and gaseous sulfides for a range of initial dissolved sulfide concentrations (287 mg/L–1997 mg/L). However, using ORP as a controlling parameter could be unreliable, as each AD system is different and a set standard for an ORP increase may not be appropriate [19]. Addition of O_2/air into the digester liquid could also lead to degradation of organics in the digestate, and therefore, a higher dose of air/O_2 may be required for adequate H_2S removal [22].

Another factor that could have affected the desulfurization efficiency is the excess formation of sulfur mats in the digester headspace. The digester headspace was never cleaned, and therefore, large-sized elemental sulfur particles would drop back into the digester, along with the formation of sulfur laden biofilms on the liquid surface [18]. Sulfate reducing bacteria are also known to use elemental sulfur as an energy source for H_2S production [23]. The accumulation of oxidized sulfates and elemental sulfur can be reduced again by SRB and can lead to increased H_2S concentrations in the biogas [24]. External vessels used by Ramos et al. (2013) and Mulbry et al. (2017) that can be cleaned on a regular basis have been suggested as a better alternative to prevent reduction of the accumulated sulfates and sulfur [17,18], which resulted in a steady CH_4 production rate within the range for mesophilic digesters (1.5 m^3/cow·day to 3.9 m^3/cow·day) [16]. The farm averaged 2003 m^3/d of biogas flow through the generator (1125 m^3/d or 1.73 m^3/cow·day CH_4 yield) and produced 689,656 kWh of electricity in 10 months at a rate of 1.95 kWh/m^3 CH_4 combusted. The average rate of electricity production was 2196 kWh/d. The average biogas flow rate was affected by the generator malfunction during the last 3 weeks of data collection (Figure 6).

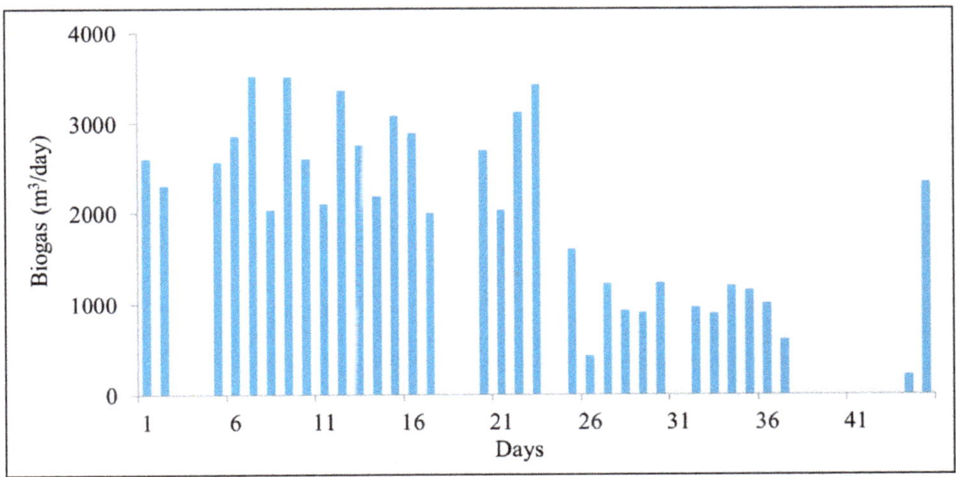

Figure 6. Daily biogas production (m^3/d) through the generator, operating on the farm with S_{BDS}, for electricity production.

3.3. Economic Analysis

The total cost of the scrubber systems was calculated using data provided by the farm owners. The total capital cost of the iron oxide scrubber system (S_{IOS}) was approximately $525 based on the reactor vessel and piping costs, as this was a homemade system. All the maintenance was conducted by the farm owner, and the labor costs were considered negligible. Additionally, scrap iron ($25 cost) was added by the farmer once during the study. Steel wool media cost $80 to fill the space within the scrubber. The replacement media for the scrubber was calculated to be $650/year with original iron scrap based on 26 media replacements per year and $960/year with grade 000 steel wool based on 12 media replacements per year. Approximately, $450/year was required for oil changes as one liter of oil was added to the generator every other day (183 L/yr). The total cost to own and operate the scrubber was $1100 (with iron scrap media) and $1410 (with grade 000 steel wool). Generator maintenance and repair can add significant costs as well, but no information was available for generator repair costs.

The total capital cost of S_{BDS} was approximately $450 for the air pump for air injection into the digester headspace. Scrubber maintenance was carried out by cleaning out the air injection connection into the digester on a weekly basis. This was estimated to take 20 min per week and cost the farm $120/year in labor costs (estimated to be $10/week at $30/h.). Oil change costs ranged from $1190 to $1795 per month and additional costs during a month were for generator repairs. The farm owner spent $10,798 for oil changes and repairs to the EGS engine head in April 2017. One of the primary reasons for the lower costs of oil change for S_{IOS} was the lower average H_2S concentrations (430 ppm) compared to S_{BDS} (1938 ppm).

Zicari (2003) tabulated data for different proprietary iron-oxide based adsorbents, where the capital costs ranged from $8000 to $43,600 and the operating costs ranged from $8290 to $23,840 for a biogas stream with 4000 ppm of H_2S and a gas flow rate of 1350 m^3/d, which is comparable to the average daily biogas flow rates for both farms in this study [2]. These cited costs were much lower than the costs associated with owning and operating the BTF units in the study conducted by Shelford et al. (2019) [4]. The operational, maintenance, and utilities costs for BTF systems in their study ranged from $17,050 for farm 2 to $32,563 for Farm 1, which are comparable to the operational costs of iron oxide scrubbers, but the capital costs were at least four times higher. The proprietary iron-oxide scrubbers examined by Zicari (2003) had high H_2S removal efficiencies and low H_2S output concentrations (up to 100% and less than 1 ppm) compared to the lower efficiencies (80.1% and 94.5%) and higher H_2S output concentrations (450 and 150 ppm) seen in the study by Shelford et al. (2019) [4,12]. However, on larger farms, the operating costs associated with iron oxide scrubbers may be much higher due to the larger volume of biogas to be treated and the higher handling and disposal costs of the spent media [12]. When the costs were normalized on the basis of volume of biogas treated, the costs were comparable, with iron-based adsorbents costs ranging from $0.024 to $0.046 per m^3 of biogas treated and BTF systems costs ranging from $0.012 to $0.03 per m^3 of biogas treated [2,4,12].

Shelford et al. (2019) also calculated the economic benefits of having a BTF scrubbing system by calculating the savings associated with less frequent oil changes after scrubber installation [4]. The farms reported a net annual loss of $61,593 for BTF 1 and $30,093 for BTF 2, which may be economically infeasible for smaller farms, especially during low milk price cycles in the US.

The results and observations from this study and Shelford et al. (2019) study showed that even though H_2S scrubbing system existed on all four farms studied, consistent performance was lacking in the inexpensive systems analyzed in our study. Both S_{BDS} and S_{IOS} had significantly lower capital and operating costs than the two BTF systems, but it is unclear if the farmers realized any economic or social benefits from these two H_2S scrubbing systems during the study period. It is also difficult to calculate monetary benefits of having the scrubbing systems, since there was no information available on oil changes prior to scrubber installation and the highly inefficient performance of the scrubbing systems. Table 2 shows the cost information of the BTF units from Shelford et al. (2019) in comparison to the scrubbing systems monitored in this study.

Table 2. Capital and operating cost summary of different scrubbing technologies in Northeast US.

Scrubber Type	Iron Oxide Scrubber (S_{IOS})	In-Vessel Biological Desulfurization (S_{BDS})	Bio-Trickling Filter 1 *	Bio-Trickling Filter 2 *
Farm Size	750 cows	650 cows	4200 cows	1500 cows
Generator Capacity	110 kW	140 kW	1000 kW	500 kW
Scrubber System Capital Cost	$525	$450	$342,000	$185,000
Annual Labor, Cleanout Costs	N/A	$0	$10,323	$4340
Annual Generator Maintenance Costs	$450	$28,708	N/A	N/A
Annual Scrubber Maintenance Costs	$960 #	$120	$8900	$9400

Notes: * data obtained from [4]; # Annual scrubber maintenance costs with steel wool as the scrubbing media.

3.4. Scrubber Management

An important factor to consider for efficient scrubber operation is scrubber management by farm or AD operators. H_2S management on agricultural digesters has lagged behind municipal and industrial digesters due to limited funding [18]. Hiring full-time operators for ensuring efficient scrubber performance can lead to unaffordable operating and labor costs, especially for farm owners with AD systems.

Changing the iron-oxide media after saturation is a labor-intensive process due to a need for careful handling of the saturated media [12]. Without proper monitoring of biogas quality, it is also impossible for farmers to know when to replace the saturated media or ascertain if biological conversion of H_2S is occurring in the digester headspace. Portable biogas quality monitoring equipment used in the study cost $17,000 and required technical expertise for regular calibration and H_2S sensor replacements every 3–6 months for accurate data collection. The farm with in-vessel biological desulfurization (S_{BDS}) had previously installed an external BTF to work in conjunction with the in-vessel micro-aeration. The BTF unit was abandoned for several years after the farmers encountered operational issues that they could not troubleshoot. It is important for manufacturers to provide on-field assistance for the maintenance of these systems for several years after they are purchased. In addition, one of the farms in the Shelford et al. (2019) study had a dedicated operator, and the H_2S scrubbing efficiency was 94.5%, whereas, the other farm had multiple personnel acting as temporary operators for the BTF unit, which contributed to the H_2S scrubbing efficiency dropping to 80.1% (Table 3) [4]. S_{IOS} and S_{BDS}, in this study, did not have dedicated operators maintaining the scrubbing systems, and monitoring H_2S concentrations in the scrubbed biogas. As a result, the scrap iron media for S_{IOS} was not replaced upon saturation, and it was impossible to determine the effectiveness of the media, leading to poor performance of the system (3% removal efficiency). In the case of S_{BDS}, regular maintenance of the air flow lines to prevent flow obstruction and appropriate modification of the air flow rates could have resulted in a lower H_2S concentration in the biogas.

Table 3. Performance summary of two different scrubbing technologies in Northeast US.

Scrubber Type	Iron Oxide Scrubber (S_{IOS})	In-Vessel Biological Desulfurization (S_{BDS})	Bio-Trickling Filter 1 *	Bio-Trickling Filter 2 *
Average Untreated H_2S (ppm)	450 ± 42	N/A	2640 ± 5.85	2350 ± 5.67
Average Treated H_2S (ppm)	430 ± 41	1938 ± 65	150 ± 1.84	450 ± 3.42
Overall removal Efficiency (%)	3.0	N/A	94.5	80.1
Avg. Mass of H_2S removed (kg/h)	0.0009	N/A	2.37	0.35
Engine-Generator Set Capacity Factor	N/A	0.76	0.93	0.68

Notes: * data obtained from [4].

In a detailed report compiled by Lusk (1998), it was shown that AD operators faced a multitude of problems caused by high H_2S content in biogas [25]. Currently, managing H_2S in biogas is still an issue, as seen from our study results. Based on interaction with the participating farmers operating the AD systems, frequent EGS oil changes to reduce corrosion instead of managing the H_2S scrubbing system were considered to be a more practical solution. Libarle (2014) found that most AD technology adopters encountered operational and maintenance issues due to a lack of training and scientific understanding of the processes involved [26]. Similar issues were observed during this study, as the farm owners of the S_{IOS} and S_{BDS} systems encountered several hurdles while trying to increase the H_2S scrubbing efficiencies of their underperforming systems. In addition, the rural locations of the farms limit access to consultants and AD experts capable of aiding farmers facing challenges from elevated H_2S concentrations in the biogas. There seems to be a need for increased assistance (education and outreach workshops, free biogas monitoring services, etc.) to impart more technical knowledge to the farm owners and offset some of the costs involved in managing and maintaining these systems.

4. Conclusions

The studied in-vessel air injection system for biological desulfurization had a low capital and time investment, with positive but inconsistent H_2S removal efficiencies. The iron-oxide scrubber also had a low time and labor investment but negligible H_2S removal efficiencies over the study period. The use of the appropriate scrubbing media (commercially available iron oxide or iron sponge) for increased reactivity and contact area, instead of scrap iron and steel wool could have increased the scrubber performance. The study also showed a substantial effect of scrubber operation and management on its performance. H_2S scrubber systems that were better managed with more time and labor investment have shown more efficient and consistent scrubbing performance. Future studies should quantify and incorporate long-term costs (5 or more years) associated with engine overhauls, down-times, repairs, etc., undertaken due to H_2S related damage to better understand the economic benefits of H_2S scrubbers.

Author Contributions: Conceptualization, G.F., C.G. and S.L.; data curation, A.C.; formal analysis, A.C.; funding acquisition, G.F., C.G. and S.L.; investigation, A.C. and T.S.; methodology, A.C., T.S., G.F., C.G. and S.L.; project administration, G.F., C.G. and S.L.; resources, C.G. and S.L.; software, A.C. and T.S.; supervision, G.F., C.G. and S.L.; validation, A.C. and T.S.; visualization, C.G. and S.L.; writing—original draft, A.C.; writing—review and editing, A.C., T.S., G.F., C.G. and S.L.

Funding: This material is based upon work supported by the National Institute of Food and Agriculture at the U.S. Department of Agriculture through a Northeast Sustainable Agriculture Research and Education (SARE) grant (#LNE15-341).

Acknowledgments: The authors would also like to thank the participating farmers on both farms for their enthusiasm and assistance with data collection for the project.

Conflicts of Interest: The authors declare no conflict of interest.

References

1. Occupational Safety and Health Administration. OSHA safety and health standards. 29 CFR 1910.1000. *Occup. Saf. Health Adm. Revis.* **1997**, *1*, 444.
2. Steven, M. *Zicari Removal of Hydrogen Sulfide from Biogas Using Cow Manure Compost*; Cornell University: Ithaca, NY, USA, 2003.
3. Krich, K.; Augenstein, D.; Batmale, J.P.; Benemann, J.; Rutledge, B.; Salour, D. *Biomethane from Dairy Waste*; Western United Dairymen: Modesto, CA, USA, 2005.
4. Shelford, T.J.; Gooch, C.A.; Lansing, S.A. Performance and Economic Results for Two Full-Scale Biotrickling filters to remove H_2S from Dairy Manure-derived Biogas. *Appl. Eng. Agric.* **2019**, *35*, 283–291. [CrossRef]
5. Gooch, C.A.; Hogan, J.S.; Glazier, N.; Noble, R. *Use of Post-Digested Separated Manure Solids as Freestall Bedding: A Case Study*; Cornell University: Ithaca, NY, USA, 2006.
6. Key, N.; Sneeringer, S. *Climate Change Policy and the Adoption of Methane Digesters on Livestock Operations*; U.S. Department of Agriculture: Washington, DC, USA, 2012.

7. Allegue, L.B.; Hinge, J. *Biogas Upgrading Evaluation of Methods for H_2S Removal*; Danish Technology Institute: Taastrup, Denmark, 2014; p. 31.
8. Petersson, A.; Wellinger, A. *Biogas Upgrading Technologies–Developments and Innovations*; IEA Bioenergy: Paris, France, 2009.
9. Oliver, J.P.; Gooch, C.A. *Microbial Underpinnings of H_2S Biological Filtration*; Cornell University: Ithaca, NY, USA, 2016.
10. Schieder, D.; Quicker, P.; Schneider, R.; Winter, H.; Prechtl, S.; Faulstich, M. Microbiological removal of hydrogen sulfide from biogas by means of a separate biofilter system: Experience with technical operation. *Water Sci. Technol.* **2003**, *48*, 209–212. [CrossRef] [PubMed]
11. Shelford, T.; Gooch, C. *Hydrogen Sulfide Removal from Biogas: Iron Sponge Basics*; Cornell University: Ithaca, NY, USA, 2017.
12. Abatzoglou, N.; Boivin, S. A review of biogas purification processes. *Biofuels Bioprod. Biorefin.* **2009**. [CrossRef]
13. Weather Underground. Available online: https://www.wunderground.com/history/daily/us/md/rising-sun/KILG (accessed on 17 October 2019).
14. Cord-Ruwisch, R.; Kleinitz, W.; Widdel, F. Sulfate-reducing Bacteria and Their Activities in Oil Production. *J. Pet. Technol.* **1987**, *39*, 97–106. [CrossRef]
15. Kohl, A.L.; Nielsen, R.B. *Gas Purification*; Elsevier: Amsterdam, The Netherlands, 1997; Volume 10, ISBN 978-0-88-415220-0.
16. Mehta, A. The Economics and Feasibility of Electricity Generation Using Manure Digesters on Small and Mid-Size Dairy Farms. 2002. Available online: https://papers.ssrn.com/sol3/papers.cfm?abstract_id=2638078 (accessed on 23 August 2019).
17. Ramos, I.; Pérez, R.; Fdz-Polanco, M. Microaerobic desulphurisation unit: A new biological system for the removal of H2S from biogas. *Bioresour. Technol.* **2013**, *142*, 633–640. [CrossRef] [PubMed]
18. Mulbry, W.; Selmer, K.; Lansing, S. Effect of liquid surface area on hydrogen sulfide oxidation during micro-aeration in dairy manure digesters. *PLoS ONE* **2017**, *12*, e0185738. [CrossRef] [PubMed]
19. Krayzelova, L.; Bartacek, J.; Díaz, I.; Jeison, D.; Volcke, E.I.P.; Jenicek, P. Microaeration for hydrogen sulfide removal during anaerobic treatment: A review. *Rev. Environ. Sci. Biotechnol.* **2015**, *14*, 703–725. [CrossRef]
20. Ramos, I.; Fdz-Polanco, M. Microaerobic control of biogas sulphide content during sewage sludge digestion by using biogas production and hydrogen sulphide concentration. *Chem. Eng. J.* **2014**, *250*, 303–311. [CrossRef]
21. Khanal, S.K.; Huang, J.-C. Online Oxygen Control for Sulfide Oxidation in Anaerobic Treatment of High-Sulfate Wastewater. *Water Environ. Res.* **2006**, *78*, 397–408. [CrossRef] [PubMed]
22. Khanal, S.K.; Huang, J.C. ORP-based oxygenation for sulfide control in anaerobic treatment of high-sulfate wastewater. *Water Res.* **2003**, *37*, 2053–2062. [CrossRef]
23. Speece, R.E. *Anaerobic Biotechnology and Odor/Corrosion Control for Municipalities and Industries*; Nashville (Ten.); Archae Press: Nashville, TN, USA, 2008; ISBN 9781578430529.
24. Greer, D. Biogas conditioning and upgrading in action. *Biocycle* **2010**, *51*, 53–56.
25. Lusk, P. *Methane Recovery from Animal Manures: A Current Opportunities Casebook*, 3rd ed.; NREL/SR-25145; National Renewable Energy Laboratory: Golden, CO, USA, 1998.
26. Libarle, D.L. *Barriers to Adoption of Methane Digester Technology on California Dairies*; California Polytechnic State University: San Luis Obispo, CA, USA, 2014.

Article

Methane and Hydrogen Sulfide Production from Co-Digestion of Gummy Waste with a Food Waste, Grease Waste, and Dairy Manure Mixture

Abhinav Choudhury and Stephanie Lansing *

Department of Environmental Science and Technology, University of Maryland, University of Maryland Energy Research Center, 1429 Animal Science/Ag Engineering Bldg., College Park, MD 20742, USA; abhinavc@terpmail.umd.edu

* Correspondence: slansing@umd.edu; Tel.: +1-(301)-405-1197

Received: 1 November 2019; Accepted: 19 November 2019; Published: 23 November 2019

Abstract: Co-digestion of dairy manure with waste organic substrates has been shown to increase the methane (CH_4) yield of farm-scale anaerobic digestion (AD). A gummy vitamin waste (GVW) product was evaluated as an AD co-digestion substrate using batch AD testing. The GVW product was added at four inclusion levels (0%, 5%, 9%, and 23% on a wet mass basis) to a co-digestion substrate mixture of dairy manure (DM), food-waste (FW), and grease-waste (GW) and compared to mono-digestion of the GVW, DM, FW, and GW substrates. All GVW co-digestion treatments significantly increased CH_4 yield by 126–151% (336–374 mL CH_4/g volatile solids (VS)) compared to DM-only treatment (149 mL CH_4/g VS). The GVW co-digestion treatments also significantly decreased the hydrogen sulfide (H_2S) content in the biogas by 66–83% (35.1–71.9 mL H_2S/kg VS) compared to DM-only (212 mL H_2S/kg VS) due to the low sulfur (S) content in GVW waste. The study showed that GVW is a potentially valuable co-digestion substrate for dairy manure. The high density of VS and low moisture and S content of GVW resulted in higher CH_4 yields and lower H_2S concentrations, which could be economically beneficial for dairy farmers.

Keywords: biochemical methane potential; biogas; anaerobic digestion

1. Introduction

Anaerobic digestion (AD) of organic substrates with dairy manure, also known as co-digestion, can increase biogas production and result in higher return on investment for dairy farmers [1]. Biogas produced from AD is a combination of 50–75% methane (CH_4) and 25–50% carbon dioxide (CO_2), with trace levels (0.01–1%) of hydrogen sulfide (H_2S) that can be used as a source of renewable energy for heat and power generation [2]. Limitations from mono-digestion of organic materials arise from substrate properties, such as unbalanced C:N ratios, recalcitrance in the feedstock, high concentrations of long chain fatty acids, and deficiency in trace minerals required for the growth of methanogens [1,3]. These limitations can lead to unfavorable economics for dairy farmers using AD to generate energy on-farm [1,4]. Furthermore, positive synergy from co-digestion of a mixture of substrates can lead to more CH_4 production than the addition of CH_4 produced from mono-digestion of each individual substrate. A review by Mata-Alvarez et al. (2014) reported that co-digestion of carbon (C)-rich organic matter with cattle and poultry manure resulted in up to 3.5 times more CH_4 production than the CH_4 potential of the individual substrates [3]. Lisboa and Lansing (2013) reported a maximum of 29.4 times more CH_4 yield when dairy manure was co-digested with chicken processing waste compared to mono-digestion of dairy manure [5]. Moody et al. (2011) determined the biomethane potential of a wide range of food waste substrates and concluded that co-digestion of manure and organic waste has the potential to increase biogas production, and in turn, increase energy generation from AD [6]. However, often studies are only applied to individual substrates due to differences in organic waste composition and collection.

Previous research on co-digestion of food waste and dairy manure has primarily focused on the CH_4 production potential of co-substrates [7–9], with limited data on the effects of co-digestion substrate selection on the production of H_2S [10]. The production H_2S in biogas occurs when sulfur-containing compounds, such as sulfates, sulfites, and thiosulfate, in AD substrates are reduced by sulfate-reducing bacteria (SRB) under anaerobic conditions [11]. High H_2S concentrations in biogas (0.05–1% by vol.) can become a major problem when utilizing the biogas due to health concerns and corrosion of biogas equipment [12]. Combined heat and power (CHP) systems usually require H_2S concentrations to not exceed 500 ppm to prevent reduced performance from corrosion, and H_2S concentrations over 100 ppm can lead to severe adverse human health impacts [10]. Most dairy farms use CHP systems to generate energy for on-farm use and lower H_2S concentrations can lead to improved energy generation efficiencies and reduced maintenance. Corro et al. (2013) observed a reduction in H_2S concentrations when coffee waste was co-digested with dairy manure compared to digestion of dairy manure only, but there was no discussion of the cause for the observed H_2S differences [13]. Research has shown that co-digestion of organic matter with higher C:N ratios in manure-based digesters can reduce ammonia inhibition and enhance methane production [3]. Co-digestion of carbon-rich organic matter with a low sulfur (S) content may also reduce the H_2S concentration in the biogas when compared to the mono-digestion of dairy manure and prevent sulfide inhibition.

Industrial food waste comprises 5% of the total food waste generated globally [14]. Although the fraction of industrial food waste is significantly less than food waste from other sources, it has logistical and economic advantages due to its high-volume generation at specific points and homogenous nature. Valorization of these industrial food waste streams can help mitigate disposal costs in landfills, while providing a source of tipping fees for dairy farmers with AD systems. The waste produced from gummy vitamin industries is high in degradable C compared to dairy manure. Production of gummy vitamin waste (GVW) from a single manufacturing facility can be up to 10% of the total weight of the product produced [15]. For example, one multi-national gummy vitamin manufacturing company produces approximately 100 million gummy vitamins daily, with a daily production of 500 tons of gummy product (5 g per gummy vitamin), resulting in approximately 50 tons/day of GVW produced [16]. Most of this waste product is landfilled, with some composting and incineration being practiced in the EU [15,17]. The GVW material can contain up to 70% sugar and gelatin, with starch or pectin-based gels that create the unique structure that is characteristic of gummy candies [18]. Due to its high sugar content, GVW can be a valuable resource for AD, yet the dense jelly-like consistency may lead to issues, such as a slow degradation rate, increased hydraulic retention time, or possible pipe clogging within the AD system. It is also possible that GVW with a high C:S ratio could reduce the H_2S concentration in the biogas when co-digested with dairy manure.

The main goal of the project was to evaluate a GVW product as a co-digestion substrate for AD. The specific objective was to evaluate the CH_4 and H_2S production and VS degradation of a GVW substrate when co-digested with a dairy manure (DM), food waste (FW), and grease waste (GW) mixtures (DM.FW.GW). A co-digestion mixture was used for testing, as many on-farm digesters incorporate multiple waste streams and to highlight the benefits of testing co-substrates as both mixtures and single substrates. Co-digestion of the tested mixtures was expected to produce a significantly higher amount of CH_4 and lower H_2S compared to the mono-digestion of DM.

2. Materials and Methods

2.1. Sample Collection

Anaerobic digester effluent (inoculum source) and the GVW product were collected from a Northeastern US farm. The farm co-digested dairy manure from heifers with gummy vitamin waste, food waste, and grease waste at a 64% DM, 9% GVW, 16% FW, and 11% GW ratio, by mass. The AD effluent sample was utilized as an inoculum source, as it had been pre-acclimated to the GVW material used at the farm. The GW and FW were collected from a local supermarket. Un-separated dairy manure from the USDA Beltsville Agricultural Research Center (BARC) in Beltsville, MD, USA, was utilized as the DM substrate. Field samples were collected and brought back to lab on ice. The mean total solids (TS) and volatile solids (VS) data for the substrates used in the experiment are shown in Table 1.

Table 1. Total and volatile solids content of the individual substrates (gummy vitamin waste, food waste, grease waste, dairy manure) and digester effluent (inoculum) used for the experiment.

Parameters	Gummy Vitamin Waste	Food Waste	Grease Waste	Dairy Manure	Inoculum
Total Solids (g/kg)	464 ± 2.0	91.0 ± 1.0	673 ± 4.5	94.5 ± 3.6	64.8 ± 0.9
Volatile Solids (g/kg)	463 ± 2.1	83.1 ± 1.1	645 ± 1.5	81.7 ± 3.6	47.5 ± 0.8

2.2. Experimental Design

The GVW product was added to individual batch digesters at four inclusion levels (0%, 5%, 9%, and 23% on a wet mass basis) to a co-digestion substrate mixture of dairy manure (DM), food-waste (FW), and grease-waste (GW) and compared to mono-digestion of the GVW, DM, FW, and GW substrates, with an inoculum control. The 9% GVW treatment (64% DM, 16% FW, 11% GW by mass) represented the mixture that was used at the farm during the time of AD effluent collection. An inoculum-to-substrate ratio (ISR) of 1:1 (VS basis) was used for the experiment. Table 2 shows the experimental design and the descriptions of the treatment levels for the experiment, with each treatment conducted using triplicate AD reactors. All mass data are expressed on a wet mass basis.

Table 2. Experimental design using a 1:1 inoculum-to-substrate ratio, with the calculated initial total solids (TS) and volatile solids (VS) of the treatment mixtures. The percent of gummy vitamin waste (GVW) inclusion was based on mass. All treatments were conducted in triplicate.

Digestion Substrate and Inoculum	Inoculum (g)	DM (g)	FW (g)	GW (g)	GVW (g)	TS (g/L)	VS (g/L)
Inoculum control	31.9	-	-	-	-	64.1	47.0
Dairy manure (DM)	31.9	18.3	-	-	-	71.7	59.5
Food waste (FW)	31.9	-	18.1	-	-	74.2	60.0
Grease waste (GW)	31.9	-	-	2.3	-	105	87.6
Gummy vitamin waste (GVW)	31.9	-	-	-	3.2	101	85.5
DM.FW.GW (0% GVW)	23.9	5.2	1.4	0.9	-	86.3	71.5
GVW.DM.FW.GW (5% GVW)	28.1	5.2	1.4	0.9	0.4	88.2	73.5
GVW.DM.FW.GW (9% GVW)	31.9	5.2	1.4	0.9	0.8	89.5	74.5
GVW.DM.FW.GW (23% GVW)	47.9	5.2	1.4	0.9	2.4	93.1	78.0

2.3. Biochemical Methane Potential (BMP) Test Procedures

The batch laboratory testing followed the biochemical methane potential (BMP) protocol, which is a laboratory batch study used to characterize CH_4 production potential [6]. Substrate and inoculum were added into 300 mL serum bottles, purged with N_2 gas to establish anaerobic conditions, capped, and incubated at 35 °C in an environmental chamber. Biogas, CH_4, and H_2S concentrations were monitored at regular intervals for 67 days, at which point the daily biogas production was less than 1% of the cumulative biogas production for the treatments, indicating biogas production had largely ceased. The mass of substrate and inoculum in each bottle ranged from 31.4 to 58.8 g (Table 2) to keep the ISR at 1:1 for all treatments.

The quantity of biogas produced was measured using a graduated, gas-tight, wet-tipped 50 mL glass syringe inserted through the septa of the digestion reactors and equilibrated to atmospheric pressure. Biogas samples were collected in 0.5 mL syringes and tested on a gas chromatograph (Agilent 7890) using a thermal conductivity detector (TCD) at a detector temperature of 250 °C for CH_4

and H_2S concentrations. The average CH_4 and H_2S production in the triplicates from the inoculum control was subtracted from the other treatments to present the total CH_4 production from the waste substrates only.

2.4. Analytical Methods

The treatment mixtures were analyzed for pH before and after digestion using an Accumet AB15 pH meter. Triplicate samples were tested for TS and VS, according to Standard Methods for the Examination of Water and Wastewater (APHA-AWWA-WEF, 2005) within 24 h of collection. For TS analysis, triplicate 10.0 mL samples were pipetted into pre-weighed porcelain crucibles. The samples were then dried at 105 °C until a constant mass was obtained for the TS concentration. The crucibles were then placed in a furnace at 550 °C until a constant weight was obtained to determine VS concentration. The gummy waste, dairy manure, and inoculum (digester effluent) were tested for total metals (iron, zinc) and sulfur using ICP-MS (inductively coupled plasma mass spectrometry), and total nitrogen using A3769 Methods for Manure Analysis at Agrolabs Inc., Harrington, DE, USA, [19]. The C:N ratio was calculated using the conversion factor from Adams et al. (1951) stating that 55% of the VS content is carbon [20]. The calculated C value and the measured N value were used to derive the C:N ratio.

2.5. Statistical Analysis

Collected data were reviewed in accordance with QA/QC procedures and analyzed for significant differences in biogas quantity, CH_4, H_2S, TS, VS, and pH using ANOVA and Tukey-Kramer post-hoc multiple mean comparison tests of the reviewed data using SAS ® statistical software package. Tests of significance were conducted with an alpha value set at 0.05. Data are reported as averages with standard errors (SE).

3. Results

3.1. Methane (CH_4) Production

The co-digestion mixtures 0–23% GVW.DM.FW.GW had a significantly higher percent CH_4 in the biogas compared to the mono-DM digestion (p-value < 0.0001; Table 3). However, there were no significant differences in the percent CH_4 among the co-digestion mixtures, with a non-significant trend in increasing percent CH_4 as the percent of GVW increased (Table 3). The cumulative CH_4 production over the 67 day AD period was normalized using two methods: (1) the total mass of the substrate added (mL CH_4/g substrate), as this normalization provides an estimate of CH_4 production that can be readily used by farmers, and (2) the VS of the substrate (mL CH_4/g VS added) for comparison with other studies [5].

Table 3. Methane (CH_4) and hydrogen sulfide (H_2S) production data from the batch digestion testing.

Treatment	CH_4 (%) *	mL CH_4/g VS	mL CH_4/g Substrate	mL H_2S/kg VS	mL H_2S/kg Substrate
Dairy manure (DM)	53.7 ± 0.5	149 ± 11	12.2 ± 0.1	212 ± 17	17.4 ± 1.4
Food waste (FW)	14.8 ± 1.1	0 #	0 #	99.7 ± 8.8	8.3 ± 0.7
Grease waste (GW)	25.7 ± 3.0	10 ± 4.5	6.3 ± 2.9	33.1 ± 30.4	21.4 ± 19.6
Gummy vitamin waste (GVW)	6.98 ± 0.9	0 #	0 #	7.0 ± 0.1	3.2 ± 0.1
DM.FW.GW (0% GVW)	67.4 ± 0.2	373 ± 6	56.0 ± 0.8	35.1 ± 2.2	5.3 ± 0.3
GVW.DM.FW.GW (5% GVW)	66.6 ± 1.6	374 ± 12	62.5 ± 2	71.9 ± 13.7	12.0 ± 2.3
GVW.DM.FW.GW (9% GVW)	68.3 ± 1.2	355 ± 3	64.1 ± 0.5	70.4 ± 5.2	12.7 ± 0.9
GVW.DM.FW.GW (23% GVW)	71.1 ± 1.0	336 ± 12	76.3 ± 2.7	68.3 ± 16.6	15.5 ± 3.8

* The % CH_4 shown is the average value from Days 53–67 of the experiment. # The CH_4 production from the inoculum was subtracted from all treatments, resulting in zero values when the inoculum outperformed the treatment.

As expected, the co-digestion treatments (with and without GVW addition) produced 359–524% more CH_4 compared to mono-DM digestion, when normalized by the mass of substrate added (Table 3).

Normalized CH_4 production in co-digestion without GVW (DM.FW.GW-only) was 11.6% lower than the 5% GVW.DM.FW.GW mixture, 14.5% lower than 9% GVW.DM.FW.GW mixture, and 36.3% lower than the 23% GVW.DM.FW.GW mixture (Table 3; Figure 1). The CH_4 production in the 23% GVW.DM.FW.GW mixture was the highest among all treatments, as expected. The total normalized volume of CH_4 increased linearly with the mass percent of GVW added ($r^2 = 0.9866$) (Figure 2).

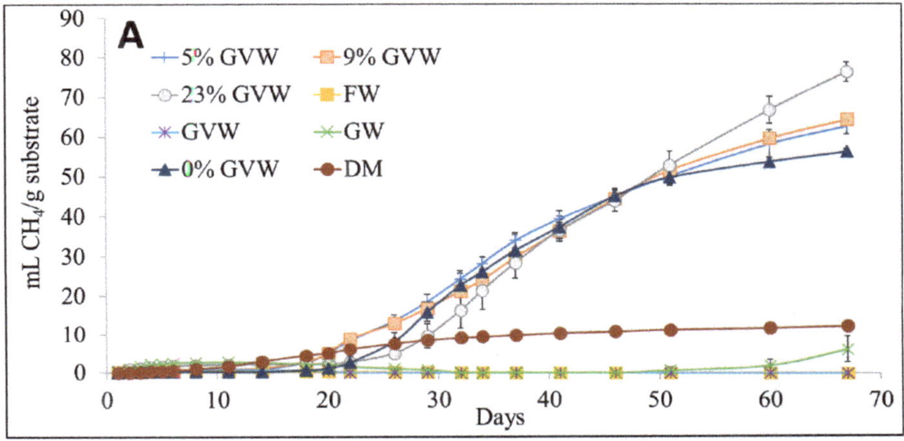

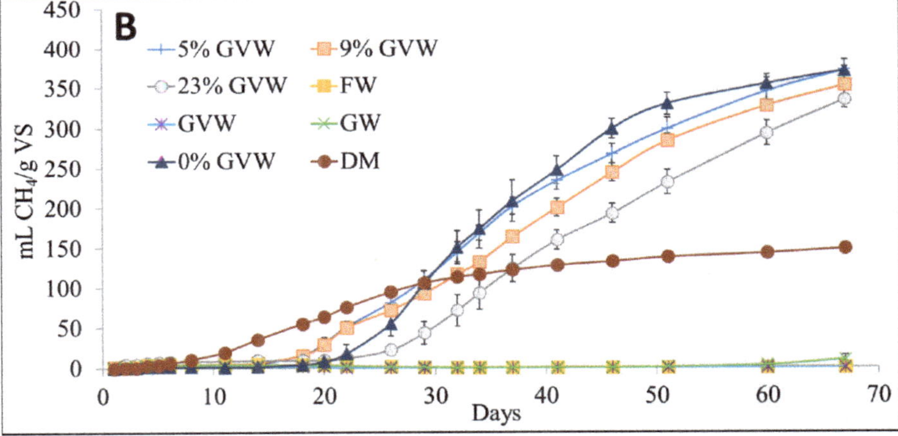

Figure 1. Methane (CH_4) production normalized by gram of substrate (mL CH_4/g substrate) ((**A**), top) and by gram of volatile solids (mL CH_4/g VS) ((**B**), bottom) in the batch digestion testing of gummy vitamin waste (GVW), grease waste (GW), food waste (FW), and dairy manure (DM) digested singularly and as a mixture (DM.FW.GW), with the percent inclusion of GVW shown for the co-digestion mixtures.

When the total CH_4 produced was normalized by the quantity of organic material added (mL CH_4/g VS), the 23% GVW.DM.FW.GW mixture was significantly lower than the DM.FW.GW mixture with 0% GVW (p-value = 0.0156) and 5% GVW.DM.FW.GW mixtures (p-value = 0.0122) (Table 3), with no significant differences between the other co-digestion treatment groups. Mono-GVW digestion resulted in negligible CH_4 production (0 mL CH_4/g VS) over 67 days of digestion due to subtraction of inoculum CH_4 production from each treatment, and higher CH_4 production values in the triplicate inoculum reactors compared to the triplicate GVW-only AD reactors. Both treatments with negligible CH_4 production (mono-GVW and mono-FW) had low final pH levels in the digestion vessels (under pH 7) (Table 4).

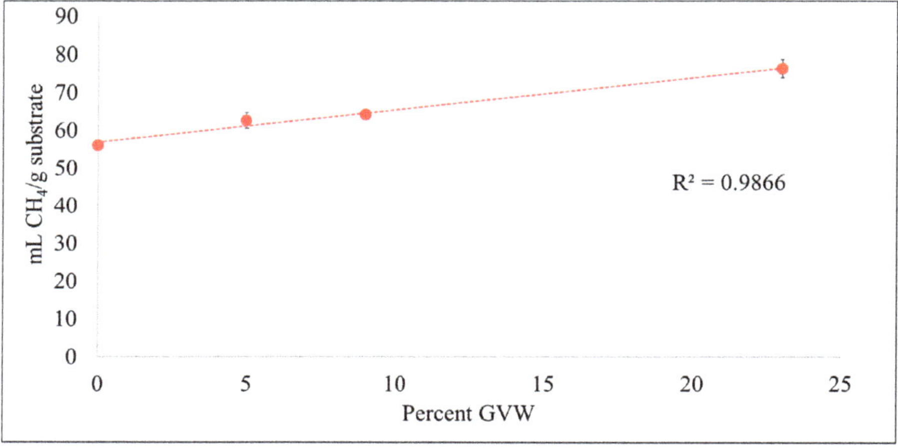

Figure 2. Linear regression of normalized methane (CH_4) production per gram of added substrate and percent gummy vitamin waste (GVW) within the co-digestion mixture.

Table 4. Average pH and volatile solids (VS) in all treatment mixtures pre-digestion (initial) and post-digestion (final). Initial VS data was calculated theoretically, and final VS data was determined experimentally.

Treatment	Initial VS (g/L)	Final VS (g/L)	Decrease in VS (%)	Initial pH	Final pH
Dairy manure (DM)	59.5	48.0 ± 1.8	19.3%	7.64	7.75
Food waste (FW)	60.0	42.0 ± 2.5	30.0%	7.11	6.24
Grease Waste (GW)	87.5	79.5 ± 1.1	9.1%	7.79	7.21
Gummy vitamin waste (GVW)	85.5	53.0 ± 0.5	38.0%	7.75	6.24
DM.FW.GW (0% GVW)	71.5	49.4 ± 0.8	30.9%	7.92	7.97
GVW.DM.FW.GW (5% GVW)	73.5	47.6 ± 3.0	35.2%	7.84	7.95
GVW.DM.FW.GW (9% GVW)	74.5	49.2 ± 1.3	34.0%	7.87	7.95
GVW.DM.FW.GW (23% GVW)	78.0	51.0 ± 2.6	34.6%	7.77	7.88

3.2. *Hydrogen Sulfide (H_2S) Production*

The DM treatment produced biogas with a peak concentration of 2145 ppm H_2S after 3 days of digestion (Figure 3). After this time, H_2S levels decreased and no H_2S was detected in the biogas by the 60th day of the experiment. The treatment with the next highest peak H_2S concentration in the biogas was the 9% GVW.DM.FW.GW mixture (804 ppm H_2S), which was 63% less than the DM treatment and 23% greater than the next highest treatment (DM.FW.GW-only mixture with 0% GVW) at 576 ppm H_2S. The peak H_2S concentrations for all treatments were observed within the first 2–3 days before peak CH_4 production. The 23% GVW.DM.FW.GW treatment, DM and FW had detectable H_2S concentrations in the biogas for the longest period (51 days). The mono-GVW treatment did not produce a measurable amount of CH_4, but it had the shortest period of detectable levels of H_2S (5 days). This is likely due to lowered microbiological activity within the digester due to the low pH levels, which led to low biogas production.

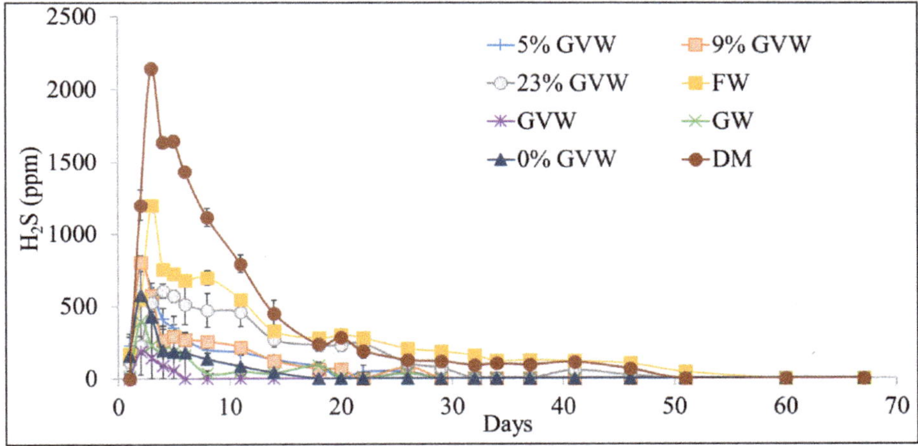

Figure 3. Hydrogen sulfide (H_2S) concentration (ppm) in the biogas over time in the batch digestion testing of gummy vitamin waste (GVW), grease waste (GW), food waste (FW), and dairy manure (DM) digested singularly and as a mixture, with the GVW inclusion shown for each co-digestion mixture tested.

The quantity of H_2S produced showed an increasing trend with increases in the percent of GVW inclusion (0–23%) when normalized by kilograms of substrate addition (5.3–15.5 mL H_2S/kg substrate; Table 3, Figure 4). The H_2S production in the DM treatment (17.4 mL H_2S/kg substrate) was significantly higher than the treatments co-digested with GVW (p-value = 0.0046). However, in the DM.FW.GW treatment (0% GVW), the normalized H_2S production was the lowest among the co-digested treatments (5.3 mL H_2S/kg substrate), and significantly lower than 23% GVW.DM.FW.GW (p-value = 0.0106) and DM (p-value = 0.0023) treatments. However, there were no significant differences for normalized H_2S production between the 5–23% GVW inclusion (p-value = 0.633) treatments.

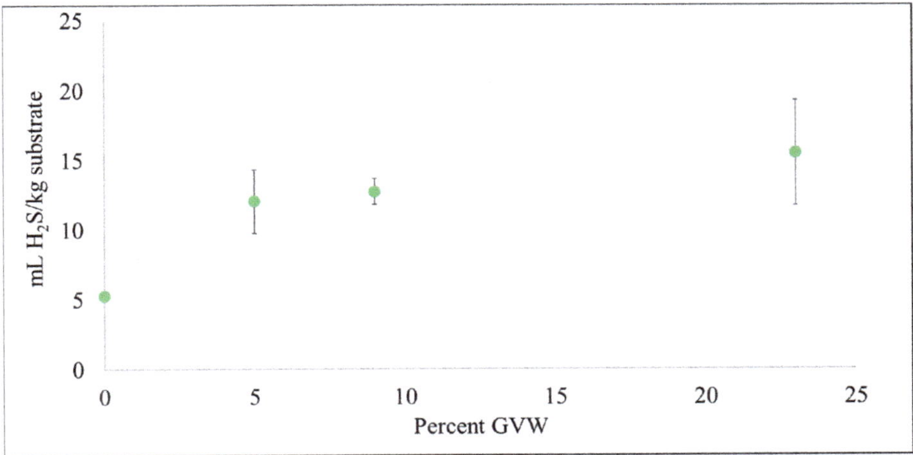

Figure 4. Normalized hydrogen sulfide (H_2S) production per kilogram of added substrate and percent gummy vitamin waste (GVW) within the co-digestion mixture.

When the total H_2S was normalized by the amount of VS added, the DM treatment (212 mL H_2S/kg VS) produced a significantly larger amount of H_2S compared to all co-digestion treatments (p-value < 0.0001) (Table 3). The addition of GVW (68–72 mL H_2S/kg VS) showed a significant increase in H_2S production compared to the DM.FW.GW (0% GVW) treatment (35 mL H_2S/kg VS; p-value = 0.0003). However, there were no significant differences within the 5–23% GVW.DM.FW.GW treatments (p-value = 1.000).

3.3. *Effect of Retention Time and Solids Degradation*

The percentage of CH_4 in the biogas of the DM treatments rose above 25% on the 11th day of digestion, while the treatments containing additional substrates (FW, GW, and GVW) had a longer lag phase and started producing higher quantities of CH_4 after 20 days of digestion (Figure 1), which is a relatively long lag-time for BMP analyses. The DM treatment produced 43% of its total cumulative CH_4 within the first 20 days, while all other treatments had less than 10% of the total cumulative CH_4 production during this time (Table 5). By the 41st day of the experiment, 89% of the total cumulative CH_4 from the mono-DM treatment had been produced, but the percent of total cumulative CH_4 from the GVW.DM.FW.GW and DM.FW.GW treatments by Day 41 varied from 57–80% of the cumulative CH_4 after 67 days of digestion. The effect of the longer retention times on GVW degradation was seen, as the CH_4 production rate for co-digestion was highest when no GVW was added (DM.FW.GW), with a maximum CH_4 production rate of 16.8 mL CH_4/VS.day). The maximum CH_4 production rate decreased with increasing GVW inclusion (10.6–11.6 mL CH_4/VS.day). The maximum CH_4 production rate was the lowest for DM (6.0 mL CH_4/VS.day) for the treatments with CH_4 generation.

Table 5. Normalized methane production (mL CH_4/g VS) after 20, 46, and 67 days, with the percentage of the cumulative CH_4 (Day 67) by Days 20 and 46 shown in parentheses.

Treatment	Day 20 (mL CH_4/g VS)	Day 46 (mL CH_4/g VS)	Day 67 (mL CH_4/g VS)
Dairy manure (DM)	64 (43%)	133 (89%)	149
DM.FW.GW (0% GVW)	7 (2%)	299 (80%)	373
GVW.DM.FW.GW (5% GVW)	30 (8%)	268 (72%)	374
GVW.DM.FW.GW (9% GVW)	29 (8%)	245 (69%)	355
GVW.DM.FW.GW (23% GVW)	10 (3%)	193 (57%)	336

The C:N ratios of the GVW (196:1) was high due to the high C (255 g C/kg GVW) and low N content (1.3 g N/kg GVW), which was much higher than the dairy manure (7.7:1) and inoculum (8.0:1) utilized. The TS and VS concentrations of the GVW showed that the VS comprised 99.7% of the total solids content (46.4% of the wet GVW). While a high percentage of the GVW was degradable, there was only a 34–35.2% degradation of VS during digestion (Table 4). While there was no CH_4 production from the mono-FW and mono-GW treatments, there was a decrease of >30% of the initial VS content, which can be attributed to the initial breakdown of the organic matter, resulting in CO_2-enriched biogas production. Biogas volume for these treatments was over 200 mL during the first two days, with less than 0.5% CH_4 and over 35% CO_2 for mono-FW and over 50% CO_2 for mono-GVW treatments.

4. Discussion

Increasing the amount of GVW during digestion did increase CH_4 production, as expected. The GVW appeared to completely hydrolyze during digestion, with no visible trace of solid GVW in the post-BMP samples after 67 days of digestion. The GVW accounted for 5–23% of the total mass of substrate added, corresponding to 15–50% of the VS inclusion. The GVW product could be beneficial for farmers interested in co-digestion waste substrates that increase CH_4 production, but the longer retention time of the GVW compared to DM digestion should be taken into consideration.

The negligible CH_4 production and low pH values in the mono-GVW, FW, and GW treatments compared to the higher CH_4 production (336–374 mL CH_4/g VS) and pH range (7.88–7.95) in treatments that co-digested GVW, FW, GW, and DM showed that the buffering capacity of the added co-substrates is important to mitigate accumulation of volatile fatty acids (VFA) and lowered pH [3,21]. Carbon-rich substrates can have a poor buffering capability, leading to an increased rate of VFA production and methanogenesis inhibition [3]. The mono-GW treatment had an initial pH of 7.79 but did not produce significant amounts of CH_4, possibly due to the slow degradation rate of lipids in the grease waste. Previous studies have also shown that digestion of lipids without co-digestion required the use of lime as a pH stabilizer [22]. The use of a buffer for pH control in the experiment was avoided since the study was originally conducted to emulate the source farm conditions. The AD system on farm did not use any pH stabilizers, as the manure provided sufficient buffering capacity for the digestion process. Generally, the high alkalinity of manure increases digester resistance to acidification for high-fat and sugar content wastes and adds a nitrogen source for micro-organisms [23]. Another important parameter that likely resulted in negligible CH_4 production in the mono-GVW treatment was the high C:N ratio of GVW (196:1). High C:N ratios have been shown to result in low pH values during the digestion process and high VFA production [24]. As DM had a C:N ratio of 7.7:1 in this study, which is typical for DM, the resulting mixture in the co-digestion treatments likely increased the C:N ratio within the ideal range of 20–30 for AD, resulting in large increases in CH_4 yield for the co-digestion mixtures compared to the mono-digestion treatments [25].

All treatments produced large amounts of biogas during the first two days of digestion (ranging from 39 mL for DM to 379 mL for 23% GVW.DM.FW.GW), mostly composed of CO_2. The biogas volume dropped sharply for all treatments (<10 mL per day) after Day 2, and the mono-DM treatment recovered the earliest (Day 11) and started producing > 50 mL biogas per day. The reduction in VS in the treatments with negligible CH_4 production for FW, GVW, and GW (Table 4) can be attributed to this initial burst of CO_2 enriched biogas production due to the initial breakdown of complex organic molecules. Bujoczek et al. (2000) showed that high organic loading rates may initially lead to large amounts of biogas, composed mainly of CO_2, after which biogas production slows down [26]. In their study, the biogas production recovered after 30 days of digestion with CH_4 as the main component, similar to the results seen in this experiment. The authors also reported that the highest TS content for feasibility of digestion was 10%, while the shortest lag phase was obtained for 2.7% TS. The TS content in our experiment varied from 7.1% for DM to 11.6% for FW and showed similar CH_4 production trends to their study. The longer lag phase associated with a high TS content could be due to either high VFA concentrations or high ammonia concentrations or a combination of the two factors [26]. The CH_4 production in this study recovered after the lag phase, indicating acclimatization of the methanogenic bacteria to the initial inhibitory conditions, but the quantity of CH_4 generated from the DM treatment (149 ± 11 mL CH_4/g VS) was 38–44% lower than the results obtained by Moody et al. (2011) for dairy manure (239–264 mL CH_4/g VS) [6]. Witarsa and Lansing (2015) showed that the normalized CH_4 production on a VS basis is often lower for unseparated dairy manure due to the recalcitrant nature of the manure solids, leading to lower VS conversion efficiency [27].

It was expected that CH_4 production normalized by VS in the GVW co-digested treatments would be similar, but a decreasing trend with increasing percent GVW was observed. Normalization by VS illustrates the efficiency of organic material conversion to CH_4. As GVW is a dense substrate in terms of grams of VS per gram of substrate, the increase in GVW inclusion decreased the efficiency and rate of converting the VS to CH_4. The longer lag phase and the larger CH_4 production rates in the GVW treatments compared to DM.FW.GW and DM-only, from Days 41 to 67, suggests that long retention times would be needed to receive the full increase in expected CH_4 production. This effect was also seen by Kaparaju et al. (2002) when black candy, chocolate, and confectionary by-products were digested with dairy manure for 160 days in order to obtain a complete cumulative CH_4 value, with similar normalized CH_4 production for the confectionary waste (320–390 mL CH_4/g VS) compared to the GVW.DM.FW.GW treatments (336–374 mL CH_4/g VS) [28].

In all treatments, the VS degradation was low compared to studies conducted by Lisboa and Lansing (2013) and Li et al. (2013), where the VS degradation rates ranged from 48–93% [5,29]. Only 19.3% of the initial VS content of the mono-DM treatment was degraded at the end of the experiment, illustrating recalcitrance in the manure feed. The VS degradation was consistent with co-digestion studies of forage radish and dairy manure by Belle et al. (2015b), which used the same manure source as this study with a 21.3% reduction in VS concentration in the mono-DM treatment [30]. The VS degradation of our study (30.9–35.2%) was also comparable to the aforementioned study (30.8–39.7%), with 50–80% co-digestion substrate with dairy manure.

In a review conducted by Xie et al. (2018), it was reported that addition of a carbon-rich substrate to sewage sludge digestion may lower the H_2S concentration due to a dilution effect [31]. This dilution effect can be attributed to a proportionally higher biogas yield compared to the additional H_2S produced from the co-digested substrates. The S concentration for GVW (212 ppm S) was lower than the inoculum source (368 ppm S), and unseparated dairy manure slurries with a TS content of 7% (average 400 ppm S) [32]. The low sulfur concentrations combined with the high VS content (46.3%) of GVW, in comparison to DM (8.2 % VS), provide more evidence to the dilution effect observed in the study, as previously hypothesized. Since more biogas was produced in the GVW treatments compared to the DM treatments, the relative percent of the biogas attributed to manure in the mixed substrate treatments was lowered, and thus, the relative contribution of H_2S from the manure substrate also decreased. Furthermore, the contribution of H_2S from GVW was comparatively lower due to its low sulfur content, leading to the overall decrease in H_2S concentrations in the biogas. However, it should be noted that the GVW addition as a co-digestion substrate increased total normalized H_2S production when compared to co-digestion with 0% GVW addition (DM.FW.GW). A co-digestion substrate with negligible S content could have led to further decreases in H_2S concentrations and total yield. Some gummy vitamins are fortified with Fe, but the concentrations seen in this study (4.3 ppm Fe) was lower than the Fe concentrations in food waste (4800 ppm) and unlikely to have affected H_2S production in our study [33].

The sulfurous compounds in the feedstock were primarily utilized during the initial phase of digestion as most of the H_2S was produced within the first 20 days, after which the CH_4 percentage started rising for all treatments. Similar results were also observed by Belle et al. (2015b) when co-digesting different mass fractions of forage radish with dairy manure in BMP experiments [30]. Forage radish has a high sulfur content and increasing the forage radish percentage led to an expected increase in H_2S production initially, but all the treatments had lowered and similar H_2S production by the end of the study. Belle et al. (2015a) also conducted a pilot-scale study on the same substrates and showed an increased rate of H_2S production during the first two weeks of digestion, after which, the concentration decreased by >75% of the maximum H_2S concentration for the remainder of the digestion period (33 days total) [10]. These observations can be attributed to increased SRB activity during the initial digestion phase, as SRBs can outcompete methanogens when the availability of biodegradable sulfur is higher.

5. Conclusions

Results from the BMP study suggested that gummy waste is a potentially valuable co-digestion substrate with dairy manure. The mixture of substrates containing gummy waste, food waste, grease waste, and dairy manure enhanced CH_4 yields compared to digestion of dairy manure alone. The high density of VS and low moisture content of the gummy waste results in high CH_4 yields per gram of the substrate, but due to the slower degradation rate of the GVW, higher retention times may be needed to yield these higher CH_4 potentials. Co-digestion of GVW with dairy manure lowered the H_2S yield and maximum H_2S concentration compared to mono-digestion of dairy manure due to its low sulfur content. The research highlighted the significance of testing co-digestion mixtures in conjunction with single substrates for both CH_4 and H_2S to provide beneficial information for researchers and AD practitioners. Co-digestion of industrial byproducts and food waste mixtures in farm-scale biogas digesters could provide economic incentives for farmers through tipping fees and increased biogas production while redirecting valuable waste products from the landfills.

Author Contributions: Conceptualization, A.C. and S.L.; Data curation, A.C.; Formal analysis, A.C.; Funding acquisition, S.L.; Investigation, A.C.; Methodology, A.C. and S.L.; Project administration, S.L.; Resources, S.L.; Software, A.C.; Supervision, S.L.; Validation, A.C.; Visualization, S.L.; Writing—Original draft, A.C.; Writing—Review & editing, A.C. and S.L.

Funding: This material is based upon work supported by the National Institute of Food and Agriculture at the U.S. Department of Agriculture through a Northeast Sustainable Agriculture Research and Education (SARE) grant (# LNE15-341).

Acknowledgments: The authors would also like to thank the participating farmer for his enthusiasm and assistance with the project.

Conflicts of Interest: The authors declare no conflict of interest.

References

1. El-Mashad, H.M.; Zhang, R. Biogas production from co-digestion of dairy manure and food waste. *Bioresour. Technol.* **2010**, *101*, 4021–4028. [CrossRef] [PubMed]
2. Weiland, P. Biogas production: Current state and perspectives. *Appl. Microbiol. Biotechnol.* **2010**, *85*, 849–860. [CrossRef] [PubMed]
3. Mata-Alvarez, J.; Dosta, J.; Romero-Güiza, M.S.; Fonoll, X.; Peces, M.; Astals, S. A critical review on anaerobic co-digestion achievements between 2010 and 2013. *Renew. Sustain. Energy Rev.* **2014**, *36*, 412–427. [CrossRef]
4. Zhang, C.; Xiao, G.; Peng, L.; Su, H.; Tan, T. The anaerobic co-digestion of food waste and cattle manure. *Bioresour. Technol.* **2013**, *129*, 170–176. [CrossRef]
5. Lisboa, M.S.; Lansing, S. Characterizing food waste substrates for co-digestion through biochemical methane potential (BMP) experiments. *Waste Manag.* **2013**, *33*, 2664–2669. [CrossRef]
6. Moody, L.B.; Burns, R.T.; Bishop, G.; Sell, S.T.; Spajic, R. Using biochemical methane potential assays to aid in co-substrate selection for co-digestion. *Appl. Eng. Agric.* **2011**, *27*, 433–439. [CrossRef]
7. Braguglia, C.M.; Gallipoli, A.; Gianico, A.; Pagliaccia, P. Anaerobic bioconversion of food waste into energy: A critical review. *Bioresour. Technol.* **2018**, *248*, 37–56. [CrossRef]
8. Xu, F.; Li, Y.; Ge, X.; Yang, L.; Li, Y. Anaerobic digestion of food waste—Challenges and opportunities. *Bioresour. Technol.* **2018**, *247*, 1047–1058. [CrossRef]
9. Zhang, C.; Su, H.; Baeyens, J.; Tan, T. Reviewing the anaerobic digestion of food waste for biogas production. *Renew. Sustain. Energy Rev.* **2014**, *38*, 383–392. [CrossRef]
10. Belle, A.J.; Lansing, S.; Mulbry, W.; Weil, R.R. Anaerobic co-digestion of forage radish and dairy manure in complete mix digesters. *Bioresour. Technol.* **2015**, *178*, 230–237. [CrossRef]
11. Lupitskyy, R.; Alvarez-Fonseca, D.; Herde, Z.D.; Satyavolu, J. In-situ prevention of hydrogen sulfide formation during anaerobic digestion using zinc oxide nanowires. *J. Environ. Chem. Eng.* **2018**, *6*, 110–118. [CrossRef]
12. Pipatmanomai, S.; Kaewluan, S.; Vitidsant, T. Economic assessment of biogas-to-electricity generation system with H2S removal by activated carbon in small pig farm. *Appl. Energy* **2009**, *86*, 669–674. [CrossRef]
13. Corro, G.; Pal, U.; Bañuelos, F.; Rosas, M. Generation of biogas from coffee-pulp and cow-dung co-digestion: Infrared studies of postcombustion emissions. *Energy Convers. Manag.* **2013**, *74*, 471–481. [CrossRef]
14. RedCorn, R.; Fatemi, S.; Engelberth, A.S. Comparing End-Use Potential for Industrial Food-Waste Sources. *Engineering* **2018**, *4*, 371–380. [CrossRef]
15. Rusín, J.; Kašáková, K.; Chamrádová, K. Anaerobic digestion of waste wafer material from the confectionery production. *Energy* **2015**, *85*, 194–199. [CrossRef]
16. ASSOCIATED PRESS. Sweet news: Germany's Haribo to produce gummy bears in Wisconsin. Available online: https://www.chicagotribune.com/business/ct-haribo-gummy-bear-factory-kenosha-wisconsin-20170324-story.html (accessed on 17 November 2019).
17. Cecilia, J.A.; García-Sancho, C.; Maireles-Torres, P.J.; Luque, R. Industrial Food Waste Valorization: A General Overview. In *Biorefinery*; Springer: New York, NY, USA, 2019; pp. 253–277.
18. Delgado, P.; Bañón, S. Determining the minimum drying time of gummy confections based on their mechanical properties. *CYTA - J. Food* **2015**, *13*, 329–335. [CrossRef]

19. Peters, J.; Combs, S.; Hoskins, B.; Jarman, J.; Kovar, J.; Watson, M.; Wolf, A.; Wolf, N. *Recommended Methods of Manure Analysis*; University of Wisconsin Cooperative Extension Publishing: Madison, WI, USA, 2003.
20. Adams, R.C.; Bennett, F.M.; Dixon, J.K.; Lough, R.C.; Maclean, F.S.; Martin, G.I. The utilization of organic wastes in NZ. *New Zeal. Eng.* **1951**, *6*, 396.
21. Lansing, S.; Martin, J.F.; Botero, R.B.; da Silva, T.N.; da Silva, E.D. Methane production in low-cost, unheated, plug-flow digesters treating swine manure and used cooking grease. *Bioresour. Technol.* **2010**, *101*, 4362–4370. [CrossRef]
22. Ugoji, E.O. Anaerobic digestion of palm oil mill effluent and its utilization as fertilizer for environmental protection. *Renew. Energy* **1997**, *10*, 291–294. [CrossRef]
23. Angelidaki, I.; Ahring, B.K. Codigestion of olive oil mill wastewaters with manure, household waste or sewage sludge. *Biodegradation* **1997**, *8*, 221–226. [CrossRef]
24. Wang, X.; Yang, G.; Feng, Y.; Ren, G.; Han, X. Optimizing feeding composition and carbon-nitrogen ratios for improved methane yield during anaerobic co-digestion of dairy, chicken manure and wheat straw. *Bioresour. Technol.* **2012**, *120*, 78–83. [CrossRef] [PubMed]
25. AgStar Increasing Anaerobic Digester Performance with Codigestion. Available online: http://www.epa.gov/agstar/documents/codigestion.pdf (accessed on 17 November 2019).
26. Bujoczek, G.; Oleszkiewicz, J.; Sparling, R.; Cenkowski, S. High Solid Anaerobic Digestion of Chicken Manure. *J. Agric. Eng. Res.* **2000**, *76*, 51–60. [CrossRef]
27. Witarsa, F.; Lansing, S. Quantifying methane production from psychrophilic anaerobic digestion of separated and unseparated dairy manure. *Ecol. Eng.* **2015**, *78*, 95–100. [CrossRef]
28. Kaparaju, P.; Luostarinen, S.; Kalmari, E.; Kalmari, J.; Rintala, J. Co-digestion of energy crops and industrial confectionery by-products with cow manure: Batch scale and farm scale evaluation. *Water Sci. Technol.* **2002**, *45*, 275–280. [CrossRef]
29. Li, Y.; Zhang, R.; Liu, X.; Chen, C.; Xiao, X.; Feng, L.; He, Y.; Liu, G. Evaluating methane production from anaerobic mono- and co-digestion of kitchen waste, corn stover, and chicken manure. *Energy and Fuels* **2013**, *27*, 2085–2091. [CrossRef]
30. Belle, A.J.; Lansing, S.; Mulbry, W.; Weil, R.R. Methane and hydrogen sulfide production during co-digestion of forage radish and dairy manure. *Biomass Bioenergy* **2015**, *80*, 44–51. [CrossRef]
31. Xie, S.; Higgins, M.J.; Bustamante, H.; Galway, B.; Nghiem, L.D. Current status and perspectives on anaerobic co-digestion and associated downstream processes. *Environ. Sci. Water Res. Technol.* **2018**, *4*, 1759–1770. [CrossRef]
32. Chastain, J.P.; Camberato, J.J. Dairy manure production and nutrient content. In *Confined Animal Manure Managers Certification Program Manual B Dairy Version 1*; Clemson University Cooperative Extension Service, 2004; Volume 1, pp. 1–16. Available online: https://www.clemson.edu/extension/camm/manuals/dairy/dch3a_04.pdf (accessed on 22 November 2019).
33. Tampio, E.; Ervasti, S.; Paavola, T.; Heaven, S.; Banks, C.; Rintala, J. Anaerobic digestion of autoclaved and untreated food waste. *Waste Manag.* **2014**, *34*, 370–377. [CrossRef]

Article

Energy and Nutrients' Recovery in Anaerobic Digestion of Agricultural Biomass: An Italian Perspective for Future Applications

Federico Battista *, Nicola Frison and David Bolzonella

Department of Biotechnology, University of Verona, Strada Le Grazie 15, I-37134 Verona, Italy
* Correspondence: federico.battista@univr.it

Received: 3 June 2019; Accepted: 22 August 2019; Published: 26 August 2019

Abstract: Anaerobic digestion (AD) is the most adopted biotechnology for the valorization of agricultural biomass into valuable products like biogas and digestate, a renewable fertilizer. This paper illustrates in the first part the actual situation of the anaerobic digestion sector in Italy, including the number of plants, their geographical distribution, the installed power and the typical feedstock used. In the second part, a future perspective, independent of the actual incentive scheme, is presented. It emerged that Italy is the second European country for the number of anaerobic digestion plants with more than 1500 units for a total electricity production of about 1400 MW_{el}. More than 60% of them are in the range of 200 kW–1 MW installed power. Almost 70% of the plants are located in the northern part of the Country where intensive agriculture and husbandry are applied. Most of the plants are now using energy crops in the feedstock. The future perspectives of the biogas sector in Italy will necessarily consider a shift from power generation to biomethane production, and an enlargement of the portfolio of possible feedstocks, the recovery of nutrients from digestate in a concentrated form, and the expansion of the AD sector to southern regions. Power to gas and biobased products will complete the future scenario.

Keywords: anaerobic digestion; biomethane potential tests; Italy; biogas; manure; energy crops; agriculture residues; digestate

---

1. Introduction

Anaerobic digestion (AD) technology is widely present in the European rural context as it enables the bioconversion of organic matter present in manure and other agro-waste (residual crops or residual streams of food processing) while recovering biogas for electricity or biomethane production [1,2] and a renewable fertilizer, digestate [3]. Recent studies showed how biogas from agro-waste allows for the production of biofuels with a relatively low environmental impact because of their reduced emission in terms of greenhouse gases (GHG) [4].

Because of its intrinsic benefits and a generous program of incentives in several countries, AD is largely diffused in Europe [5,6]. The European Biogas Association reports that the anaerobic digestors in operation in Europe numbered more than 17,200 units, with installed electrical capacity of 8000 MW_{el}, while biomethane upgrade units number more than 400 (data from the Annual Report of the European Biogas Association [7]).

These anaerobic plants are mainly farm-based (around 80%) and are fed with agricultural biomasses like energy crops, livestock effluents, and other agro-waste [8]. Sometimes, the necessity to maximize the energy production (i.e., incomes), and an erroneous designing and business planning approach, determined a distorted situation where energy crops, and maize silage in particular, are massively used as feedstock, determining a strong local impact [9]. Corn (*Zea mays* L.) is a typical example of concern because of its use in the food and feed sectors as well bioenergy with a consequent increase in prices of this crop.

To solve these controversial situations Germany, for example, revised the Renewable Energies Act in 2012, 2014 and 2016/2017 and introduced the so-called maize cap, that is a limit of 60% from 2014 on and 50% since 2016 for energy crops in the feedstock [10].

Moreover, some studies demonstrated how subsidies for bioenergy generation determined the displacement of grasslands and other crops. On the other hand, some energy crops, such as switchgrass (*Panicum virgatum*) can impact favorably on soil properties (erosion prevention) and carbon sequestration [11].

Considering the depicted scenario, it is believed that agro-wastes are the best substrates for anaerobic digestion, as they are not in competition with food/feed production [12].

Beside biogas, an energy vector, digestate, a so called renewable fertilizer, is produced [13]. Moreover, digestate allows for the supplementation of stable carbon on fields thus increasing the carbon sink capability of soils [14].

Digestate is particularly rich in nitrogen (N), phosphorus (P), and potassium (K), vital elements for intensive agriculture. Moreover, it is important to emphasize that P and K are typically mined and are present in defined geographical regions at a global level [15]. Interestingly, these nutrients can be recovered from in concentrated forms: livestock manure in particular, can be considered a mine for these elements. In caft, during anaerobic digestion, the organic backbone of molecules is (at least partially) destroyed while N, P, and K are made available: N and K will be found in soluble forms while is mainly bound to particulate matter. Therefore, agricultural digestate can be used as it is on fields [16,17] or further treated to recover concentrated nutrients to be then transferred in other agricultural areas. The excessive presence of nutrients is a typical problem of some European regions [18]. Today, commercial technologies like stripping, drying, evaporation and membranes technology are available to recover nutrients from digestate [19–21].

Therefore, it is obvious to imagine anaerobic digestion at the center of a future biorefinery approach where agro-waste are converted into high added-value biobased products other than biofuels. This new bioeconomy approach is crucial for the rural renaissance of Europe [22].

Italy is an important actor in this scenario: with its 1500 AD plants, mainly in rural areas, it represents the second European market after Germany and the third in the world after China [7].

In this paper we will report in the first part of the manuscript a picture of the actual Italian scenario for the agricultural biogas sector and will critically analyse the actual situation, then we will expose our vision of the future development of the sector, considering in particular modification of the feedstock recipes based on territorially available biomass, especially in the southern part of the country, and report some full scale experience about nutrients recovery.

2. Materials and Methods

The most abundant substrates from Italian agricultural and farming activities have been tested to determine their biomethane potential (BMP). In particular, the substrates tested along this work have been selected considering their abundance in rural area of some administrative regions in northern, central and southern Italy. Moreover, the portrait of the distribution of the anaerobic digestion plants and their energetic capabilities along the Italian territory have been discussed. Lastly, to close the circular economy approach, the conventional and the more innovative tendencies for the digestate valorization in valuable fertilizers have been analyzed.

2.1. Data Analysis

Data analysis considered the number of AD plants and their installed capacity as well as the main Italian crop production.

The number of the biogas plants located in the different Italian administrative regions, their power capabilities and the relative electrical power production have been obtained combining official data from the official annual reports of the Consorzio Italiano Biogas (CIB), the Gestore Servizi Energetici (GSE) and the data available through the European Project ISAAC (Increasing Social Awareness and ACceptance of biogas and biomethane) [23–25].

The amounts of the most diffused cultivations in Italy have been taken from official web site of the Italian National Institute of Statistics for agricultures and food activities, Agristat [26]. In particular, data of the most cultivated crops, vegetables and fruits have been reported for the two Italian regions with the highest number of biogas plants in North Italy (Lombardia and Veneto), Central Italy (Toscana and Lazio) and South Italy (Campania and Puglia).

2.2. Analytical Methods

To avoid the degradation, the substrates were kept at −18 °C until the experimental campaign started. The substrates considered by this work, were physically and chemically characterized. In particular, the concentrations of dry matter (TS), volatile solids (TVS), chemical oxygen demand (COD), total Kjeldahl nitrogen (TKN) and total phosphorus (TP) were determined according to the standard methods [27]. For the measurement of TKN and TP contents, a high-performance Ethos-One microwave digestion system by Milestone (Italy) and the UDK 129 distillation unit by Velp Scientifica (Italy) were used.

2.3. Biomethane Potential of Substrates

The BMP tests of the most abundant substrates in Italy were conducted according the methods by Angelidaki et al. [28]. They were fed in triplicate in 1 L sealed bottles, with 0.5 L working volume. The duration of the tests was established by a more recent protocol (Holliger et al. [29]) which decided to stop theBMP tests when the daily biogas production is lower than 1% of the cumulative amount, at least for three consecutive days. Inoculum, was taken from a full scale reactor operating in mesophilic condition and treating a mixture of cow and chicken manures and energy corps residues (maize silage, sorghum silage, triticale silage). Before its utilization, the inoculum was filtered at 2 mm to remove coarse material, diluted two-fold with the digestate and, then, kept at the operative mesophilic temperature (37 °C) for one week to assure the endogenous methane production. Microcrystalline cellulose BMP tests were used as positive control [28,29]. All the reactors were manually stirred once a day. The inoculum was also characterized in terms of TS and TVS contents. The average solids content of inoculum was 26.5 ± 12.8 $g \cdot kg^{-1}$, while its volatile content was 63 ± 4% on TS. The volume of biogas generated during the batch trials was determined by water displacement method, while the methane content was determined using a portable biogas analyser (Geotech Biogas 5000 by GeoTech, London, UK).

2.4. Definition of the Hydrolysis Rate

To gain an indication of the degradability index of each substrates, te hydrolysis rate constant, K_h, was determined following the first order model described in Angelidaki et al. [28]. In particular, the biogas production derived from the first 5 days after beginning the experiment was considered. The first order equation, reported below, was recognized a kinetic model describing adequately the methane yield by a recent work [13]:

$$-k_h\, t = \ln \frac{B_\infty - B}{B_\infty} \tag{1}$$

where "B" is the cumulative methane yield (L $CH_4 \cdot$kg VS_{fed}^{-1}) at digestion time "t" days and "B_∞" is the ultimate methane potential of the substrate L $CH_4 \cdot$kg VS_{fed}^{-1} which is obtained at the end of the tests.

Another indication of the degradation kinetic of the substrates is provided by T-50, which is the required time, in days, to produce half of the total cumulative methane production. It was calculated considering the daily cumulated methane production from each BMP test.

3. Results and Discussion

3.1. The Actual Italian Scenario for Biogas Production

During the period 2008–2012 Italy benefited of one of the most generous incentive schemes for power generation from biogas thanks to the so-called "all inclusive" tariff of 280 €/MWh for plants able to generate up to 999 kW_{el}. Once granted, the incentive is valid for a period of 15 years. As a consequence, the AD sector grew up considerably, invested more than €4 billion euros and installed a total energy capacity of some 1000 MW_{el} in the period 2008–2012 [30]. In the following years, up today, the tariff system was modified and substantially decreased determining a reduced rate of new plants and installation. In particular, three different ranges for installed power were identified, and different corresponding tariffs were introduced which depend on the feedstock fed in the bioreactor. For agro by products and energy crops the tariffs are: 180 €/MWh up to 300 kW_{el}, 160 €/MWh from 300 to 600 kW_{el}, and 140 € per MWh produced for AD plants with an installed capacity larger than 600 kW_{el}. However, the tariffs are higher if the substrates are represented by livestock effluents: 236 €/MWh up to 300 kW_{el}, 206 €/MWh from 300 to 600 kW_{el}, and 178 € per MWh produced for AD plants with an installed capacity larger than 600 kW_{el}. The intention to incentivise the adoption of anaerobic digestion for agro wastes and by-products' valorisation in Italian rural area is clear [31].

Today, after 10 years from the first incentive scheme, the total number of AD plants operating in the agricultural sector in Italy is around 1500 units for an installed capacity of some 1400 MW_{el} (average electrical capacity of 700 kW_{el} per AD plant): these represent the 90% of the total AD plants in operation [24]. More than 62% of the Italian biogas plants are represented by a power class in the range 200 kW–1 MW. Only 5% and 15% of AD plants are classified as lower than 50 kW and within 50–200 kW, respectively. The remaining biogas plants have power capacity higher than 1 MW but lower than 10 MW [24]. These numbers make Italy the second biogas producer in Europe and third at global scale after China and Germany [7]. However, the number of installed plants is very different in the 20 different administrative regions: biogas generation is concentrated in the northern part of the country (Po valley) where intensive agriculture and husbandry are present while in the South other alternative energetic sources, such as wind and solar power, are present. Figure 1 reports both plants and their installed capacity in the 20 administrative regions (data elaborated from ISAAC Project [25]).

It should be also considered that the northern part of the country is in proximity of Austria and Germany where the biogas experience in Europe originated: it was therefore easy to transfer technologies and knowledge to the southern side of the Alps.

As a consequence, about 500 of the AD plants operating in the agricultural sector are placed in Lombardy, 220 in Veneto, while 180 are in Emilia Romagna and Piemonte, respectively. All the other regions reported less than 100 AD units on their administrative territories. In total, 67% of the plants and 75% of the installed power are based in the northern part of the country.

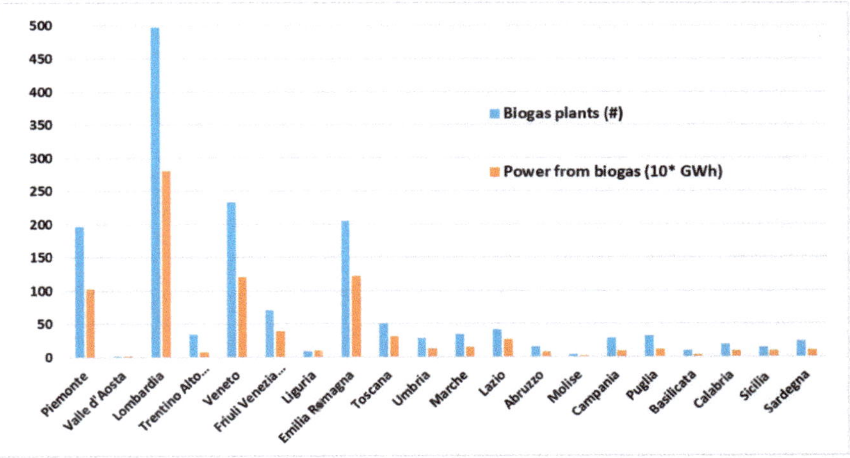

Figure 1. The number and the power capabilities of biogas plants in the 20 Italian administrative regions (data source ISAAC Project [25], modified).

3.2. Typical Feedstocks for the Actual Italian Scenario

According to a specific survey carried out by the Italian Biogas Association [23] and reported in the deliverables of the ISAAC project [25], the typical feedstock composition is due to livestock effluents, energy crops, and agricultural by-products. In particular, it was found that half of the biogas plants in Italy are fed by a mixture of manure and by-products and/or energy crops while the other half are fed with a mixture where energy crops are predominant.

As a consequence, energy crops, especially maize and triticale silage, are the main and sometimes only substrate used for the feeding of AD reactors. Maize is by far the most used crop in anaerobic digestion feedstock: in fact, more than 40 tons of maize per hectare can be produced in southern Europe and its biogas yield is up to 0.35 m^3CH_4 per kg VS (after silaging). The typical cost for growth, transport and silaging in northern Italy is around 30 € per ton. In these conditions, the typical feedstock costs for producing 1 MWh is around 2000 € per day, while incentives can arrive at 5800 € per day. This difference allows for a rapid payback of plants which cost is around 4–5 million € per MW. Since energy crops growth requires for land, water and fertilizers, these substrates are not sustainable on a long term perspective and should be replaced by agro-waste.

The anaerobic codigestion of manure and agro-waste is the normal practice in Italy and Europe in general [32]. Livestock production, in fact, is one of the main activities in rural areas within the European Union. Italy, especially in its northern part (Po valley), is one of these areas: as a consequence, livestock effluents are the typical substrates treated in anaerobic digestion plants. For example, the amount of liquid and solid manure produced in the Veneto Region in 2010 accounted for 6 and 5 million cubic meters, respectively, cattle manure being the dominant (67%) part [26]. A similar scenario is observed also for Lombardia, Piemonte and Emilia Romagna Regions [26].

On the other hand, the very high tariff for the production of renewable energy, leads to the use of substrates characterized by high organic content, energy density, and biogas yields like energy crops (especially maize silage) and agro-waste [32]. These co-substrates present similar characteristics in terms of total and volatile solids, thus COD, lower nutrients content and higher biogas yields.

3.3. Actual Use of Digestate

Nitrogen and phosphorus concentrations in livestock effluents are in the range 5–15 kgN/ton and 0.1–1 kgP/ton, respectively, while concentrations in energy crops and other biodegradable by-products are typically lower [26]. The European production of digestate is estimated in 56 Mtons per year [33]: this can be a real renewable mine for nutrients recovery for the European agricultural sector: the new directive on fertilizers will probably act as a driver in this sense [34].

Digestate is usually valorised because of its nutrients content into fertilizers or soil improvers production, considering its high nitrogen and phosphorus contents not mentioning potassium. Interestingly, N and K will be mainly present in the liquid form after organic substrates undergo the anaerobic digestion process while P is mainly present in the solid form. As a consequence, digestate can be used as it is nearby the farm [16,17].

The efficacy of digestate as a fertilizer [35] was proved for example by Grigatti et al. [36], which conducted pot tests using phosphorous salts from different digestates. ^{31}P nuclear magnetic resonance (NMR) showed how orthophosphate was the main form determining different fertilization potentials. In particular, anaerobic digestates from livestock effluents and energy crops demonstrated to be good alternatives to fertilization with chemical P.

4. Future Perspective for the Biogas Sector in Italy

When considering the future perspective for the biogas sector in Italy one should consider that in 2018 a new decree came into force with the aim to incentive the biogas valorisation into biomethane after biogas up-grading. Biomethane can be injected in the national grid or adopted as automotive fuel [37]. The decree introduced particular tariffs for the biomethane originated from biogas produced from agricultural feedstocks like manure rather than dedicated energy crops.

The fact that the biogas sector is already developed in the North while the South is still waiting for the implementation of infrastructures and that new incentives for biomethane are coming into force together with the necessity to decarbonize the industrial sector together determine the necessity to develop anaerobic digestion sector in the southern part of the country: it should be emphasized here that there are important potentials for the development of the biogas sector in rural areas in Campania, Apulia and Sicily in particular, as will be analyzed in more detail in the following paragraph.

The estimated biogas production is around 5 billion cubic meter per year, making Italy the fourth country in the world for biogas production [23]. Because of the most recent regulatory framework part of this biogas will be converted into biomethane in the next future. However, the estimation by SNAM, the Italian company responsible for methane net and distribution, puts future biogas production at 10 billion cubic meters [38].

4.1. Future Feedstocks

The necessity to make the biogas sector sustainable and to respond to specific requests for biomethane production, open the doors to the use of several by-products which can be used in the feedstock instead of energy crops [39]. Because of their abundance in the Italian territory, agro-wastes from agriculture and animal farm activities can be considered ideal substrates for the co-digestion process instead of energy crops.

Tables 1 and 2 show the most abundant cultivations and farmed animals, respectively in the selected regions of North, Central and South Italy (Agristat [26]). Table 3 shows the main characteristics of the considered substrates.

Table 1. The most diffused cultivations, vegetables and fruits in some administrative regions of North, Central and South Italy with the highest number of biogas plants.

	NORTH ITALY		CENTRAL ITALY		SOUTH ITALY		
Crops, Fruits (Tons)	**Veneto**	**Lombardia**	**Toscana**	**Lazio**	**Puglia**	**Campania**	**Italy**
Wheat (common + durum)	709,795	410,952	318,658	194,560	1,026,600	246,863	7,054,799
Chickpea	843	3828	7580	1190	3032	534	47,438
Beans and string beans	6018	6830	1698	3220	7670	46,893	151,452
Onions	30,592	10,227	5757	2190	39,650	34,155	382,634
Carrots	30,021	Low	1143	88,180	33,370	4620	480,824
Fennel	557	88	2430	17,260	146,400	63,819	537,444
Lettuce	9390	18,493	1547	16,750	100,480	33,139	349,017
Fresh fruits (apples, pears, apricots, cherries)	405,819	66,154	35,744	13,528	72,274	172,199	3,516,837
Olive oil	24,371	4987	120,364	88,434	565,100	112,926	1,867,662
Wine (DOP *, IGP **, table wine)	1,015,801	148,833	270,830	131,961	955,257	132,749	5,043,610
Citrus fruits (oranges, tangerines, lemons)	Low	Low	122	3411	115,023	49,700	2,080,377

* DOP stays for the Italian "Denominazione Origine Protetta", that means "Protected Designation of Origin; ** IGP stays for the Italian "Indicazione Geografica Protetta, that means "Protected Geographical Indication".

Table 2. The number of animals farmed in the selected Italian regions grouped in different categories.

Substrates (Tons)	NORTH ITALY		CENTRAL ITALY		SOUTH ITALY		
Farm Animals' Categories (#)	**Veneto**	**Lombardia**	**Toscana**	**Lazio**	**Puglia**	**Campania**	**ITALY**
Ovines	11,178	81,356	199,300	511,088	170,950	151,369	2,984,336
Bovines	546,171	542,209	36,484	26,203	60,867	195,862	2,651,010
Swines	437,428	4,265,523	272,445	93,999	70,698	195,383	11,380,546
Poultries and Rabbits	108,841,000	66,043,108	6,145,918	414,186	18,770,251	114,747	606,062,235
Equines	12,382	3497	225	2589	31,144	679	67,005

Table 3. Summary of chemical-physical characteristics of the organic biomass more abundant in the Italian context. Energy crops.

	Total Solids (%)	**Total Volatile Solids (%)**	**TVS/TS (%)**	**COD ($g \cdot kg^{-1}$)**	**TKN ($g \cdot kg^{-1}$)**	**TP ($g \cdot kg^{-1}$)**
Energy Crops						
Millet—*Panicum Miliaceum* L.	21.8	20.1	92.0	-	-	-
Barley—*Hordeum distichon* L.	25.8–66.3	25.1–59.1	89.3–97.2	517	7.0–19.9	0.8–3.9
Maize—*Zea mays* L.	40–66.50	38.3–64.0	90.7–96.5	293–304	4.0–4.8	0.3–0.6
Sorghum—*Sorghum* spp.	28.6–39.6	25.5–35.4	89.3–94.0	302–353	3.2–13.0	0.5
Triticale—*Triticum aestivum* L.	30–30.8	27.9	90.4–93.1	296	13.5	0.7
Vegetables and fruits by products						
Carrot Leaves—*Daucus carota* L.	14.8	12.3	83.6	258	3.1	-
Radicchio Leaves—*Red Cichorium* L.	10.4	9.3	89.0	38.1	0.9	0.3
Potato Peels—*Solanum tuberosum* L.	11.8	10.7	90.6	186	4.8	0.5
Apple Pomace—*Malus domestica* L.	50.2	47.9	95.6	580	4.2	-
Tomato Pomace—*Solanum lycopersicum* L.	30.1	29.0	96.1	380	7.7	-
Grape Marcs—*Vitis vinifera.* L.	29.6–36.7	27.8–34.3	93.1–93.7	312–347	5.7–9.2	2.8–3.3
Grape Vinasse—*Vitis vinifera.* L.	35.6–64.2	28.5–53.1	80.0–82.7	178–324	17.6–37.4	-
Lemon Pomace—*Citrus lemon* L.	12.6–85.5	12.0–64.3	75.1–95.3	127–692	1.7–6.1	0.3–0.4
Livestock effluents						
Bovine Slurry	4.9–14.5	3.6–12.2	72.5–100	48.0–128	2.1–6.2	0.3–1.2
Bovine Manure	15.6–47.7	13.5–32.1	48.7–99.8	135–291	3.2–7.1	0.2–1.5
Pig Manure	36.1	35.9	99.3	381	-	-
Pig Slurry	0.7–6.4	0.5–5.3	75.0–82.7	5.2–46.6	0.2–5.0	0.1–1.5
Poultry Manure	31.5–78.3	21.3–51.7	44.7–84.1	235–586	2.3–38.9	5.2–15.3

Several authors reported in recent years the possibility to use different by-products, typically originated from the food-processing industry instead of dedicated energy crops [34]. Table 4 shows the great methane potentials from agro-waste byproducts which is comparable, and in some cases higher (radicchio and carrots leaves and potatoes and onions peels) than energy crops.

Table 4. K_h, T-50 and methane yields of biomasses considered in this study.

	K_h (d^{-1})	T-50 (d)	CH_4 Yield ($LCH_4 \cdot kg\ TVS^{-1}$)
Energy Crops			
Millet—*Panicum miliaceum* L.	0.080	13.1	253
Barley—*Hordeum distichon* L.	0.097	8.1	290 ± 83
Maize—*Zea mays* L.	0.135 ± 0.06	6.2	289 ± 86
Sorghum—*Sorghum* spp.	0.091 ± 0.06	10.6	313 ± 73
Triticale—*Triticum aestivum* L.	0.154 ± 0.07	10.3	351 ± 5
Vegetables and Fruits by-Products			
Carrot Leaves—*Daucus carota* L.	0.096	7.0	312
Radicchio Leaves—*Red Cichorium* L.	0.185	3.0	431
Potato Peels—*Solanum tuberosum.* L.	0.063	3.9	446
Onion Peels—*Allium cepa* L.	0.213	3.0	455
Apple Pomace—*Malus domestica* L.	0.148	0.0	204
Tomato Pomace—*Solanum lycopersicum* L.	0.068	10.9	239
Grape Marcs—*Vitis vinifera.* L.	0.103 ± 0.04	11.4	248 ± 48
Grape Vinasse—*Vitis vinifera.* L.	0.162	5.3	274 ± 123
Lemon Pomace—*Citrus lemon* L.	0.226 ± 0.06	4.3	355 ± 10
Livestock Effluents			
Bovine Slurries	0.039 ± 0.02	14.1	35.2 ± 4.3
Bovine Manure	0.038 ± 0.02	12.3	97.5 ± 9.3
Pig Manure	0.090	8.0	128
Pig Slurries	0.120	8.0	187 ± 89
Poultry Manure	0.098 ± 0.03	6.8	208 ± 103

Schievano et al. [39] reported operating with different mixtures, where municipal organic waste, waste molasses, fruits waste, can substitute energy crops but guarantee the same biogas production while lowering the feedstock costs.

Giuliano et al. [40] demonstrated how, in thermophilic conditions, rotten onions and potatoes can substitute maize silage maintaining the same operational conditions (Organic Load Rate and Hydraulic Retention Time) of the anaerobic digester.

De Menna et al. [41] investigated the BMP potential of five different varieties of artichokes, whose cultivation is particular important in Sardinia. They found a methane yield of 292 LCH_4/kg_{VS}. Considering the regional availability of artichokes by-products, this means that about 20×10^6 Nm^3 CH_4 could be produced.

In the last few years, several studies dealt with the definition of the potential for biogas production in the southern regions of Italy. In fact, even if it was already remarked that the national biogas production is concentrated in North Italy, and in particular in Lombardia, Veneto, and Emilia Romagna (Figure 1), Tables 1 and 2 show that agricultural and animal farm activities are very strong in the Central and South regions of Italy, with a consequent by-products production which can be exploited by biorefinery for biogas production. Southern Italian regions are leaders in unique agriculture products, which are exported around the world, such as extra virgin olive oil [42] and citrus productions [43].

With almost 1.9 Mtons of olives per year (Table 1), Italy is a large producer of olive oil. The cultivation of olive trees and olive oil extraction are mainly concentrated in southern Italy, especially in Puglia where it was estimated to be located about the 40% of the national olive oil production. Battista et al. [42] realized a study for biogas production on a pilot reactor working in continuous mode with a feed represented by 75% *v/v* of olive oil byproducts and 25% *v/v* by cheese whey. The daily biogas was of 1.4 L/L day for a potential annual production of 55 GJ. Taking into account the annual amounts of olive oil and dairy productions' residues, this biogas production rate would assure the generation of about 375,000 GJ, able to cover approximately the 0.015% of Puglia's energy demand.

Valenti et al. [43] focused their attention on the most abundant agro-wastes in Sicily: the by-products of lemon (citrus pulp), olive oil (olive pomace), poultry manure, Italian sainfoin and nopals of prickly pears (opuntia indica). The BMP tests investigated different mix of these substrates and showed that all these substrates can be used as feedstock in biogas plants, with methane production between 240 and 260 LCH4/kgVS. It was estimated that agro-waste and by-products from the agro-food sector could produce 562 million Nm^3/year of biomethane in Sicily in 2030. This is equivalent to 8% of the total Italian generation [44]. Although these encouraging estimations, the effective biogas production in Central and South Italy is far from the North Italy situation, where there is already an adequate level of valorisation of the agro-wastes byproducts. Instead, in the southern and central Italian regions the major part of the substrates remain unused in the better cases, and often are simply disposed of on the soil or burned in the open air with negative consequences on human health and increasing contamination of the air, soil and aquifers [42].

In these years our lab characterized several substrates in terms of chemico-physical characteristics, BMP and tendency to biodegradability, showed in Table 4. By this way, it is possible to estimate the great potential of the most abundant Italian agro-waste byproducts.

To have a complete scenario of the Italian situation one should consider that because of the specific climatic conditions two harvesting shift are normally possible during the summer season: this concept is at the base of the "biogas done right" model, where on the same land both crops for food/feed and dedicated energy crops for biogas production are cultivated.

4.2. Recovery of Nutrients from Digestate

In perspective, digestate can be separated in two distinct streams, one liquid and the other solid, with different fertilizing characteristics. These two streams can be further processed to obtain concentrated nutrient forms so as to minimize the transport costs to different agricultural areas [19,20].

However, there are now several technological options for digestate treatment available on the market. Drying of the whole digestate or of its solid fraction, evaporation of the whole digestate or its fractions, membrane filtration of the liquid fraction or stripping of ammonia from the liquid fraction are examples of different options [18–20].

Drying consists in removing water from digestate using hot air generated from the engines burning biogas. Vapors produced within the process are treated to recovery volatilize ammonia. The two main outputs are, therefore, a solid dried fraction and a liquid phase rich in nitrogen [19,20].

Ammonia nitrogen in digestate can be displaced using vapor or can be blocked in an acidic environment after adding mineral acids. Once evaporated, gaseous ammonia can be recovered by means of scrubbing or osmosis. Since the digester sludge is diluted (<10% total solids), the amount of heat recovered from the CHP unit is insufficient to treat all the digestate produced [45].

In the stripping systems, digestate is previously sent to a solid/liquid separation. Then, the liquid phase is fed to a packed bed tower where gaseous ammonia (NH_3) is stripped, passing from the aqueous to the gas phase. The gas stream, rich in ammonia, is then sent to a second tower where NH_3 is absorbed in an acidic media, typically sulfuric acid, producing ammonium sulfate at 25–35% [18–20].

In membrane filtration systems digestate is separated from coarse solids and then the liquid phase of digestate is treated in ultrafiltration (UF) and reverse osmosis (RO) systems: here most of the nutrients are concentrated and separated from water: the obtained concentrated stream is normally from 20% to 30% of the initial treated volume [19,20,46].

Recently, Battista and Bolzonella [13] reported the use sof olar energy for simultaneous digestate drying and ammonium sulfate recovery. In particular, they tested four digestates, different for origin and characteristics in a transparent greenhouse exposed to solar irradiation. The liquid phase evaporation was favored by three solar air fans, which also addressed the ammonia rich vapors to a Drechsle trap filled with 38% w/w sulfuric acid solution. In this way, ammonia reacted with sulfuric acid, forming a solution of ammonium sulfate to be used as fertilizer. It was found that substrates rich in proteins (thus nitrogen and ammonia), like animal manure and food wastes were indicated for ammonium sulfate recovery. The solution in the Drechsle trap reached concentrations up to 2 M.

4.3. Biogas, Power to Gas and Added-Value Biobased Compounds

Another interesting perspective for the biogas sector is the transformation of AD farm-based plants into biorefineries. In fact, there is a growing interest in the production of biobased chemicals like volatile fatty acids, lactic acid, succinic acid, poly-hydroxy-alkanoates, and single cell proteins [22,47].

In this approach mixed cultures fermentative processes are applied to produce high added products like carboxylic acids [48] or bioplastics [22] while anaerobic digestion for biogas production is the last process of the biorefinery train and is dedicated to eventual energy recovery (thermal or power).

The strength of this vision is mainly related to the fact that farm-based AD plants are already in existence and have been paid for, and therefore infrastructure is already available without excessive capital costs expenditures.

Another important perspective for the biogas sector is the integration of biogas plants with other renewable energy technologies like solar and wind power to generate hydrogen from water lysis and combine then hydrogen and carbon dioxide present in biogas to further produce methane [49]. This process, known as power to gas, can be one of the future developments of the rural biogas sector when associated with photo-voltaic or wind power generation: this allows for the storage of pick power generation typically associated with sun and wind cycles into a carbon-based energy vector, which is easy to store and liquify and which can be used for several purposes, including transportation.

5. Conclusions and Perspectives

Anaerobic digestion is widely applied in the European rural scenario; in fact, this technology is complementary to other renewable energy technologies like photovoltaic, wind and hydro and is, therefore, a fundamental piece of the energy puzzle.

By contrast with the other technologies, power generated from AD can be modulated (biogas can be stored) while biogas can be upgraded to biomethane and used as biofuel.

The biogas sector within the European Union is still largely dependent on energy crops like maize silage, thus opening the competition with the food and feed sector, but it is rapidly changing to the treatment of livestock effluents and other agro-waste, thus participating in the reduction of the environmental burden associated with these streams. Residual digestate can be used as it is, directly in the farm or nearby, while in some situations it can be necessary to apply a technique which allows for the recovery of nutrients in concentrated forms easy to be transported and used in other rural areas.

Author Contributions: Conceptualization, D.B. and F.B.; Methodology, all the authors; Software, all the authors; Validation, all the authors; Writing—Original Draft Preparation, F.B. and D.B.; Writing—Review & Editing, F.B. and D.B.; Supervision, D.B.

Funding: Part of the results of this research paper is financed by Eranet Med Biogasmena (project number 72-026), which contemplates the optimization of technologies for biogas production and nutrients recovery from digestate in the MENA (Middle East and North Africa) regions.

Conflicts of Interest: The authors declare no conflict of interest. The funders had no role in the design of the study; in the collection, analyses, or interpretation of data; in the writing of the manuscript, or in the decision to publish the results.

References

1. Poeschl, M.; Ward, S.; Owende, P. Evaluation of energy efficiency of various biogas production and utilization pathways. *Appl. Energy* **2010**, *87*, 3305–3321. [CrossRef]
2. Wall, D.M.; O'Kiely, P.; Murphy, J.D. The potential for biomethane from grass and slurry to satisfy renewable energy targets. *Bioresour. Tech.* **2014**, *149*, 425–431. [CrossRef] [PubMed]
3. Arthurson, V. Closing the global energy and nutrient cycles through application of biogas residue to agricultural land–Potential benefits and drawbacks. *Energies* **2009**, *2*, 226–242. [CrossRef]
4. Czyrnek-Deletre, M.M.; Smyth, B.M.; Murphy, J.D. Beyond carbon and energy: The challenge in setting guidelines for life cycle assessment of biofuel systems. *Renew. Energy* **2017**, *105*, 436–448. [CrossRef]
5. Del Pablo-Romero, M.; Sánchez-Braza, A.; Salvador-Ponce, J.; Sánchez-Labrador, N. An overview of feed-in tariffs, premiums and tenders to promote electricity from biogas in the EU-28. *Renew. Sustain. Energy Rev.* **2017**, *73*, 1366–1379. [CrossRef]
6. Scheftelowitz, M.; Becker, R.; Thran, D. Improved power provision from biomass: A retrospective on the impacts of German energy policy. *Biomass Bioenergy* **2018**, *111*, 1–12. [CrossRef]
7. European Biogas Association. Annual Report. 2015. Available online: http://european-biogas.eu/2015/12/16/biogasreport2015/ (accessed on 28 May 2019).
8. Gruda, N.; Bisbis, M.; Tanny, J. Impacts of protected vegetable cultivation on climate change and adaptation strategies for cleaner production—A review. *J. Clean. Prod.* **2019**, *225*, 324–339. [CrossRef]
9. Hamelin, L.; Naroznova, I.; Wenzel, H. Environmental consequences of different carbon alternatives for increased manure-based biogas. *Appl. Energy* **2014**, *114*, 774–782. [CrossRef]
10. Theuerl, S.; Herrmann, C.; Heiermann, M.; Grundmann, P.; Landwehr, N.; Kreidenweis, U.; Prochnow, A. The Future Agricultural Biogas Plant in Germany: A Vision. *Energies* **2019**, *12*, 306. [CrossRef]
11. Cobuloglu, H.I.; Büyüktahtakin, I.E. Food vs. biofuel: An optimization approach to the spatio-temporal analysis of land-use competition and environmental impacts. *Appl. Energy* **2015**, *140*, 418–434. [CrossRef]
12. Garcia, N.H.; Mattioli, A.; Gil, A.; Frison, N.; Battista, F.; Bolzonella, D. Evaluation of the methane potential of different agricultural and food processing substrates for improved biogas production in rural areas. *Renew. Sustain. Energy Rev.* **2019**, *112*, 1–10. [CrossRef]
13. Battista, F.; Bolzonella, D. Exploitation of Solar Energy for Ammonium Sulfate Recovery from Anaerobic Digestate of Different Origin. *Waste Biomass Valoris.* **2019**. [CrossRef]
14. Lal, R. Digging deeper: A wholistic perspective of factors affecting Soil Organic Carbon sequestration. *Glob. Change Biol.* **2018**, *24*, 3285–3301. [CrossRef] [PubMed]
15. Mehta, C.M.; Khunjar, W.O.; Nguyen, V.; Tait, S.; Batstone, D.J. Technologies to recover nutrients from waste streams: A critical review. *Crit. Rev. Environ. Sci. Technol.* **2015**, *45*, 385–427. [CrossRef]
16. Moller, K.; Müller, T. Effects of anaerobic digestion on digestate nutrient availability and crop growth: A review. *Eng. Life Sci.* **2012**, *12*, 242–257. [CrossRef]
17. Vaneeckhaute, C.; Darveau, O.; Meers, E. Fate of micronutrients and heavy metals in digestate processing using vibrating reversed osmosis as resource recovery technology. *Sep. Purif. Technol.* **2019**, *223*, 81–87. [CrossRef]
18. Bernet, N.; Beline, F. Challenges and innovations on biological treatment of livestock effluents. *Bioresour. Technol.* **2009**, *100*, 5431–5436. [CrossRef] [PubMed]
19. Fuchs, W.; Drosg, B. Assessment of the state of the art of technologies for the processing of digestate residue from anaerobic digesters. *Water Sci. Technol.* **2013**, *67*, 1984–1993. [CrossRef]
20. Drosg, B.; Fuchs, W.; Al Seadi, T.; Madsen, M.; Linke, B. *Nutrient Recovery by Biogas Digestate Processing*; IEA Bioenergy: Paris, France, 2015; ISBN 978-1-910154-15-1.
21. Bolzonella, D.; Fatone, F.; Gottardo, M.; Frison, N. Nutrients recovery from anaerobic digestate of agro-waste: Techno-economic assessment of full scale applications. *J. Environ. Manag.* **2018**, *216*, 111–119. [CrossRef]
22. Gontard, N.; Sonesson, U.; Birkved, M.; Majone, M.; Bolzonella, D.; Celli, A.; Angellier-Coussy, H.; Jang, G.W.; Verniquet, A.; Broeze, J.; et al. A research challenge vision regarding management of agricultural waste in a circular bio-based economy. *Crit. Rev. Environ. Sci. Technol.* **2018**, *48*, 614–654. [CrossRef]
23. Consorzio Italiano Biogas. 2019. Available online: https://www.consorziobiogas.it/ (accessed on May 2019).

24. Gestore Servizi Energetici. Rapporto Statistico: Energia da Fonti Rinnovabili in Italia. Anno 2016. 2018. Available online: https://www.gse.it/documenti_site/Documenti%20GSE/Rapporti%20statistici/Rapporto%20statistico%20GSE%20-%202016.pdf (accessed on 28 May 2019).
25. ISAAC. Increasing Social Awareness and Acceptance of Biogas and Biomethane. Available online: http://www.isaac-project.it/ (accessed on 28 May 2019).
26. Agristat, Istituto Nazionale di Statistica per l'Agricoltura e la Zootecnia. Available online: http://agri.istat.it/ (accessed on 28 May 2019).
27. American Public Health Association, American Water Works Association and Water Pollution Control Federation. *APHA, AWWA & WPCF Standard Methods for the Examination of Water and Wastewater*, 17th ed.; American Public Health Association, American Water Works Association and Water Pollution Control Federation: Washington, DC, USA, 1989.
28. Angelidaki, I.; Alves, M.; Bolzonella, D.; Borzacconi, L.; Campos, J.L.; Guwy, A.J.; Kalyuzhnyi, S.; Jenicek, P.; van Lier, J.B. Defining the biomethane potential (BMP) of solid organic wastes and energy crops: A proposed protocol for batch assays. *Water Sci. Technol.* **2009**, *59*, 927–934. [CrossRef] [PubMed]
29. Holliger, C.; Alves, M.; Andrade, D.; Angelidaki, I.; Astals, S.; Baier, U.; Bougrier, C.; Buffière, P.; Carballa, M.; De Wilde, V.; et al. Towards a standardization of biomethane potential tests. *Water Sci. Technol.* **2016**, *74*, 2515–2522. [CrossRef] [PubMed]
30. Benato, A.; Macor, A. Italian Biogas Plants: Trend, Subsidies, Cost, Biogas Composition and Engine Emissions. *Energies* **2019**, *12*, 979. [CrossRef]
31. Ragazzoni, A. Analisi della Redditività Degli Impianti per la Produzione di Biogas alla Luce delle Nuove Tariffe Incentivanti 2013. Available online: http://www.crpa.it/media/documents/crpa_www/Convegni/20130314_BiogasBiometano_RA/Ragazzoni_RA_14-3-2013.pdf (accessed on 28 May 2019).
32. Weiland, P. Biogas production: Current state and perspectives. *Appl. Microbiol. Biotechnol.* **2010**, *2010. 85*, 849–860. [CrossRef]
33. Beggio, G.; Schievano, A.; Bonato, T.; Hennebert, P.; Pivato, A. Statistical analysis for the quality assessment of digestates from separately collected organic fraction of municipal solid waste (OFMSW) and agro-industrial feedstock. Should input feedstock to anaerobic digestion determine the legal status of digestate? *Waste Manag.* **2019**, *87*, 546–558. [CrossRef] [PubMed]
34. Council of the European Union. Proposal for a Regulation of the European Parliament and of the Council Laying Down Rules on the Making Available on the Market of CE Marked Fertilising Products and Amending Regulations (EC) No 1069/2009 and (EC) No 1107/2009. 2018. Available online: http://data.consilium.europa.eu/doc/document/ST-15103-2018-INIT/en/pdf (accessed on 28 May 2019).
35. Tambone, F.; Scaglia, B.; D'Imporzano, G.; Schievano, A.; Orzi, V.; Salati, S.; Adani, F. Assessing amendment and fertilizing properties of digestates from anaerobic digestion through a comparative study with digested sludge and compost. *Chemosphere* **2010**, *81*, 577–583. [CrossRef]
36. Grigatti, M.; Boanini, E.; Bolzonella, D.; Sciubba, L.; Mancarella, S.; Ciavatta, C.; Marzadori, C. Organic wastes as alternative sources of phosphorus for plant nutrition in a calcareous soil. *Waste Manag.* **2019**, *93*, 34–46. [CrossRef] [PubMed]
37. Gazzetta Ufficiale della Repubblica Italiana. Decreto 2 marzo 2018. Promozione Dell'uso del Biometano e Degli Altri Biocarburanti Avanzati Nel Settore dei Trasporti. 2018. Available online: https://www.gazzettaufficiale.it/eli/id/2018/03/19/18A01821/SG (accessed on 28 May 2019).
38. SNAM. Available online: http://www.snam.it/en/Natural-gas/green-energy/biomethane (accessed on 28 May 2019).
39. Schievano, A.; D'Imporzano, G.; Adani, F. Substituting energy crops with organic wastes and agro-industrial residues for biogas production. *J. Environ. Manag.* **2009**, *90*, 2537–2541. [CrossRef] [PubMed]
40. Giuliano, A.; Bolzonella, D.; Pavan, P.; Cavinato, C.; Cecchi, F. Co-digestion of livestock effluents, energy crops and agro-waste: Feeding and process optimization in mesophilic and thermophilic conditions. *Bioresour. Technol.* **2013**, *128*, 612–618. [CrossRef]
41. De Menna, F.; Malagnino, R.A.; Vittuari, M.; Molari, G.; Seddaiu, G.; Deligios, P.A.; Solinas, S.; Ledda, L. Potential Biogas Production from Artichoke Byproducts in Sardinia, Italy. *Energies* **2016**, *9*, 92. [CrossRef]
42. Battista, F.; Ruggeri, B.; Fino, D.; Erriquens, F.; Rutigliano, L.; Mescia, D. Toward the scale-up of agro-food feed mixture for biogas production. *J. Environ. Chem. Eng.* **2013**, *1*, 1223–1230. [CrossRef]

43. Valenti, F.; Porto, S.M.C.; Selvaggi, R.; Pecorino, B. Evaluation of biomethane potential from by-products and agricultural residues co-digestion in southern Italy. *J. Environ. Manag.* **2018**, *223*, 834–840. [CrossRef] [PubMed]
44. Selvaggi, R.; Pappalardo, G.; Chinnici, G.; Fabbri, C.I. Assessing land efficiency of biomethane industry: A case study of Sicily. *Energy Policy* **2018**, *119*, 689–695. [CrossRef]
45. Vaneeckhaute, C.; Lebuf, V.; Michels, E.; Belia, E.; Vanrolleghem, P.A.; Tack, F.M.G.; Meers, E. Nutrient recovery from digestate: Systematic technology review and product classification. *Waste Biomass Valor.* **2017**, *8*, 21–40. [CrossRef]
46. Kertesz, S.; Beszedes, S.; Laszlo, Z.; Szabo, G.; Hodur, C. Nanofiltration and reverse osmosis of pig manure: Comparison of results from vibratory and classical modules. *Desalin. Water Treat.* **2010**, *14*, 233–238. [CrossRef]
47. Battista, F.; Frison, N.; Pavan, P.; Cavinato, C.; Gottardo, M.; Fatone, F.; Eusebi, A.L.; Majone, M.; Zeppilli, M.; Valentino, F.; et al. Food wastes and sewage sludge as feedstock for an urban biorefinery producing biofuels and added-value bioproducts. *J. Chem. Technol. Biotechnol.* **2019**, in press. [CrossRef]
48. Agler, M.T.; Wrenn, B.A.; Zinder, S.H.; Angenent, L.T. Waste to bioproduct conversion with undefined mixed cultures: The carboxylate platform. *Trends Biotechnol.* **2011**, *29*, 70–78. [CrossRef] [PubMed]
49. Hahn, H.; Krautkremer, B.; Hartmann, K.; Wachendorf, M. Review of concepts for a demand-driven biogas supply for flexible power generation. *Rrenew. Sustain. Energy Rev.* **2014**, *29*, 383–393. [CrossRef]

Article

Development of a Modified Plug-Flow Anaerobic Digester for Biogas Production from Animal Manures

Daniel Gómez [1,*], Juan Luis Ramos-Suárez [1,2], Belén Fernández [3], Eduard Muñoz [3], Laura Tey [3], Maycoll Romero-Güiza [3,4] and Felipe Hansen [1]

1 ProCycla SL, Pont de Vilomara 140, 2-1, 08241 Manresa, Spain
2 Departamento de Ingeniería Agraria, Náutica, Civil y Marítima, Universidad de La Laguna, Carretera de Geneto 2, 38071 San Cristóbal de La Laguna, Spain
3 IRTA, Torre Marimon, E-08140 Caldes de Montbui, Spain
4 Aqualia I + D, Av. Camino de Santiago, 40, Edificio 3, 4ª Planta, E-28050 Madrid, Spain
* Correspondence: dgomez@procycla.com; Tel.: (+34)-649-989-971

Received: 13 May 2019; Accepted: 27 June 2019; Published: 8 July 2019

Abstract: Traditional plug-flow anaerobic reactors (PFRs) are characterized by lacking a mixing system and operating at high total solid concentrations, which limits their applicability for several kinds of manures. This paper studies the performance of a novel modified PFR for the treatment of pig manure, characterized by having an internal sludge mixing system by biogas recirculation in the range of 0.270–0.336 m^3 m^{-3} h^{-1}. The influence on the methane yield of four operating parameters (recirculation rate, hydraulic retention time, organic loading rate, and total solids) was evaluated by running four modified PFRs at the pilot scale in mesophilic conditions. While the previous biodegradability of organic matter by biochemical methane potential tests were between 31% and 47% with a methane yield between 125 and 184 L_{CH4} $kgVS^{-1}$, the PFRs showed a suitable performance with organic matter degradation between 25% and 51% and a methane yield of up to 374 L_{CH4} $kgVS^{-1}$. Operational problems such as solid stratification, foaming, or scum generation were avoided.

Keywords: plug-flow reactor; anaerobic digestion; animal manures; biogas; unconfined gas injection mixing; mixing recirculation

1. Introduction

Anaerobic digestion (AD) is a biological process in which organic matter breaks down naturally in the absence of oxygen to produce biogas [1]. The most common forms of large-scale anaerobic digesters are the continuously stirred tank reactor (CSTR) and plug-flow reactor (PFR) [2,3]. In a CSTR system, microorganisms are suspended in the digester through intermittent or continuous mixing, which offers good substrate–sludge contact with slight mass transfer resistance, but higher energy is required [4]. In an ideal CSTR, the concentration in any point of the reactor is identical. In contrast, in an ideal PFR, there is no lengthwise mixing of the substrates under digestion as they move through the PFR. Therefore, the concentration distribution is not uniform throughout the reactor [5]. Actually, PFR is characterized by the fact that the flow or fluid through the reactor is orderly with no element of fluid overtaking or mixing with any other element ahead or behind [6]. A PFR system is simple, economical [7], and attractive in terms of efficiency and overall bioconversion compared to CSTR [4].

The expected volatile solid (VS) conversion to gas both in CSTR and PFR with high loads is in the range of 35–45% [8], depending on the substrate. For a similar hydraulic retention time (HRT), PFR usually reaches similar or higher removal efficiencies [9,10] and a better utilization of the volume requirements due to hydraulics and the high solid concentrations [11–14] with lower initial investment and running costs than CSTR systems [4]. Moreover, no short-circuiting can happen in PFRs, and the operational energy demand is lower than CSTRs due to the mixing and heating requirements [15]. PFRs are increasingly being used, particularly in North America, for high-solid manure digestion systems [12], and these have become an industry standard in the USA for scraped-manure treatments [16]. Among this, there are different examples of efficient pilot and industrial-scale PFRs reported in the literature for AD and the co-digestion of organic wastes [10–13,17–23]. The PFRs are designed for manure with a high solid content in the range of 11–14% total solids (*TS*). Typical operational parameters are an HRT between 20 and 50 days, an organic loading rate (OLR) of 1–6 kgCOD m^{-3} d^{-1} and biogas production between 0.4 and 0.8 m^3 m^{-3} d^{-1} [13,24].

However, PFRs' drawbacks are a lower mass transfer due to lack of mixing, a lower efficiency when treating low *TS* content substrates (<10%), thermal stratification and solid sedimentation, or floating/scum-formation problems [13,15,25]. These problems have generated unsuccessful fermentations when treating manure feedstock in conventional biogas plants [26]. It must be noted that neither dung or manure are homogeneous liquids which contain floating solids; in the case of PFRs for such materials, a length to width ratio in the range of 2.0–2.5 has been suggested, although conventional liquid-based PFRs usually have a larger length to width ratio to avoid mixing.

The most common option to prevent solid stratification in PFRs is a partial mixing of the inner content, using mechanical mixers or biogas blower mixers, combined with a recirculation rate (RR) of the effluent [27–29]. Biogas blower mixers by biogas reinjection were found to be the best solution for treating diluted manures (*TS* of 5%, 10% and 15%) in AD for dairy cattle [27,30]. Appels et al. [30] recommended a mixing biogas flow range of 0.27–0.30 m^3 m^{-3} min^{-1} for unconfined systems.

Another aspect regarding mixing is the intensity and pattern, but significant differences of values can be found. In general, mixing with an intermediate intensity and an intermittent pattern was concluded to be the most optimal in terms of biogas production [27,31]. In [32], typically, mixing power inputs between 5 and 8 W m^{-3} were reported; Wu [33] found that 0.5–4 W m^{-3} may be sufficient; and both recommendations are below the proposed range of 40–100 W m^{-3} by Couper et al. [34].

This work describes the assessment of a modified PFR equipped with a biogas reinjection system in order to avoid operational problems such as solid stratification, foaming, or scum generation during the anaerobic digestion of animal slurries. The effect of the effluent *RR*, the *HRT*, and the inlet *TS* on the methane yield (*MY*) was evaluated, four modified PFRs running in parallel at the pilot scale for biogas production from animal manure (pig slurry) in mesophilic conditions.

2. Materials and Methods

2.1. Biomass Sources: Animal Manure and Inoculum

Fresh pig manure (FM), selected as an example of animal slurries, was collected in an intensively rearing pig farm (Girona, Spain) five times for the experiment. FM samples were characterized and used as the influent of four PFR digesters by applying different dilutions with tap water to adjust the inlet total solid content, depending on the experimental conditions (see next section). Both fresh and diluted manure were periodically characterized; Table 1 shows the characterization of FM samples. The net methane yield (L_{CH4} $kgVS^{-1}$), at 273 K and 1013 hPa, and the biodegradability (% COD) were calculated through biochemical methane potential assays at 35 °C on FM samples [35,36]. Inoculum (5 gVS L^{-1}), bicarbonate (1g $gCOD^{-1}$), deionized water (to accomplish a 0.5 L of medium per vial), and FM (5 gCOD L^{-1}) were added to 1.2 L glass vials for 30 days at 35 °C. In parallel, controls were prepared to determine the residual biogas production of the inoculum. In order to calculate the methane production rate I and the lag phase (λ) of the fresh manure (see Table 1), the modified Gompertz Equation (1) was used [37]:

$$R(t) = R_0 \cdot exp\left\{-exp\left[\frac{R_{max} \cdot e}{R_0}[\lambda - t] + 1\right]\right\} \quad (1)$$

The inoculum used for the start-up of the digesters and for the biochemical methane potential assays was sampled in a mesophilic CSTR digester of a municipal wastewater treatment plant (WWTP) in Barcelona (Spain).

Table 1. Characterization of fresh pig manure. (FM, fresh manure. COD, chemical oxygen demand. TS, VS, total and volatile solids. TKN, TAN, total Kjeldahl and ammonia nitrogen. *R*, methane production rate. λ, lag phase. Nd, not determined).

Fresh Manure (FM)	FM1	FM2	FM3	FM4	FM5	Averages
COD (gO_2 kg^{-1})	120.5	96.1	90.4	83.1	113.5	100.7 ± 14.1
TS (g kg^{-1})	105.1	105.6	99.3	104.4	113.0	105.5 ± 4.4
VS (g kg^{-1})	75.2	84.3	79.3	75.7	80.0	78.9 ± 3.3
TKN (gN kg^{-1})	6.0	6.1	6.0	6.8	6.3	6.2 ± 0.3
TAN (gN kg^{-1})	3.7	3.8	4.0	4.5	4.2	4.0 ± 0.3
Biodegrability (%COD)	33%	nd	31%	47%	nd	37 ± 7.0
CH_4 Yield ($m^3{}_{CH4}$ $kg_{VS}{}^{-1}$)	0.184	nd	0.125	0.181	nd	0.163 ± 0.027
R ($m^3{}_{CH4}$ $kg_{COD}{}^{-1}$ d^{-1})	5.8	-	6.9	14.0	-	8.9 ± 4.1
λ (d)	1.9	-	14.0	-	-	

2.2. Pilot-Scale Biogas Plant

A pilot-scale plant comprised of four horizontal PFRs (R1, R2, R3, and R4), a feeding system, a heating system, a biogas flow meter, and a programmable logic controller (PLC) to automate the plant equipment (timing and data acquisition system). All recorded data (temperature profiles of each reactor, biogas flows, inlet flows, etc.) were downloaded periodically as Excel files directly from the panel. Figure 1 shows a scheme of the pilot plant.

Once collected, FM samples were stored at ambient temperature (10–40 °C) in a 1 m^3 tank that was diluted with tap water daily just before the feeding of each digester. The weight of tap water and FM added to the dilution tank were registered by a weight cell BL-7 (Sensocar S.A., Terrassa, Spain). The diluted FM was stored in another 1 m^3 tank, also at ambient temperature, which was periodically stirred with a waterproof pump GR BluePRO (Zenit Europe, Bascharage, Luxembourg) or a vertical rotor (dilute manure tank) during the storage.

Each PFR (2830 mm length, 646 mm width) consisted of a horizontally oriented U-shaped container of stainless steel and a methacrylate cover, which allowed periodic visual revisions of the inner material, with a working volume of 160 L (total volume of 235 L). Each PFR had its own external gas holder (flexible balloon of 100 L) located on the digester´s cover that, along with the gas headspace of the stainless tank, led to a total volume of 175 L per PFR for gas storage. A flowmeter (TG5, Ritter, Bochum, Germany) recorded the biogas flow, once the biogas passed through a silica filter. The biogas composition was measured off-line daily with a gas analyzer, equipped with electrochemical (H_2S, O_2) and dual-beam infrared (CH_4, CO_2) sensors BIOGAS5000 (Geotech Ltd., Conventry, UK). Each digester was heated with individual electric blankets For-Flex Super (Electricfor S.A., Rubí, Spain) and insulated with polyurethane boards. The working temperature was set up at 34 °C and monitored with three temperature probes Pt-100 PR-24-3-100-A-G1/4-6-150 (OMEG, Connecticut, USA) that were distributed regularly along the length of each reactor, controlled by PLC.

The biogas reinjection system was used between the feeding and effluent withdrawal operations. This system includes for each PFR, a silica filter, a gasholder, a compressor V-DTN16 (Elmo Rietschle, Gardner Denver Iberia S.L., Madrid, Spain), and inner 45 polyethylene gas diffusers of 8 mm diameter Tee quick connection (Ningmao Hydraulic Pneumatic Components Factory, Zhejiang, China), which were distributed along the digester floor, near the dividing wall. The gas pipes were made of polyamide. The diffuser distribution fitted well with recommended configurations reported in the literature [38]. The reinjected biogas, introduced perpendicularly to the PFRs, generated turbulence to mix the reactor's content. The stored biogas was compressed to attain a specific flow range of 0.270–0.336 m^3 m^{-3} h^{-1} and intermittently released (2 min h^{-1} and 8 times d^{-1}), with the gas flow range being 4–5 m^3 h^{-1}.

Safety valves were located along the cover of each digester in order to keep the internal pressure of each PFR ≤20 mbar (differential pressure).

Feeding, effluent recirculation and effluent withdrawal were done manually once per day, from Monday to Friday, dividing the loading equally over the 5 days. Daily flows were recorded directly in the PLC.

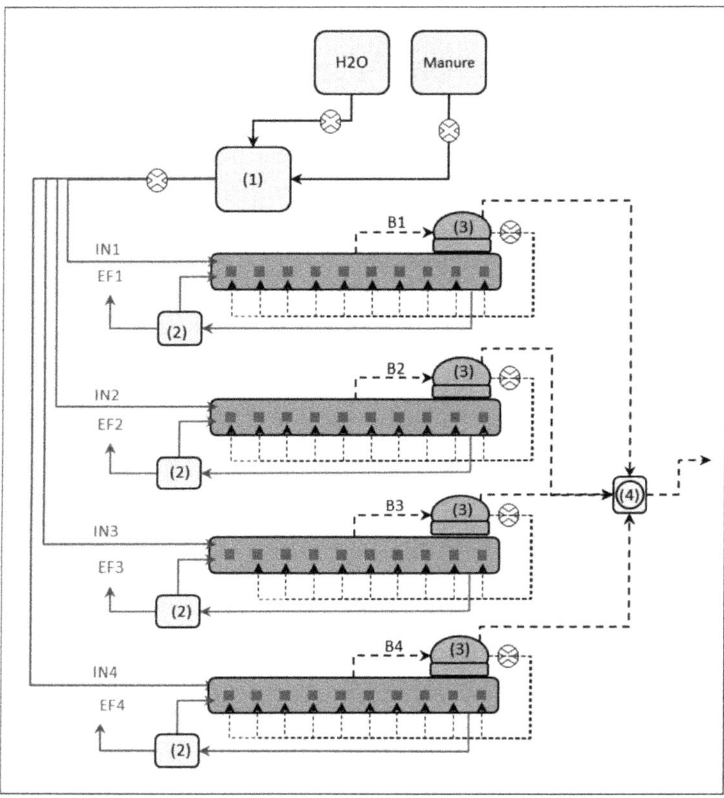

Figure 1. Scheme of the pilot plant: Numbers: (**1**) influent tank; (**2**) effluent tanks; (**3**) gas holder; (**4**) biogas flow-meter. Nomenclature: IN, influent; EF, effluent; B, biogas.

Sampling ports were located in the cover and in the bottom of the tank, allowing the collection of sludge samples from the initial, intermediate, and final points of the digesters. Influents and effluents of each PFR were characterized once per week by their content of total chemical oxygen demand (COD), total solids (TS), volatile solids (VS), total Kjeldahl nitrogen (TKN), and total ammonium nitrogen (TAN). Effluents were also characterized by pH, total and partial alkalinities (TA, PA), and volatile fatty acid (VFA) concentration once a week. All parameters were determined following standard methods [39], except for COD, which was determined as per Noguerol-Arias et al. [40] and VFA profile (acetic, propionic, i-butyric, n-butyric, i-valeric, n-valeric) which was determined by gas chromatography determined as per Rodríguez-Abalde et al. [35].

2.3. Experimental Conditions

Table 2 shows the recorded experimental conditions, which are also shown in Figure 2, in order to improve the understanding of the different conditions in each period and the dependence between the different experimental parameters (*inlet-TS, HRT, RR* and *OLR*) shown. The experiment was developed in four periods (P1, P2, P3, and P4), being evaluated in terms of *MY* (L_{CH4} kgVS^{-1}, expressed at 273 K and 1013 hPa) and organic matter removal efficiency (*VS removal*). The length of each period was at least 2.5 × HRT or the minimum amount of time to achieve a "stable condition", which was defined as that moment in which the biogas production and COD concentration in the effluent were inside 15% of the average value [27].

Table 2. Operational conditions. Average values ± Standard Error. Nomenclature: inlet total solids (*inlet-TS)*, hydraulic retention time (*HRT*), recirculation rate (*RR*), organic loading rate (*OLR*). Abbreviations: period (P), average (Av), volatile solids (VS).

Parameters	P1			P2			P3			P4		
Time Interval (Days)	1–56			57–120			121–186			187–255		
Digester R1	Av.		Error	Av.		Error	Av.		Error	Av.		Error
Inlet-TS (%)	3.58	±	0.01	5.30	±	0.54	5.85	±	0.68	5.57	±	1.07
HRT (d)	21.08	±	4.26	24.64	±	2.84	25.37	±	3.60	19.94	±	2.68
RR (%effluent)	31.17	±	6.23	30.52	±	4.15	33.55	±	5.51	39.74	±	4.59
OLR (kgVS m^{-3} d^{-1})	1.12	±	0.22	1.39	±	0.15	1.50	±	0.24	1.92	±	0.29
Digester R2												
Inlet TS (%)	3.68	±	0.05	5.64	±	0.57	5.45	±	0.99	6.21	±	0.33
HRT (d)	20.35	±	4.08	24.54	±	3.00	30.15	±	3.82	31.44	±	2.54
RR (%effluent)	31.03	±	6.21	30.53	±	3.94	26.37	±	4.38	32.71	±	4.93
OLR (kgVS m^{-3} d^{-1})	1.14	±	0.23	2.40	±	0.13	2.30	±	0.60	2.90	±	0.50
Digester R3												
Inlet TS (%)	3.57	±	0.00	5.60	±	0.86	6.81	±	1.35	3.43	±	0.18
HRT (d)	20.39	±	4.09	28.51	±	4.41	27.06	±	4.37	22.41	±	2.24
RR (%effluent)	31.21	±	6.24	15.17	±	16.15	40.57	±	7.99	38.73	±	4.89
OLR (kgVS m^{-3} d^{-1})	1.11	±	0.22	1.21	±	0.18	1.61	±	0.33	1.41	±	0.17
Digester R4												
Inlet TS (%)	3.10	±	0.00	5.10	±	0.40	5.70	±	0.69	6.12	±	0.31
HRT (d)	20.39	±	4.09	25.76	±	3.57	32.01	±	4.07	33.19	±	3.07
RR (%effluent)	31.23	±	6.25	24.46	±	4.17	46.08	±	9.81	26.50	±	4.78
OLR (kgVS m^{-3} d^{-1})	1.60	±	0.20	1.60	±	0.30	1.60	±	0.10	2.00	±	0.20

Period P1 consisted in the start-up of all four digesters (R1, R2, R3, and R4), which were inoculated with 150 L inoculum obtained from a CSTR operating at a WWTP in Barcelona (Spain). Conservative conditions were established for 25 days. Different conditions were applied for each reactor in the next periods, P2 to P4, by changing *inlet-TS, HRT,* and *RR* values. The *inlet-TS* content was in the range of 3.0% and 7.0%, as the representative TS range of FMs. The *HRT* was between 10 and 40 days, as these values were reported as the minimum for anaerobic digesters in the literature [10,12]. The *RR* was fixed between 20% and 50% of the influent flow to ensure a stable process through the anaerobic biomass recirculation into the reactors.

Digesters R1 and R2 were used to evaluate the effect of changing the inlet flow and/or the inlet *TS* concentration. In this way, in R1 a progressive increase of the inlet flow and *inlet-TS* content through the periods were done, which increased the *OLR* and decreased the *HRT* from period P2 to P4.

In R2, the *OLR* was promptly doubled by pulse additions in each period. Two additions of dilute FM were done in period P2 reaching 2.5 kgVS m^{-3} d^{-1} these days. Three pulses of different VFA (acetic acid 30 g pulse^{-1}, propionic acid 20 g pulse^{-1}, or butyric acid 16 g pulse^{-1}) were done in period P3 reaching 2.5 kgVS m^{-3} d^{-1} these days. Finally, three glucose additions (30 g pulse^{-1}) in period P4 were done reaching 2.5 kgVS m^{-3} d^{-1} these days. These increases in *OLR* were performed to evaluate resilience of the different group of microorganisms involved in the anaerobic digestion process: FM pulses help to evaluate reactor response focusing mainly on hydrolytic and acidogenic bacteria, VFA pulses help to evaluate reactor response focusing on methanogenic bacteria, whereas glucose pulses help to evaluate reactor response focusing on acidogenic and acetogenic bacteria.

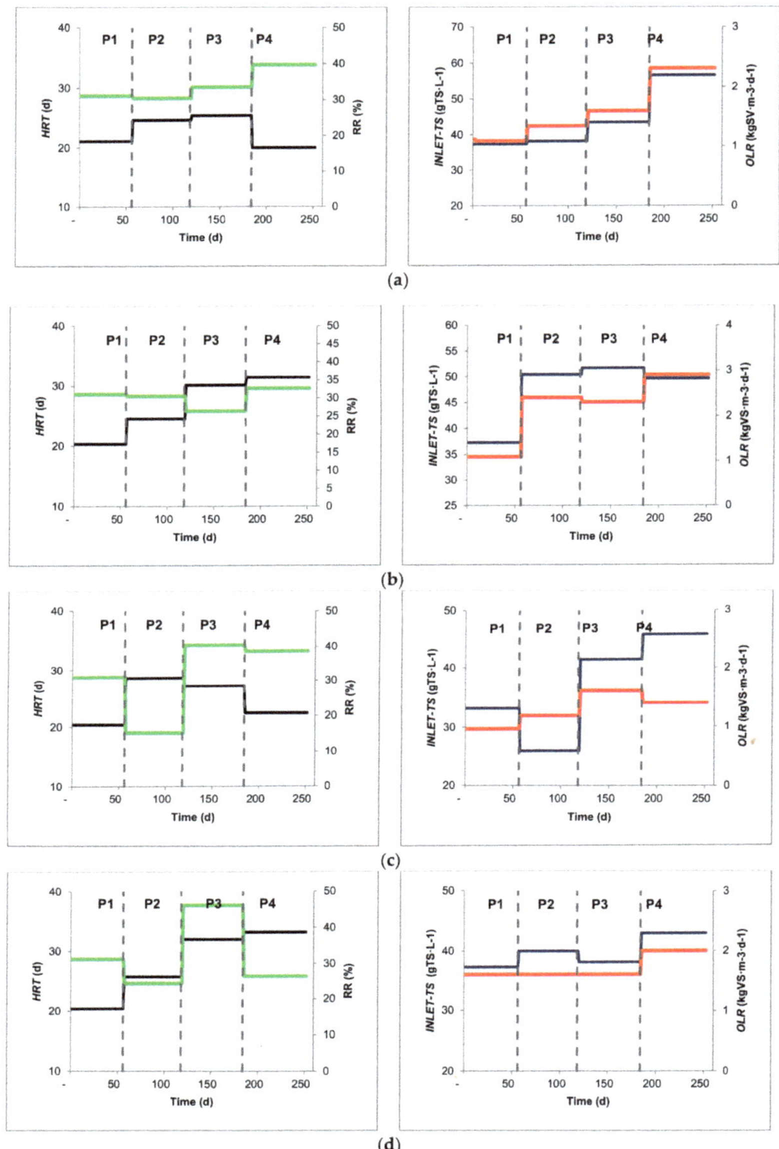

Figure 2. Operational parameter: Hydraulic retention time (*HRT*), recirculation rate (*RR*), inlet total solids content (*inlet-TS*), and organic loading rate (*OLR*) per reactor and period. (**a**) R1; (**b**) R2; (**c**) R3; (**d**) R4. Note: Vertical grey lines denote periods; black lines denote average *HRT* per period; green lines denote average *RR* per period; blue lines denote average *inlet-TS* per period; red lines denote average *OLR* per period.

Digester R3 was a used to evaluate possible problems at full-scale plants, such as pumping failures or feed blockages. In this way, during period P2, both feeding and recirculation of the effluent were stopped for 1 week and *RR* was stopped for another week at the end of this period.

Finally, digester R4 was used as a control by maintaining stable conditions, with an *OLR* of 1.6 kgVS m^{-3} d^{-1} and a *HRT* of 24–36 days. The only parameter changed was the effluent *RR*, varying from 50% to 25% depending on the experimental period (see Table 2).

Periodically, a mass balance calculation was done in terms of total N and COD. For that purpose, an accumulation term was estimated based on total N balance (since total N concentration is conservative in AD processes assuming a constant growth of the involved microorganisms). This accumulation was included in the COD mass balance in order to calculate both methane and biogas production. In this regard, Kinyua et al. [41] included N related compounds, such as struvite, and the presence of a dead volume inside tubular anaerobic digesters. Similarly, Jagadish et al. [26] discovered a floating layer that remained during the entire fermentation period inside horizontal PFRs when digesting a chopped blend of herbaceous terrestrial weeds and leaf biomass.

2.4. Statistical Analysis

The statistical analysis was done with the SPSS Statistics v.20.0.0 (IBM) software (Armonk, N.Y., USA). An ANOVA and HSD Tukey post-hoc analyses were performed to determine significant differences ($p < 0.05$) for variables whose data were normally distributed and with equal variances. For data of variables with heterogeneous variances, the Welch statistic and Games–Howell post hoc analyses were used to determine which groups were significantly different ($p < 0.05$). Linear correlations between control parameters and operation conditions during the assays were studied using the Pearson correlation coefficient (r). Correlations were considered significant at $p < 0.05$.

3. Results and Discussion

3.1. Substrate Characterization

The collected FM (Table 1) was representative of typical pig slurry or liquid manure in terms of *TS* content, as reported in [42]. The TS (9.9–11.3% wet weight) and TKN (6.0–6.8 gN kg^{-1}) contents were almost constant among all experiments. The biodegradability of FM ranged between 31–47% COD with a MY between 125–184 L_{CH4} $kgVS^{-1}$. These values were lower than the reported by Møller et al. [43] and Hansen et al. [44] when treating slurries from fattening pigs, with ranges of 47–78% COD and MY of 300–356 L_{CH4} $kgVS^{-1}$. However, the VS and TAN contents of fresh FM varied, which could explain the different qualities regarding the biodegradability. This result is coherent with other references in the literature where the storage time in the farm facility was related with the quality and methane potential of animal slurries [45].

3.2. General Plant Performance

Pilot plant operation showed several challenges due to the complex automation and the operation of four reactors simultaneously.

During P2, P3, and P4, some problems regarding the recorded biogas flow with the flowmeter were observed, and the biogas production was lower than the calculated values by COD balance (Figure 3). Two main reasons for this were identified. The first one was caused by the compressor operation for the biogas reinjection. Because of it, a regular overpressure in the digesters was observed, causing the opening of the safety gas valves in order to keep the internal pressure below 20 mbar and losing some biogas without being recorded. Secondly, the pipe length between the gas extraction points and the gas flowmeter changed depending on the digester, leading also to some biogas losses and condensate generation inside of these pipes. For these reasons, the recorded biogas flows in the beginning (P1) were well measured in all reactors, but immediately lower in the next periods.

Reactor R4 had an extra time operation of 30 days. Before this period, a maintenance in gas piping was performed in order to avoid problems regarding the recorded biogas flow. After this maintenance, the *MY*s in both reactors suddenly increased (Figure 3), following a similar trend to the calculated values by COD balance. This showed the difficulty of operating an automated pilot biogas plant comprised of four different reactors, but also confirmed that the COD balance performed during the periods in which the biogas flow meter was not working appropriately, was correct and exact enough to accept methane yields obtained through this method.

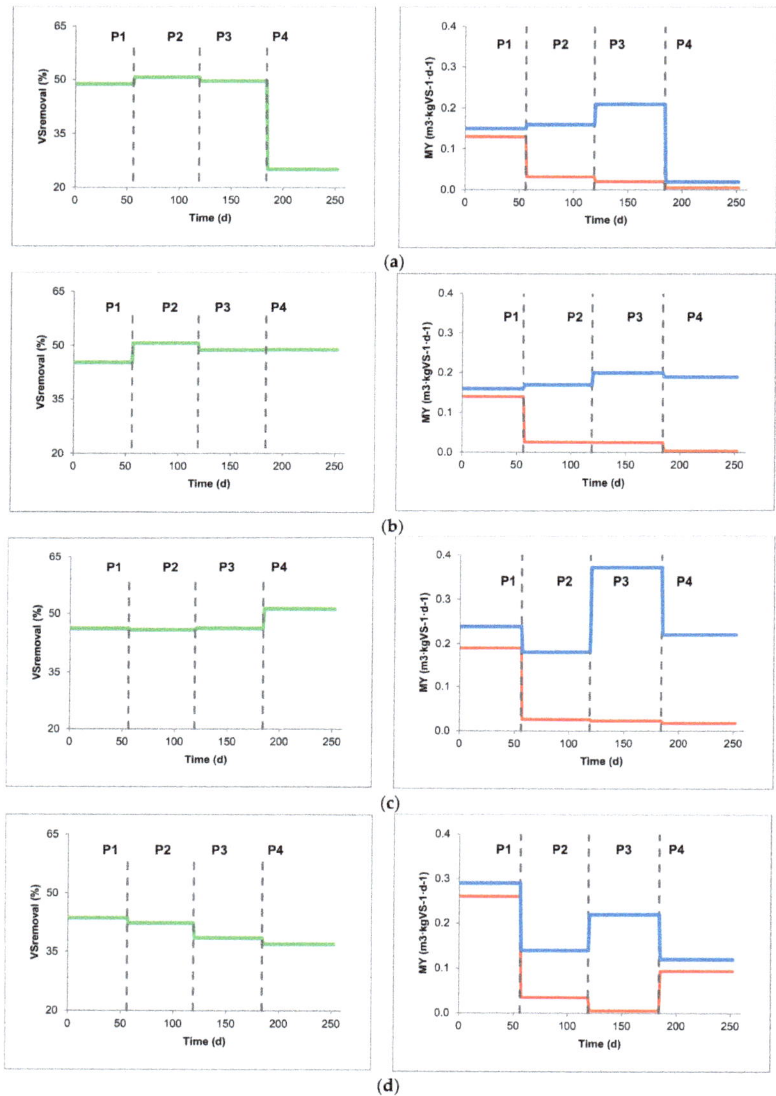

Figure 3. Evolution of organic matter removal (*VS removal*) and methane yield (*MY*) per reactor and period: (**a**) R1, (**b**) R2, (**c**) R3, (**d**) R4. Note: Vertical grey lines denote periods; green lines denote average values of *VS removal* efficiency for each whole period; red and blue lines denote average values of *MY* registered by flowmeter and COD balance, respectively, for each whole period.

The biogas reinjection flow was well adapted to the literature [38] in the range of 0.270–0.336 m^3 m^{-3} h^{-1}. Some of the 45 valve diffusers per reactor were clogged (10–15%), as reported in the literature [32], but no foaming or scum problems were observed in the digesters. No corrosion of gas piping and equipment was observed.

In general, the reactors showed appropriate performance, avoiding common problems associated with the operation of plug flow reactors at low *inlet-TS* concentrations, such as solid stratification (sedimentation or crust formation), foaming or scum generation.

3.3. Pilot Plant Monitoring

3.3.1. Biogas Composition and Organic Matter Removal

Table 3 shows the results obtained throughout the experiment regarding methane yield and methane volumetric yield, organic matter removal (*VS removal*) and biogas composition.

Table 3. Control parameters (methane productivity and yield, and organic matter removal). Nomenclature: methane yield (*MY*), volumetric methane yield (*VMY*), period (P), average (Av).

Parameters	*P1*			*P2*			*P3*			*P4*		
Time Interval (Days)	**1–56**			**57–120**			**121–186**			**187–255**		
Digester R1	**Av.**		**Error**	**Av.**		**Error**	**Av.**		**Error**	**Av.**		**Error**
MY (m^3_{CH4} $kgVS^{-1}$)	0.15	±	0.02	0.16	±	0.10	0.21	±	0.10	0.02	±	0.02
VMY (m^3_{CH4} m^{-3} d^{-1})	0.19	±	0.01	0.21	±	0.12	0.30	±	0.14	0.04	±	0.04
CH_4 (%)	63.01	±	1.52	62.97	±	3.34	61.52	±	1.62	57.22	±	2.02
O_2 (%)	0.60	±	1.12	0.26	±	0.32	0.71	±	0.71	1.14	±	1.15
VS removal (%VS inlet)	48.86	±	16.53	50.77	±	18.08	49.68	±	16.70	25.12	±	30.01
Digester R2												
MY (m^3_{CH4} $kgVS^{-1}$)	0.16	±	0.02	0.17	±	0.04	0.20	±	0.06	0.19	±	0.04
VMY (m^3_{CH4} m^{-3} d^{-1})	0.20	±	0.01	0.22	±	0.06	0.25	±	0.10	0.23	±	0.06
CH_4 (%)	61.32	±	2.10	62.06	±	3.75	62.90	±	1.49	58.81	±	3.32
O_2 (%)	0.80	±	0.54	0.31	±	0.37	0.43	±	0.42	1.04	±	1.40
VS removal (%VS inlet)	45.34	±	9.44	50.72	±	10.05	48.78	±	13.83	48.91	±	14.08
Digester R3												
MY (m^3_{CH4} $kgVS^{-1}$)	0.24	±	0.03	0.18	±	0.12	0.37	±	0.05	0.22	±	0.11
VMY (m^3_{CH4} m^{-3} d^{-1})	0.30	±	0.02	0.21	±	0.13	0.57	±	0.04	0.42	±	0.13
CH_4 (%)	60.12	±	2.21	59.26	±	3.29	62.51	±	2.33	59.38	±	4.89
O_2 (%)	0.41	±	0.76	0.45	±	0.57	0.23	±	0.32	1.06	±	2.50
VS removal (%VS inlet)	46.38	±	20.11	46.08	±	29.33	46.47	±	20.03	51.38	±	26.01
Digester R4												
MY (m^3_{CH4} $kgVS^{-1}$)	0.29	±	0.03	0.14	±	0.11	0.22	±	0.04	0.12	±	0.04
VMY (m^3_{CH4} m^{-3} d^{-1})	0.23	±	0.01	0.16	±	0.13	0.26	±	0.05	0.16	±	0.05
CH_4 (%)	56.49	±	1.76	57.54	±	1.95	55.01	±	5.01	57.23	±	4.34
O_2 (%)	1.52	±	0.81	0.43	±	0.44	3.13	±	3.34	0.63	±	1.24
VS removal (%VS inlet)	43.62	±	20.75	42.36	±	32.29	38.57	±	14.66	37.01	±	18.53

Biogas composition was similar for each period and reactor, with the highest methane content of 63.0 ± 3.3% in R1 during the P2 period and a lowest methane content of 55.0 ± 5.0% in R4 during the P3 period (Table 2). Biogas produced from animal slurries typically has a methane content of 60% [46,47] and it was shown that R1, R2, and R3 accomplish that during the P2 and P3 periods, but not during the P4 period, which is also true for the R4 reactor in any period. The unconfined gas injection mixing introduced O_2 as a consequence of air inlet in the digesters. The maximum and minimum average amounts of O_2 detected were 3.13 ± 3.34% in R4 during the P3 period and 0.23 ± 0.32% in R3 during the P3 period, respectively. In general, the O_2 concentration was higher during the previous hours at 10:00 a.m., in accordance with the reinjection mixing cycles, with the exception of R3, in which a higher H_2S concentration was evaluated during P3 and P4. It was assumed that R3 mixing was lower than other reactors, due to several mixing stoppings for different reasons such as leakage verifications, O_2 concentration determination, or compressor maintenance. Although there is no correlation between O_2 and H_2S concentrations with experimental data (see Table 4), a decrease in H_2S concentration is observed in the reactors after an increase of O_2 concentration. Figure 4 shows

O_2 and H_2S concentrations in each reactor before and after 10:00 a.m., the time in which mixing was stopped and the feeding and discharge operations were performed. This shows an average difference decrease before and after 10:00 a.m. of 602, 263, 473, and 589 ppm of H_2S in reactors R1, R2, R3, and R4, respectively.

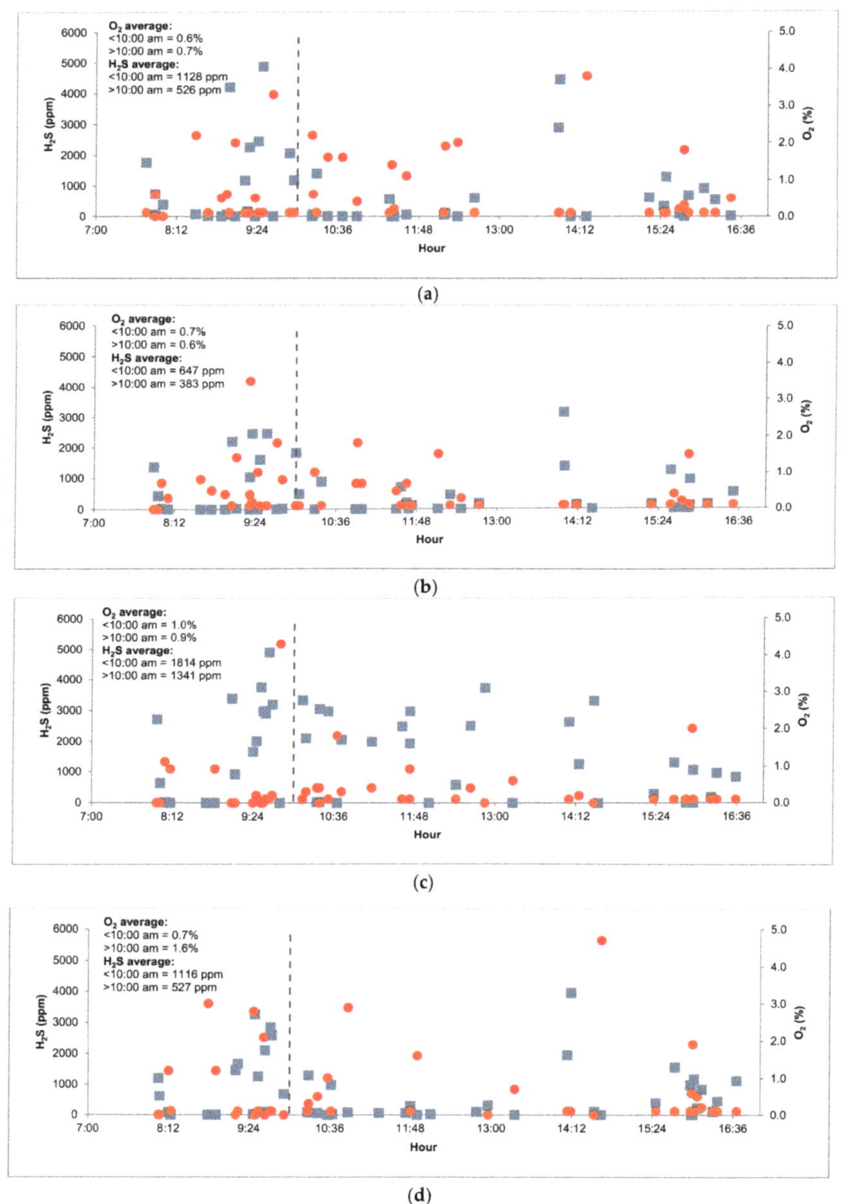

Figure 4. Evolution of H_2S and O_2 daily concentrations per reactor: (**a**) R1; (**b**) R2; (**c**) R3; (**d**) R3. Note: Vertical grey lines denote measurements done before and after 10:00 am; blue squares denote H_2S concentrations; red circles denote O_2 concentrations.

Table 4. Correlation coefficients (Pearson coefficients) for parameters. Note: * Significant correlation at 0.05 levels (dark grey). ** Significant correlation at 0.01 levels (light grey). Notes: (1) *OLR* in gVS m^{-3} d^{-1}. (2) Methane yield (*MY*) in L_{CH4} gVS^{-1}. (3) Volumetric methane yield (*VMY*) in $m^3{}_{CH4}$ m^{-3} d^{-1}. (4) Total Kjeldahl and ammonia nitrogen (TKN, TAN) in mgN L^{-1}. (5) Total volatile fatty acids (VFA) in $g_{eq\text{-}acetic\ acid}$ L^{-1}. (6) Intermediate alkalinity (IA) in g_{CaCO3} L^{-1}.

Parameter	Period	*OLR* (1)	*HRT* (d)	*VMY* (2)	*MY* (3)	*RR* (%)	NTK (4)	TAN (4)	pH	IA (5)	VFA (6)	*VS-Rem* (%)	CH_4 (%)	CO_2 (%)	H_2S (ppm)	COD-Rem (%)
Period	1	0.411 **	−0.086	−0.115	−0.168	0.370 **	0.306 *	0.079	−0.301*	0.461 **	0.412 **	−0.213 *	−0.356 *	0.344 *	0.535**	−0.218 *
OLR (1)		1	−0.800 **	−0.150	−0.279	0.438 **	0.636 **	0.139	0.044	0.392 **	0.684 **	−0.331 *	0.062	−0.073	0.467 **	−0.339 **
HRT (d)			1	0.190	0.311 *	−0.323 *	−0.567 **	−0.175	−0.100	−0.308 *	−0.582**	0.331 *	−0.201	0.205	−0.142	0.279
VMY (2)				1	0.972 **	0.051	−0.448 **	−0.243	−0.240	−0.191	−0.275	0.596 **	−0.001	−0.005	0.271	0.812 **
MY (3)					1	−0.053	−0.502 **	−0.299 *	−0.247	−0.233	−0.329 *	0.658 **	−0.073	0.070	0.168	0.866 **
RR (%)						1	0.236	−0.004	0.021	0.159	0.217	−0.156	0.030	−0.038	0.364 *	−0.136
NTK (4)							1	−0.204	0.028	0.259	0.515 **	−0.277	0.201	−0.200	−0.062	−0.423 **
TAN (4)								1	0.000	0.278 *	0.251	−0.623 **	−0.134	0.129	0.220	−0.474 **
pH									1	−0.277	0.018	−0.122	0.008	−0.002	−0.251	−0.343 *
IA (5)										1	0.682 **	−0.367 *	−0.283	0.277	0.219	−0.266
VFA (6)											1	−0.465 **	−0.217	0.212	0.217	−0.362 *
VS-Rem (%)												1	0.112	−0.113	0.037	0.818 **
CH_4 (%)													1	−1.000 **	−0.045	−0.019
CO_2 (%)														1	0.023	0.018
H_2S (ppm)															1	0.034
COD-Rem (%)																1

Because of operational problems regarding biogas measurements, biogas production was estimated using the COD balance. This methodology has been widely used in many publications and textbooks [48–50], and we validated the use of COD balance in the present work by comparing the *MY* values obtained during period P1 from both methods, by flowmeter and by COD balance, as in this period the flowmeter ran correctly. *MY* averages values are shown in Figure 3, and how a good correlation between them during this period P1 (0.15 ± 0.02, 0.16 ± 0.02, 0.24 ± 0.05, 0.29 ± 0.03 m^3CH_4 $kgVS^{-1}$ in reactors R1, R2, R3, and R4, respectively) can be observed. This is also shown during period P4, when gas piping maintenance was performed in R4 and flows of both methods (flowmeter and COD balance) were similar. Figure 5 shows the average *MY* values for all periods and reactors for the different operational conditions (*HRT*, *RR*, *inlet-TS*, and *OLR*). The maximum *MY* values obtained were in R3 during P3, with an average of 373.7 ± 49.7 L_{CH4} $kgVS^{-1}$, together with the lowest O_2 concentrations and an organic matter removal of 46.5 ± 20.0% VS. The minimum *MY* values obtained were in R1 during P4, with an average of 23.5 ± 23.0 L_{CH4} $kgVS^{-1}$, matching with the lowest organic matter removal of 25.1 ± 30.0% VS.

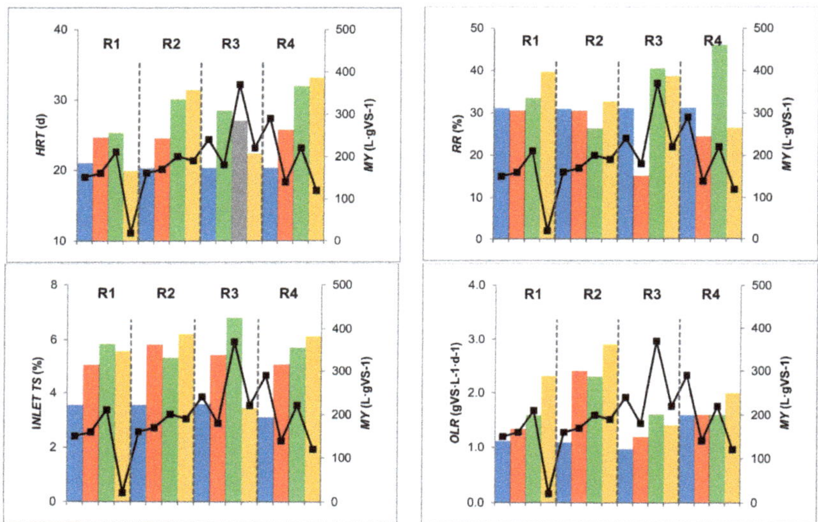

Figure 5. Comparison of methane yields (*MY*) and the hydraulic retention time (*HRT*), the recirculation rate (*RR*), the inlet total solids content (*TS*) or the organic loading rate (*OLR*). Note: Vertical grey lines denote reactors; black line denotes average values of *MY* registered in a whole period; blue, red, green and yellow bars denote average values of a whole period (P1, P2, P3, and P4, respectively) of the corresponding operational parameter per reactor.

3.3.2. VFA, pH, and Alkalinity Profiles

VFA, pH, and alkalinity are important control parameters during AD [47]. The pH average values in all reactors were slightly higher (8.1 ± 0.2, 8.0 ± 0.1, 8.0 ± 0.1 and 8.0 ± 0.2 in reactors R1, R2, R3, and R4, respectively) than the range for the normal operation of AD, between 7.5 and 7.8. This was probably due to the highest TA values registered during all periods, with averages of 11.1 ± 1.9, 10.1 ± 0.6, 10.6 ± 0.7, 10.6 ± 1.1 g_{CaCO3} L^{-1} in reactors R1, R2, R3, and R4, respectively. Moreover, changes in VFA were shown to have a significant effect on TA, with the biggest fall in R1 during P3, when a decrease of VFA from 1859 to 333 $mg{\cdot}L^{-1}$ caused a decrease of TA from 15.7 to 9.2 g_{CaCO3} L^{-1}. The percentages of acetic acid were 91 ± 10.5, 98 ± 2.5, 94 ± 9.7, and 92 ± 16.6 in reactors R1, R2, R3, and R4, respectively. These were higher than other studies reported in the literature, with ranges of 35% to 75% [51] or 60% to 75% [52]. Angelidaki et al. [53] found a common value of 1.5 g L^{-1}

was considered to be limiting for a stable operation of AD reactors. It can be seen that total VFA concentrations in effluent (Figure 6) were lower than 2 g·L^{-1} in all reactors with the exception of R1 when it was operated at high *OLR* during the P4 period. The average total VFA values were 1.0 ± 1.0, 0.4 ± 0.2, 0.4 ± 0.5, and 0.6 ± 0.4 g·L^{-1} in reactors R1, R2, R3, and R4, respectively.

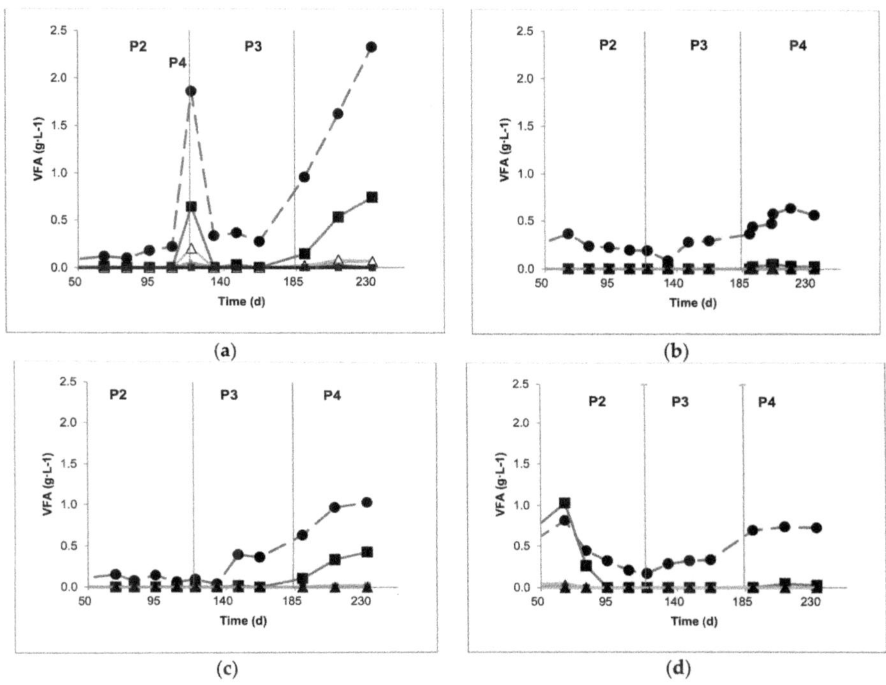

Figure 6. Evolution of volatile fatty acids (VFA) per reactor and period: (**a**) R1, (**b**) R2, (**c**) R3, (**d**) R3. Symbols: Acetic (●), propionic (■), i-butyric (▲), n-butyric (○), i-valeric (x), n-valeric (Δ), i-caproic (■), n-caproic (□).

3.3.3. Nitrogen (TKN, TAN) Profile

In general, the inhibition of the anaerobic digestion process has been reported to start at a TAN level of 1.5–2.0 gN·L^{-1}. However, in the case of pig manure, it has been reported to start at 3.1 gN·L^{-1}. The registered TAN averages values were 2.1 ± 0.2, 2.1 ± 0.1, 2.7 ± 0.2, 2.9 ± 0.3 gN·L^{-1} in reactors R1, R2, R3, and R4, respectively. The maximum amount of TAN registered (3.8 gN·L^{-1}) was in reactor R1 during P4, where a lower MY value was obtained (23.5 ± 23.0 L_{CH4} kgVS^{-1}). That would confirm the starting TAN level of 3.1 gN·L^{-1} to be inhibitory for the process.

3.3.4. Effect of Recirculation

Figure 2 shows the operational parameter evolution for each period and reactor. The reported literature has confirmed positive effects by varying *RR* in AD, such as the liquefaction and inoculation of the fresh biogas substrate, the stabilization of biogas synthesis and the optimization of biogas production [54].

When *RR* decreased, VFA concentration increased. That can be seen clearly in R4, when changing from P2 to P3, the RR variation from 24.5 ± 4.2 to 46.1 ± 9.8% caused a maximum decrease of 90% of VFA. Increases in *MY* values were observed with higher *RR* values, as can be seen in R3 and R4 from P2 to P3 (increases of 105.5% and 57.1% in R3 and R4, respectively). Results suggest *RR* values up to 40% should be used in order to keep stable conditions of the bacterial population inside the reactor, optimizing biogas production.

3.3.5. Effect of the Organic Loading Rate

In general, the expected trend for *HRT* and *OLR* was confirmed in all reactors, as the biogas production increased when increasing *HRT* or *OLR*, with some exceptions: R1 during P4, probably due to the ammonia inhibition effects mentioned before. Also increases of *OLR* in R3 during P2 and in R4 during P2 and P4 did not increase the *MY*, probably due to the decreased *RR* in these reactors and periods. An increase of acetic and propionic acid concentrations was also observed with *OLR* increases, such as in R1 and R2 during both P3 and P4 periods.

3.3.6. Effect of Total Solid Content

As the *OLR* depends on *inlet-TS* concentration, similar performances were evaluated. At higher *inlet-TS* concentrations, higher *MY* values were obtained. The maximum *MY* obtained was in R3 during P3 (373.7 ± 49.7 L_{CH4} $kgVS^{-1}$) with 6.81 ± 1.35 % TS. H.M. El-Mashad et al. [51] concluded that a higher TS content resulted in higher VFA concentrations. Although there was not a clear correlation in our results between *inlet-TS* contents and VFA (see Table 4), a similar trend was exhibited as the reported study of Page et al. [55]; when evaluating dairy manure effluent at 2.4% TS, a range of 89–439 $mg \cdot L^{-1}$ of VFA was obtained.

4. Conclusions

A novel modified plug-flow anaerobic reactor for treating animal slurries was developed and tested successfully with appropriate performance. The mixing power inputs and the intermittent mixing pattern allowed for the avoidance of operational problems such as solid stratification, foaming, or scum generation. In general, there was no inhibition due to ammonia concentration, except for reactor R1 during P4. VFA generation allowed stable conditions. The methane yields were optimized during some of the periods, showing how significant the operational parameters evaluated are. However, it seems that complex automation and equipment design involves facing the important problem of the scale factor.

Author Contributions: D.G., J.L.R., B.F. and F.H. wrote the paper; J.L.R. and B.F. designed the experiments; D.G., E.M., L.T. and M.R. performed the experiments.

Funding: This research was funded by private funds from ProCycla SA, a privately held company. Research was conducted in IRTA—Institut de Recerca i Tecnologia Agroalimentàries (Spain). IRTA thanks the financial support of CERCA program of the Generalitat de Catalunya.

Conflicts of Interest: The authors declare no conflicts of interest.

References

1. Lukehurst, C.T.; Frost, P.; Al Seadi, T. Utilisation of digestate from biogas plants as biofertiliser. *IEA Bioenergy* **2010**, *2010*, 1–36.
2. De Mes, T.Z.D.; Stams, A.J.M.; Reith, J.H.; Zeeman, G. Chapter 4: Methane production by anaerobic digestion of wastewater and solid wastes. In *Bio-Methane Bio-Hydrogen: Status and Perspectives of Biological Methane and Hydrogen Production*; Dutch Biological Hydrogen Foundation: The Hague, The Netherlands, 2003; pp. 58–102.
3. Manyi-Loh, C.; Mamphweli, S.; Meyer, E.; Okoh, A.; Makaka, G.; Simon, M.; Manyi-Loh, C.E.; Mamphweli, S.N.; Meyer, E.L.; Okoh, A.I.; et al. Microbial Anaerobic Digestion (Bio-Digesters) as an Approach to the Decontamination of Animal Wastes in Pollution Control and the Generation of Renewable Energy. *Int. J. Environ. Res. Public Health* **2013**, *10*, 4390–4417. [CrossRef] [PubMed]
4. Mao, C.; Feng, Y.; Wang, X.; Ren, G. Review on research achievements of biogas from anaerobic digestion. *Renew. Sustain. Energy Rev.* **2015**, *45*, 540–555. [CrossRef]
5. Crittenden, J.C.; Montgomery, W.H. *Water Treatment Principles and Design*; John Wiley & Sons: Hoboken, NJ, USA, 2005.

6. Levenspiel, O. *Chemical Reaction Engineering*, 3rd ed.; John Wiley & Sons: New York, NY, USA, 1999.
7. Wen, Z.; Frear, C.; Chen, S. Anaerobic digestion of liquid dairy manure using a sequential continuous-stirred tank reactor system. *J. Chem. Technol. Biotechnol.* **2007**, *82*, 758–766. [CrossRef]
8. Burke, D.A. *Dairy Waste Anaerobic Digestion Handbook*; Environmental Energy Company: Olympia, WA, USA, 2001; Volume 6007, pp. 17–27.
9. Krich, K.; Augenstein, D.; Batmale, J.P.; Benemann, J.; Rutledge, B.; Salour, D. Chapter 8 Financial Analysis of Biomethane Production. In *Biomethane from Dairy Waste. A Sourcebook for the Production and Use of Renewable Natural Gas in California*; Western United Dairymen: Modesto, CA, USA, 2005; pp. 147–162.
10. Li, Y.-F.; Chen, P.-H.; Yu, Z. Spatial and temporal variations of microbial community in a mixed plug-flow loop reactor fed with dairy manure. *Microb. Biotechnol.* **2014**, *7*, 332–346. [CrossRef] [PubMed]
11. Azbar, N.; Ursillo, P.; Speece, R.E. Effect of process configuration and substrate complexity on the performance of anaerobic processes. *Water Res.* **2001**, *35*, 817–829. [CrossRef]
12. Batstone, D.J.; Puyol, D.; Flores-Alsina, X.; Rodríguez, J. Mathematical modelling of anaerobic digestion processes: Applications and future needs. *Rev. Environ. Sci. Bio/Technol.* **2015**, *14*, 595–613. [CrossRef]
13. Cantrell, K.B.; Ducey, T.; Ro, K.S.; Hunt, P.G. Livestock waste-to-bioenergy generation opportunities. *Bioresour. Technol.* **2008**, *99*, 7941–7953. [CrossRef]
14. Speece, R.E.; Duran, M.; Demirer, G.; Zhang, H.; DiStefano, T. The role of process configuration in the performance of anaerobic systems. *Water Sci. Technol.* **1997**, *36*, 539–547. [CrossRef]
15. Monnet, F. An introduction to anaerobic digestion of organic wastes. *Remade Scotl.* **2003**, *379*, 1–48.
16. Wilkie, A.C. Anaerobic digestion of dairy manure: Design and process considerations. *Dairy Manure Manag. Treat. Handl. Community Relat.* **2005**, *301*, 312.
17. Dvorak, S.; Frear, C. Commercial Demonstration of Nutrient Recovery of Ammonium Sulfate and Phosphorous Rich Fines from AD Effluent. 2014. Available online: http://hdl.handle.net/1813/36526 (accessed on 7 July 2019).
18. Lansing, S.; Martin, J.F.; Botero, R.B.; da Silva, T.N.; da Silva, E.D. Methane production in low-cost, unheated, plug-flow digesters treating swine manure and used cooking grease. *Bioresour. Technol.* **2010**, *101*, 4362–4370. [CrossRef] [PubMed]
19. Lansing, S.; Botero, R.B.; Martin, J.F. Waste treatment and biogas quality in small-scale agricultural digesters. *Bioresour. Technol.* **2008**, *99*, 5881–5890. [CrossRef] [PubMed]
20. Martin, J.H. *An Evaluation of a Mesophilic, Modified Plug Flow Anaerobic Digester for Dairy Cattle Manure*; EPA Contract No. GS 10F-0036K Work Assignment/Task Order 9; East. Res. Group, Inc.: Morrisville, NC, USA, 2005.
21. Martin, J.H. *Comparison of Dairy Cattle Manure Management with and without Anaerobic Digestion and Biogas Utilization*; Eastern Research Group, Inc.: Boston MA for AgStar Program; U.S. Environmental Protection Agency: Washington, DC, USA, 2003.
22. St-Pierre, B.; Wright, A.-D.G. Metagenomic analysis of methanogen populations in three full-scale mesophilic anaerobic manure digesters operated on dairy farms in Vermont, USA. *Bioresour. Technol.* **2013**, *138*, 277–284. [CrossRef] [PubMed]
23. Yu, Z.; Morrison, M.; Schanbacher, F.L. Production and utilization of methane biogas as renewable fuel. Biomass to biofuels Strateg. *Glob. Ind.* **2010**, 403–433. [CrossRef]
24. Beddoes, J.C.; Bracmort, K.S.; Burns, R.T.; Lazarus, W.F. An analysis of energy production costs from anaerobic digestion systems on US livestock production facilities. *USDA NRCS Tech. Note* **2007**, *1*, 7–10.
25. Chynoweth, D.P.; Isaacson, R. *Anaerobic Digestion of Biomass*; Chynoweth, D.P., Isaacson, R., Eds.; Elsevier applied Science; Springer: Dordrecht, The Netherlands, 1987; ISBN 978-1-85166-069-8.
26. Jagadish, K.; Chanakya, H.; Rajabapaiah, P.; Anand, V. Plug flow digestors for biogas generation from leaf biomass. *Biomass Bioenergy* **1998**, *14*, 415–423. [CrossRef]
27. Karim, K.; Thomas Klasson, K.; Hoffmann, R.; Drescher, S.R.; DePaoli, D.W.; Al-Dahhan, M.H. Anaerobic digestion of animal waste: Effect of mixing. *Bioresour. Technol.* **2005**, *96*, 1607–1612. [CrossRef] [PubMed]
28. Meroney, R.N.; Colorado, P.E. CFD simulation of mechanical draft tube mixing in anaerobic digester tanks. *Water Res.* **2009**, *43*, 1040–1050. [CrossRef] [PubMed]
29. Stalin, N.; Prabhu, H.J. Performance evaluation of partial mixing anaerobic digester. *ARPN J. Eng. Appl. Sci.* **2007**, *2*, 1–6.

30. Vesvikar, M. *Understanding the Hydrodynamics and Performance of Anaerobic Digesters, Chemical Reaction Engineering Laboratory*; Washingt. Univ.: St. Louis, MO, USA, 2007.
31. Kaparaju, P.; Buendia, I.; Ellegaard, L.; Angelidakia, I. Effects of mixing on methane production during thermophilic anaerobic digestion of manure: Lab-scale and pilot-scale studies. *Bioresour. Technol.* **2008**, *99*, 4919–4928. [CrossRef] [PubMed]
32. Agency, E.P. EPA Design Information Report–anaerobic digester mixing systems. *J. Water Pollut. Control Fed.* **1987**, *59*, 162–170.
33. Wu, B. Large eddy simulation of mechanical mixing in anaerobic digesters. *Biotechnol. Bioeng.* **2012**, *109*, 804–812. [CrossRef] [PubMed]
34. Couper, R.; Penney, W.R.; Fair, J.R.; Walas, S.M. *Chemical Process Equipment*, 2nd ed.; Elsevier Inc.: Amsterdam, The Netherlands, 2005; ISBN 978-0-7506-7510-9.
35. Rodríguez-Abalde, Á.; Fernández García, B.; Flotats Ripoll, X. Effect of thermal pre-treatments on biogas production potential of solid slaughterhouse wastes. In Proceedings of the 17th European Biomass Conference & Exhibition—From Research to Industry and Markets, Hamburg, Germany, 29 June–3 July 2009; pp. 252–256.
36. Soto, M.; Méndez, R.; Lema, J.M. Methanogenic and non-methanogenic activity tests. Theoretical basis and experimental set up. *Water Res.* **1993**, *27*, 1361–1376. [CrossRef]
37. Li, P.; Li, W.; Sun, M.; Xu, X.; Zhang, B.; Sun, Y. Evaluation of Biochemical Methane Potential and Kinetics on the Anaerobic Digestion of Vegetable Crop Residues. *Energies* **2019**, *12*, 26. [CrossRef]
38. Appels, L.; Baeyens, J.; Degrève, J.; Dewil, R. Principles and potential of the anaerobic digestion of waste-activated sludge. *Prog. Energy Combust. Sci.* **2008**, *34*, 755–781. [CrossRef]
39. Association, A.P.H.; Association, A.W.W.; Federation, W.P.C.; Federation, W.E. *Standard Methods for the Examination of Water and Wastewater*; American Public Health Association (APHA): Washington, DC, USA, 2005.
40. Noguerol-Arias, J.; Rodríguez-Abalde, A.; Romero-Merino, E.; Flotats, X. Determination of Chemical Oxygen Demand in Heterogeneous Solid or Semisolid Samples Using a Novel Method Combining Solid Dilutions as a Preparation Step Followed by Optimized Closed Reflux and Colorimetric Measurement. *Anal. Chem.* **2012**, *84*, 5548–5555. [CrossRef]
41. Kinyua, M.N.; Rowse, L.E.; Ergas, S.J. Review of small-scale tubular anaerobic digesters treating livestock waste in the developing world. *Renew. Sustain. Energy Rev.* **2016**, *58*, 896–910. [CrossRef]
42. Agency, E.P. *Anaerobic Digester Conservation Practice Standard Number 366*; USDA Natural Resources Conservation Service: New York, NY, USA, 2009.
43. Møller, H.B.; Sommer, S.G.; Ahring, B.K. Methane productivity of manure, straw and solid fractions of manure. *Biomass Bioenergy* **2004**, *26*, 485–495. [CrossRef]
44. Hansen, K.H.; Angelidaki, I.; Ahring, B.K. Anaerobic digestion of swine manure: Inhibition by ammonia. *Water Res.* **1998**, *32*, 5–12. [CrossRef]
45. Flotats, X.; Bonmatí, A.; Fernández, B.; Magrí, A. Manure treatment technologies: On-farm versus centralized strategies. NE Spain as case study. *Bioresour. Technol.* **2009**, *100*, 5519–5526. [CrossRef]
46. Hreiz, R.; Adouani, N.; Fünfschilling, D.; Marchal, P.; Pons, M.-N. Rheological characterization of raw and anaerobically digested cow slurry. *Chem. Eng. Res. Des.* **2017**, *119*, 47–57. [CrossRef]
47. Nagy, G.; Wopera, A. Biogas production from pig slurry-feasibility and challenges. *Mat. Sci. Eng.* **2012**, *37*, 65–75.
48. Muzondiwa, R.; Kamusoko, R. Methods for determination of biomethane potential of feedstocks: A review. *Biofuel Res. J.* **2017**, *4*, 573–586. [CrossRef]
49. Sunada, N.S.; Amorim, A.C.; Previdelli, M.A.; Miranda, F.; Garófallo, R.; Mendes, A.R. Potential of biogas and methane production from anaerobic of poultry slaughterhouse effluent. *R. Bras. Zootec.* **2012**, *41*. [CrossRef]
50. Donoso-Bravo, A.; Sadino-Riquelme, C.; Gómez, D.; Segura, C.; Valdebenito, E.; Hansen, F. Modelling of an anaerobic plug-flow reactor. Process analysis and evaluation approaches with non-ideal mixing coniderations. *Bioresour. Technol.* **2018**, *260*, 95–104. [CrossRef]
51. El-Mashad, H.M.; Zhang, R.; Arteaga, V.; Rumsey, T.; Mitloehner, F.M. Volatile Fatty acids and Alcohols Production during Anaerobic Storage of Dairy Manure. *Trans. ASABE* **2011**, *54*, 599–607. [CrossRef]
52. Ratanatamskul, C.; Saleart, T. Effects of sludge recirculation rate and mixing time on performance of a prototype single-stage anaerobic digester for conversion of food wastes to biogas and energy recovery. *Environ. Sci. Pollut. Res.* **2016**, *23*, 7092–7098. [CrossRef] [PubMed]

53. Angelidaki, I.; Boe, K.; Ellegaard, L. Effect of operating conditions and reactor configuration on efficiency of full-scale biogas plants. *Water Sci. Technol.* **2005**, *52*, 189–194. [CrossRef]
54. Michele, P.; Giuliana, D.; Carlo, M.; Sergio, S.; Fabrizio, A. Optimization of solid state anaerobic digestion of the OFMSW by digestate recirculation: A new approach. *Waste Manag.* **2015**, *35*, 111–118. [CrossRef]
55. Page, L.H.; Ni, J.Q.; Heber, A.J.; Mosier, N.S.; Liu, X.; Joo, H.S.; Ndegwa, P.M.; Harrison, J.H. Characteristics of volatile fatty acids in stored dairy manure before and after anaerobic digestion. *Biosyst. Eng.* **2014**, *118*, 16–28. [CrossRef]

Article

Effect of Mixing Regimes on Cow Manure Digestion in Impeller Mixed, Unmixed and Chinese Dome Digesters

Abiodun O. Jegede [1,2,*], Grietje Zeeman [2] and Harry Bruning [2]

1 Centre for Energy Research and Development, Obafemi Awolowo University, 220282 Ile-ife, Nigeria
2 sub-Department Environmental Technology, Wageningen University and Research, 6708 WG Wageningen, The Netherlands
* Correspondence: jegedeabiodun@yahoo.com or ajegede@cerd.gov.ng

Received: 10 April 2019; Accepted: 13 June 2019; Published: 2 July 2019

Abstract: This study examines the effect of mixing on the performance of anaerobic digestion of cow manure in Chinese dome digesters (CDDs) at ambient temperatures (27–32 °C) in comparison with impeller mixed digesters (STRs) and unmixed digesters (UMDs) at the laboratory scale. The CDD is a type of household digester used in rural and pre-urban areas of developing countries for cooking. They are mixed by hydraulic variation during gas production and gas use. Six digesters (two of each type) were operated at two different influent total solids (TS) concentration, at a hydraulic retention time (HRT) of 30 days for 319 days. The STRs were mixed at 55 rpm, 10 min/hour; the unmixed digesters were not mixed, and the Chinese dome digesters were mixed once a day releasing the stored biogas under pressure. The reactors exhibited different specific biogas production and treatment efficiencies at steady state conditions. The STR 1 exhibited the highest methane (CH_4) production and treatment efficiency (volatile solid (VS) reduction), followed by STR 2. The CDDs performed better (10% more methane) than the UMDs, but less (approx. 8%) compared to STRs. The mixing regime via hydraulic variation in the CDD was limited despite a higher volumetric biogas rate and therefore requires optimization.

Keywords: mixing; Chinese dome digester; impeller mixed digester; unstirred digester; hydraulically mixed; total solids (TS) concentration

1. Introduction

About 2.5 billion people globally depend on traditional biomass, for example firewood, as their main source of energy for heating and cooking [1]. The use of firewood as cooking fuel has several negative effects on the environment, health and social life. The collection of firewood is sometimes done by women and children, and this activity can take many hours a day, which indirectly affects productive periods, education and leisure time. The use of firewood and other biomass for cooking produces hazardous particles [2,3], which are dangerous to human health. The use of firewood for cooking is one of the factors that causes deforestation, erosion, reduction of water resources, and indirectly contributes to climate change [2].

About 1.4 billion people worldwide will possibly be left without access to modern sources of cooking energy such as gas and clean stove, if sustainable energy sources are not made available [4]. The conversion of biomass such as agricultural waste, cow manure and pig manure to clean sources of energy such as biogas via anaerobic digestion could help to solve part of this energy problem in the rural areas of developing countries, and improve the standards of living, health, the local environment, and mitigate climate change [5].

Biogas consists of mainly methane (CH_4) and carbon dioxide (CO_2), with traces of ammonia (NH_3), nitrogen (N_2) and hydrogen sulphide (H_2S). In large-scale anaerobic digesters where biogas is used for electricity and heat production, or injection into the gas network, the calorific value of biogas needs to be increased by removing the unwanted components (for example CO_2 and H_2S) with the use of different technologies such as water scrubbing, physical and chemical absorption, membrane technology, in situ upgrading, etc. These unwanted components are harmful to the downstream units such as microturbines and combined heat and power (CHP) [6]. However, in household anaerobic digesters, where biogas is primarily for cooking, only H_2S is removed with the use of simple and cheap de-sulphuring units, e.g., by allowing biogas to pass through iron fillings.

Household biogas plants for the treatment of organic wastes such as cow manure are mostly popular in Asia. More than forty million household low-cost anaerobic digesters have been built across China and India [7–9], with an estimated potential of about 140 million household biogas systems in the agricultural regions in China [10]. The most popular low-cost household digesters are the Chinese dome [11–14], Indian floating drum, plug flow and Puxin digester—a prefabricated version of the Chinese dome digester [15].

These digesters are relatively inexpensive, unheated and have non-forced mixed systems, making them well suitable for farmers and people living in rural areas. However, the application and specific design of these digesters still depend on location, socio-economic context, and weather conditions of the particular location [16,17].

The anaerobic digestion process depends on mixing for distribution of the inoculum during start-up, improving contact between nutrients substrate and microorganisms, temperature equalization, removal of intermediate products and prevention of settling and floating layers [18]. The Chinese dome digester (CDD), which is the most used applied household digester is usually constructed underground with a hemispherical dome top, which serves as gas storage. Gas pressure is created as a result of biogas production, while collection in a closed environment and slurry level difference in the reactor is a result of the pressure build-up. The stored biogas pushes part of the slurry into the effluent (expansion) chamber, because water or the slurry is an incompressible liquid. During gas use, the effluent flows back into the main digester chamber creating a mixing regime [19]. Mixing in the CDD depends on the hydraulic variation in the digester during digester use and could be regarded as intermittent natural mixing. The mixing depends on the feeding regime, the gas production rate, gas use frequency, and slurry viscosity. Significant efforts have been made to evaluate the effect of mixing in mechanically mixed reactors by applying impeller mixing, slurry recirculation and biogas injection, in comparison to non-mixed reactors [20–29], but little research was done in studying mixing in naturally mixed reactors. Most of these studies focus on the effects of mixing modes and intensities on biogas yields in relation to retention time and organic loading rate. The optimum mixing mode is still a subject of debate, but most researchers found that intermittent mixing aids anaerobic digestion. Most household digesters are operated at total solid (TS) influent concentrations $<7\%$ and a long hydraulic retention time (HRT) (>40 days) [16,30–39], as compared to mechanically mixed systems, which are generally operated at an HRT ≤ 20 days at mesophilic conditions [26,40–47].

In CDDs, improved mixing might reduce applicable HRTs and therewith reactor volume and investment costs. Moreover, increasing the influent TS concentration by reduced dilution, at the same HRT, will result in an additional reduction of the reactor volume and might increase the mixing conditions due to a higher volumetric gas production. Therefore, the objective of this study was to study the performance of the Chinese dome digester and comparison with the impeller mixed digester and unmixed digester using cow manure at two different total solid (TS) concentrations at the same HRT of 30 days, to evaluate if mixing induced by hydraulic variation is sufficient to produce superior digestion efficiency.

2. Materials and Methods

2.1. Reactor Design and Setup

The study was performed in six laboratory-scale digesters consisting of two impeller mixed digesters or stirred tank reactors (STR), two unmixed digesters (UMD) and two Chinese dome digesters (CDD). The working volume of the impeller mixed and the unmixed reactors was 39 L, while the Chinese dome digesters (CDDs) have a working volume of 39 L and an additional 10 L for the extension chamber, which is not regarded as part of the working volume of the reactors. A scheme of the three types of reactors is shown in Figure 1a–c. All digesters were constructed from polyvinyl chloride (PVC). Generated biogas was collected in plastic gas bags. Biogas produced in the impeller mixed reactors and unmixed reactors was directly collected into the gas bags while the biogas produced in the Chinese dome lab scale reactors was stored in the reactor headspace, thereby creating pressure to displace some of the reactor content to the extension chamber. Pressure was released once a day before feeding and the biogas was collected in gas bags. In addition, in the CDDs, effluents were removed from the reactors through the extension chambers. In the mechanically and non-mixed reactors, influents were added from the top and effluents withdrawn from the bottom as shown in Figure 1a,b. Biogas was collected and measured with a wet gas meter before feeding was done in all reactors daily. The two STRs were mixed with an 18 cm impeller at 55 rpm for 10 min/hour throughout the study period, based on Karim et al. [22] and Hoffmann et al. [48]. A reasonable level of intermittent mixing was achieved during reactor feeding and effluent removal from all the reactors based on the reactor's geometry. The reactors set-up and arrangement are given in Table 1.

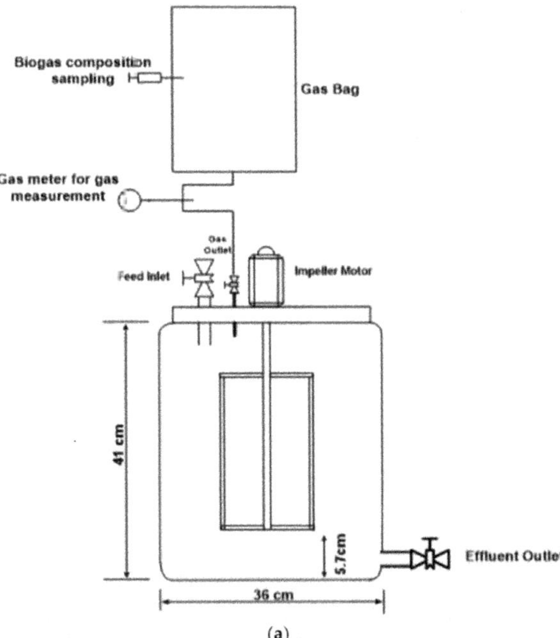

(a)

Figure 1. *Cont.*

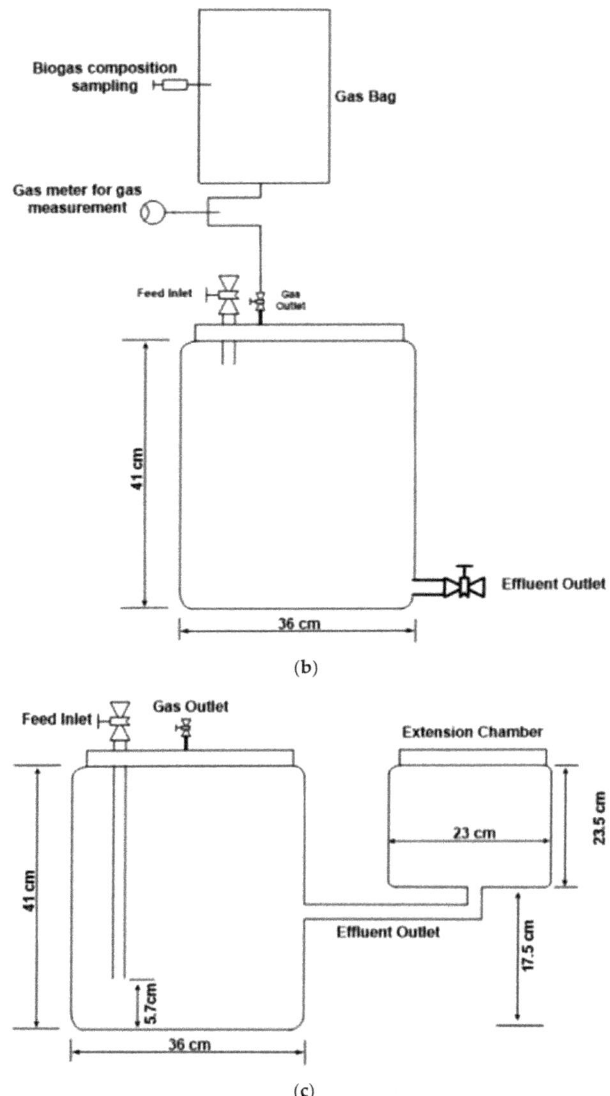

Figure 1. (**a**) Impeller mixed, (**b**) Unmixed, and (**c**) hydraulic mixed (Chinese dome) digesters.

Table 1. Reactors set-up.

Reactor Name	Reactor Design	% TS
STR 1	Impeller mixed, 55 rpm for 10 min/hour	3–7.3
STR 2	Impeller mixed, 55 rpm for 10 min/hour	6–15
UMD 1	Unmixed	3–7.3
UMD 2	Unmixed	6–15
CDD 1	Chinese dome, hydraulically mixed	3–7.3
CDD 2	Chinese dome, hydraulically mixed	6–15

2.2. Manure Collection and Preparation

The inoculum used for the reactor start-up was collected from a small-scale biogas plant treating cow manure at the Agricultural Engineering Department, Obafemi Awolowo University, Nigeria, at an average ambient temperature of 32 °C and operated with an average influent concentration of 5.5% TS. The inoculum was collected on the same day the reactors were started and occupied 23% volume of each reactor during start-up. The cow manure used for feeding the reactors was collected freshly at the Obafemi Awolowo University School farm. Each batch was stored in a refrigerator at 3 °C prior to use. The manure was prepared by manual screening for stones, blending and water dilution into two total solid concentrations. The blending of the substrates was done in a household blender at about 8000 rpm for 2 min to break large pieces of manure. The characteristics of the prepared substrate feed, including the biomethane potential (BMP) before dilution, are given in Table 2. The BMP was done in triplicate as described by Anaerobic Biodegradation, Activity and Inhibition (ABAI), task group [49]. The applied organic loading rates and period of operation for each of the reactors are given in Table 3.

Table 2. Applied feed characteristics before dilution for the feeding periods.

Feed Manure	Initial TS (g/L)	Initial vs. (g/L)	BMP (Biogas) L/g vs.	BMP (CH_4) L/g VS	NH_4^+–N (g/L)
1	284 ± 16	229 ± 14	0.25	0.17	2.2
2	262 ± 12	152 ± 2	0.23	0.15	2.1
3	316 ± 5	250 ± 1	0.22	0.15	2.3
4	339 ± 4	287 ± 7	0.24	0.16	1.9
5	200 ± 14	137 ± 2	0.23	0.15	2
6	257 ± 1	167 ± 2	0.25	0.16	1.8
7	296 ± 13	168 ± 3	0.25	0.16	2.2
8	383 ± 13	195 ± 13	0.28	0.18	2.1

Table 3. Influent concentrations and applied loading rates for the impeller mixed digesters or stirred tank reactors (STR 1 and 2), unmixed digesters (UMD 1 and 2), and the Chinese dome digesters (CDD 1 and 2), Digesters 1–6.

Feed Manure	Day	TS (g/L)		vs. (g/L)		OLR g VS/L/Day	
		1, 3, 5	2, 4, 6	1, 3, 5	2, 4, 6	1, 3, 5	2, 4, 6
1	1–35	43	86	35	70	1.16	2.33
2	36–59	40	80	23	46	0.77	1.54
3	60–78	47	94	37	74	1.22	2.44
4	79–93	52	105	44	88	1.47	2.94
	94–125		Recirculation				
4	126–133	52	105	44	88	1.47	2.94
5	134–169	30	60	20	41	0.68	1.36
6	170–198	40	80	26	52	0.86	1.72
7	199–259	44	88	25	50	0.83	1.67
8	260–319	73	147	42	84	1.39	2.79

2.3. Operation

All reactors were operated at a hydraulic retention time (HRT) of 30 days throughout the study. About 1.3 L of effluents were daily removed from the outlet port for STR 1, STR 2, UMD 1 and UMD 2, and from the extension chambers for CDD 1 and CDD 2 after which the same quantity of freshly prepared manure was added. Reactors were considered to operate in steady state when the change in biogas production was within 15% [22]. All reactors were operated at laboratory ambient temperatures between 27 and 32 °C. All the digesters were started under similar operational conditions but with different mixing schemes and loading rates. Digesters 1 and 2 (STR 1 and 2) were mixed at 55 rpm, 10 min/hour, and the Reynolds number (*Re*) and power were calculated using Equations (1) and (2):

$$R_e = \frac{N\rho D^2}{K\,\gamma^{n-1}}, \text{ and} \tag{1}$$

$$P = Np^* \rho^* N^3 * D^5, \tag{2}$$

where P is the power consumption (W), ρ is the density of manure influent ($kg{\cdot}m^{-3}$), N is impeller speed in S^{-1}, D is impeller diameter in m, K is the consistency coefficient ($P_a\ S^n$), γ is shear (S^{-1}) and n is the power-law index. Digesters 5 and 6 were mixed by hydraulic variation and the energy created and utilized consumption were calculated in the form of potential energy created as a result of slurry displacement in the extension chamber. Digesters 3 and 4 were not mixed, however all digesters could be assumed to be "mixed" intermittently once a day during effluent withdrawal and feeding. The reactors were fed, and effluents removed daily from day 1 to day 319, except for day 94 to 125 when recirculation was applied for all digesters as overloading was observed. During the recirculation period, there was no addition of new feed into the reactors but feeding of effluent. Energy consumption for STR 1 and 2 was calculated from Equation (2).

The values of parameters mentioned in the text are given in Table 4 according to Wu [50]. The power number was determined from the *Np* vs. R_e chart [51].

Table 4. Manure physical characteristic for energy calculation.

Digester	K ($P_a\ S^n$)	n	γ (S^{-1})	ρ ($kg{\cdot}m^{-3}$)	Reference
1	0.525	0.533	11	1000	[50]
2	31.3	0.3	11	~1000	[50]

During biogas production, the volume of effluent displaced in the extension chamber was stored as potential energy (P.E). During gas use, pressure would be reduced in the headspace of the CDD and would result in the follow of effluent from the extension chamber into the main digester volume in the form of kinetic energy (K.E). As a consequence, the energy created and consumed for mixing was estimated in form of potential energy and kinetic energy using Equations (3) and (4), respectively,

$$\text{P.E} = mgh, \text{ and} \tag{3}$$

$$\text{K.E} = 0.5\, mv^2, \tag{4}$$

where P.E is the potential energy (J), K.E is the kinetic energy also in J, m is the mass (kg) of the maximum volume of manure displaced during gas production (as a result of pressure build-up due to gas production) each day, g is 9.8 $m{\cdot}s^{-2}$, h (m) is the height of the extension chamber and v is the velocity ($m{\cdot}s^{-1}$). The distance is the height of extension from the base of the reactor and time is the duration it takes for the displaced slurry to flow back in the digester. These were experimentally determined as 7 s and 6 s for CDD 1 and 2, respectively. The volume of the displaced slurry in the extension chambers were calculated from the displacement to be 0.0025 and 0.0042 m^3 for CDD 5 and CDD 6, respectively. The mass of the displaced slurry was estimated from the volume and density.

The gas space of the reactors was filled with nitrogen after inoculation and gas produced in the first three days was purged and not recorded. Biomethane potential of each substrate batch was measured in triplicate according to ABAI [49].

2.4. Monitoring and Analytical Methods

The laboratory temperature was monitored using an EL-USB digital temperature logger. The pH of feeds and effluents were measured using a table-top pH meter with a probe, WTW InoLab Level 1 model. The feed and effluent samples were analysed for TS, volatile solids (VS), volatile fatty acids (VFAs, acetate, propionate and butyrate) and ammonium nitrogen (NH_4^+-N). Generated biogas was collected in gas bags, five days a week (Monday to Friday) and measured daily using a Schlumberger Lab wet gas meter. Biogas composition was determined in terms of carbon dioxide (CO_2) and methane (CH_4) content. The CH_4 content was indirectly measured by measuring the concentration of CO_2 viz. CO_2 absorption using NaOH in the gas bag once a week. TS, vs. and NH_4^+-N were analysed according to standard procedures [52]. Specific biogas and methane yields were expressed as daily methane produced, divided by the amount of vs. daily fed to the digester, and used to monitor the digestion efficiency of the digesters.

Concentration of volatile fatty acids (VFAs) in effluent samples were determined in triplicate using a 7890 B gas chromatograph (Agilent Technologies) equipped with an HP-5 column (30 m × 0.32 mm × 0.25 μm, Agilent Technologies) and a flame ionization detector (FID). The carrier gas was nitrogen with a flow rate of 6.5 mL/min. The operating conditions were as follows: Injector temperature, 120 °C (split-splitless); detector temperature, 250 °C; and an oven temperature program initiating at 40 °C, followed by three sequenced temperature increases (i) at a rate of 60 K/min up to 100 °C; (ii) at a rate of 50 K/min up to 150 °C; and finally (iii) at a rate of 90 K/min until 240 °C was reached. Calibration stock solution and sample preparation were done according to standard methods for the examination of water and wastewater [53].

Average steady-state biogas production data and the standard deviation over the period between 150 and 319 days of observations are presented in this paper. The specific methane yield, which was calculated as daily methane produced divided by the amount of vs. fed to the digester, was used to monitor the efficiency of the digesters as stated earlier. The digester performance was evaluated based on the effect of loading rate on methane production to volumetric gas production, volatile fatty acid concentration (VFA), treatment efficiency and energy consumption.

The energy consumption for mixing in the stirred digesters were estimated based on Equation (2), while the natural potential energy created by Chinese dome digesters (CDD) was estimated using Equation (3), but no power requirement or consumption for the non-mixed reactors were calculated because in the non-mixed reactors, no external or internal energy was applied. In the CDDs it was possible to estimate the mixing energy created as a result of slurry displacement, which would later be utilized as kinetic energy when slurry flows back into the digester.

2.5. Statistical Analysis

The statistical significance of the experimental data at steady state condition for all the digesters was performed using a one-way analysis of variance (ANOVA) statistical program (Microsoft Excel 2016).

3. Results

The six digesters exhibited different volumetric gas production and peaked at the highest organic loading rates. The highest volumetric biogas production rates for the steady state period are 0.34, 0.67, 0.23, 0.43, 0.29, and 0.53 L/L/day for digesters 1, 2, 3, 4, 5 and 6 at organic loading rates (OLRs) of 1.39 g VS/L/day for digesters 1, 3 and 5, and 2.79 g VS/L/day for digesters 2, 4, and 6 as shown in Table 5. The higher the loading rate the higher the observed volumetric biogas production. The specific methane production for the same type of digesters were comparable but different for different types of digesters. The maximum specific methane production was 0.17 ± 0.003 L/g vs. at an OLR of 1.39 g VS/L/day for STR 1, 0.16 ± 0.003 L/g vs. at 2.79 g VS/L/day for STR 2, 0.107 ± 0.003 L/g vs. at 1.39 g VS/L/day for UMD 1, 0.095 ± 0.002 L/g vs. at 2.79 g VS/L/day for UMD 2, 0.135 ± 0.003 L/g vs. at 1.39 g VS/L/day for CDD 1, and 0.126 ± 0.003 L/g vs. at 2.79 g VS/L/day for CDD 2. The methane production in all digesters increased slightly from day 260 till the end of the experiment, which is in agreement with the BMP of the applied substrates. The eighth substrate batch had the highest BMP of 0.18 L CH_4/g vs. compared to earlier applied substrate. As a consequence, the highest recorded specific methane production in all the digesters was achieved during the application of the eighth substrate.

The average specific methane production from highest to lowest are 0.16, 0.15, 0.13, 0.12, 0.10 and 0.09 L/g vs. for digesters 1, 2, 5, 6, 3 and 4, respectively.

Table 5. Mean volumetric, specific methane production, and effluent volatile fatty acids (VFA) concentrations, volatile solids (VS) reduction at different organic loading rates (OLRs) for 6 differently operated digesters at steady state. Hydraulic retention time (HRT) = 30 days.

Day	OLR g VS/L Day	CH_4 L/L/Day	CH_4 L/g VS	CH_4 %	VFAs (g/L)	VS red. (%)
(STR 1)						
149–169	0.68	0.09 ± 0.002	0.13 ± 0.002 (a)	67	0.83 ± 0.14	57.17 ± 0.8
170–198	0.86	0.12 ± 0.002	0.14 ± 0.002 (b)	68	0.97 ± 0.13	63.98 ± 1
199–259	0.83	0.12 ± 0.01	0.14 ± 0.01 (c)	68	0.80 ± 0.18	64.24 ± 1.7
260–319	1.39	0.23 ± 0.005	0.16 ± 0.003 (d)	68	0.86 ± 0.20	70.87 ± 1.5
(STR 2)						
149–169	1.36	0.17 ± 0.007	0.12 ± 0.004 (a_2)	66	1.0 ± 0.03	55.47 ± 1.6
170–198	1.72	0.22 ± 0.005	0.13 ± 0.006 (b_2)	66	1.2 ± 0.24	58.76 ± 48
199–259	1.67	0.22 ± 0.02	0.13 ± 0.01 (c_2)	66	0.95 ± 0.10	62.07 ± 1.6
260–319	2.79	0.43 ± 0.01	0.15 ± 0.003 (d_2)	66	1.03 ± 0.17	68.76 ± 1.6
(UMD 1)						
149–169	0.68	0.06 ± 0.001	0.08 ± 0.02 (e)	63	2.4 ± 0.24	38.35 ± 0.5
170–198	0.86	0.08 ± 0.003	0.09 ± 0.003 (f)	63	2.7 ± 0.25	40.59 ± 1.5
199–259	0.83	0.09 ± 0.004	0.09 ± 0.002 (f)	63	2.2 ± 0.15	41.74 ± 2
260–319	1.39	0.13 ± 0.007	0.10 ± 0.003 (g)	63	2.3 ± 0.28	46.80 ± 3.8
(UMD 2)						
149–169	1.36	0.12 ± 0.002	0.09 ± 0.001 (e_2 *)	61	2.5 ± 0.19	41.40 ± 0.4
170–198	1.72	0.14 ± 0.001	0.08 ± 0.004 (f_2)	61	3.0 ± 0.04	40.51 ± 4.3
199–259	1.67	0.13 ± 0.004	0.08 ± 0.002 (f_2)	61	2.8 ± 0.18	39.34 ± 1.7
260–319	2.79	0.25 ± 0.008	0.09 ± 0.002 (g_2)	61	2.81 ± 0.19	43.48 ± 1.3
(CDD 1)						
149–169	0.68	0.07 ± 0.003	0.09 ± 0.003 (h)	65	1.6 ± 0.18	43.97 ± 1.1
170–198	0.86	0.09 ± 0.004	0.10 ± 0.003 (i)	65	1.5 ± 0.12	47.58 ± 1.5
199–259	0.83	0.10 ± 0.003	0.11 ± 0.003 (j)	65	1.4 ± 0.13	52.70 ± 1.9
260–319	1.39	0.18 ± 0.005	0.13 ± 0.003 (k)	65	1.44 ± 0.10	58.76 ± 1
(CDD 2)						
149–169	1.36	0.13 ± 0.003	0.10 ± 0.002 (h_2 *)	64	1.58 ± 0.05	44.23 ± 2.8
170–198	1.72	0.17 ± 0.002	0.10 ± 0.006 (i)	64	1.7± 0.11	44.82 ± 1.9
199–259	1.67	0.18 ± 0.004	0.12 ± 0.003 (j_2 *)	64	1.65 ± 0.01	50.49 ± 1.7
260–319	2.79	0.33 ± 0.008	0.12 ± 0.003 (K_2)	64	1.7± 0.13	54.84 ± 1.2

* Letters in parentheses indicate significant difference between each type of digester at each OLR, ($p < 0.05$) for specific methane production a to k; a_2 to k_2. Values with the same alphabet means no significant difference. Alphabets with (*) mean value is higher and not lower.

At "steady state" periods, there were differences in the biogas production and methane content depending on the applied OLRs and the type of digester. The volumetric methane production increased with increasing organic loading rates (OLRs) in all digesters. The biogas production and methane composition observed during these experiments are summarized in Table 5.

The VFAs during the steady state period are 0.82 ± 0.21, 0.98 ± 0.2, 2.21 ± 0.5, 2.66 ± 0.55, 1.4075 ± 0.25 and 1.61 ± 0.26 g/L in respectively digesters 1 through 6. The observed VFA concentration in the stirred reactors (STR 1 and 2) during this period are lower compared to that in the UMD (1 and 2) and CDD (1 and 2) reactors. The average VFAs concentration in the stirred reactors are lesser or equal to 1 g/L, and they could be regarded as well-balanced digesters according to Hill et al. [54] The average of VFAs concentration in STR 2 is slightly higher but not significant ($p > 0.05$) than STR 1 because STR 2 had higher OLRs. The VFA concentrations differ in the reactors and are significantly higher in the unmixed and Chinese dome digesters compared to the stirred reactors.

The analysis of variance (ANOVA) using Microsoft Excel programme (2016) was performed on the specific methane production and VFA for the digesters in two different batches representing the two influent TS loading rates, the single (3–7.3% TS) and the double (6–15% TS). STR 1, UMD 1 and CDD 1 for the single and STR 2, UMD 2 and CDD 2 for the double concentrations represent the three types of mixing, impeller stirred, unmixed, and the hydraulic mixed (Chinese dome digester) investigated in this study. The digesters were compared based on OLR (single and double feeding), and the results show that the differences between each type of digester is significant for both specific methane production and VFA. Specifically, for the specific methane production, the significant differences are represented by a–d and a_2–d_2 for STRs, e–g and e_2–g_2 for UMDs and h–k and h_2–k_2 for CDDs, shown in Table 5.

4. Discussion

4.1. Effect of Loading Rate, Volumetric Biogas Production, VFA Concentration and Treatment Efficiency

The higher biogas production in the stirred reactors compared to the hydraulic and unstirred reactors is attributed to the impeller mixing at 55 rpm for 10 min/hour, which might have minimized stratification in the reactors. Biogas release in the liquid phase during intermittently mixed reactors has been reported to increase up to 70% during mixing in comparison to non-mixing regimes [43,55,56]. This implies gas release may be hindered in unmixed digesters, and mixing increases the chances of mass transfer from liquid phase to gas phase. This is consistent with results of Lin and Pearce, [57] and Karim et al. [22] with the conclusion that there is impact on methane production between intermittent mixing mode and unmixed systems. In addition, Stafford [58] showed that there was a gradual release of biogas from the liquid phase to the gas phase during the first minute of mixing for various intermittent mixing periods (140–1000 rpm). The Chinese dome digesters produced more methane than the unmixed digesters as seen in Table 5. The digesters also have slightly higher methane concentration compared to the unmixed digesters. Vavilin and Angelidaki [59] reported that uneven mixing in digesters can lead to the creation of initiation zones in the anaerobic digesters, where methane producing bacteria can grow and flourish and could seed the rest of the digester from these zones.

The volumetric biogas production rate increases with increasing vs. influent concentration, but the specific methane production from digesters (STR 2, UMD 2 and CDD 2) where a double loading rate was applied were lower compared to that of STR 1, UMD 1 and CDD 1. Similar observations were reported by Linke [60] and Karim et al. [22]. For the naturally mixed reactors, a higher volumetric methane production rate as a result of increased loading rate did not improve mixing in the digesters and specific methane production. For example, as seen in CDD 1 and 2, CDD 1 had a slightly higher specific methane production than CDD 2 despite the fact that it was operated at a double TS concentration and exhibited higher volumetric biogas rate. This is understandable because manure is a non-Newtonian material, the higher the solid content, the higher the apparent viscosity [61], and more force is required for mixing.

The range of specific methane production (0.10–0.16 L/g VS) in this study for all the three mixing modes is in an agreement with the BMP of the applied feeds and with specific methane gas productions measured by [62], but lower compared to many studies reported [22,28,41] with slightly different mixing modes and intensities, and a lower HRT of <20 days. However, the results from this study are higher than results (0.08–0.10 L/g VS) of Ong et al. [43] for continuous and intermittent mixing modes. The lower specific biogas production of Ong et al. [43] could be attributed to a high OLR of 7.2 g VS/L d at 10 days HRT. Lastly, difference in gas production between the CDDs and the non-mixed digesters is in agreement with the review of Lindmark et al. [63] showing that unmixed digesters will produce 10–20% lower biogas production than intermittently or mixed digesters.

The higher VFAs concentration in the UMD and CDD could be attributed to reduction of the real HRT caused by limited mixing. The limited mixing in these reactors might have created dead zones and lowered the actually working volume of the reactors. The higher levels of VFAs in the unstirred reactors (UMD 1 and 2) probably did not inhibit biogas production or biodegradation. This is coherent with the results of authors Banks et al. [64], Angelidaki et al. [65] and Ghanimeh et al. [66]. They stated that inhibition of biogas production may not occur at high VFAs (max. 4 g/L) if pH stays between 6.8 and 7.7, which is in agreement with the pH range of 6.7–7.2 in this study.

The stability exhibited by the reactors, especially the unmixed digesters (UMD 1 and 2), could be attributed to the reactor geometry in which the outlet is located at the bottom of the reactor and a relatively long HRT of 30 days. The effluent withdrawal from the bottom could have prevented strong accumulation of solids, which may have resulted in large dead zones and indirectly VFAs in the reactor beyond the tolerance level. The long HRT of 30 days applied throughout the experiments might have prevented washout of microbes. In addition, the "intermittent" mixing in the unmixed digesters (viz. feeding and effluent withdrawal) and Chinese dome digesters (viz. feeding, effluent withdrawal and hydraulic variation during gas collection) could have slowed down the fermentation processes at a higher OLR to allow large percentage of intermediates products to be consumed by methanogens and synthrophs without VFAs accumulations and toxicity effects [39].

In ideal continuously stirred tank reactors (CSTR), the removal of volatile solids (VS) should be equivalent to the methane production. Since no reactor was stirred continuously in this study, it was expected that the vs. removal may not fit completely to the methane recovery. The percentage of vs. removal in the reactors during the steady state period (day 150–318) are in Table 5 for reactors 1 to 6. Volatile solid (VS) reductions in all the reactors are different but not significant, and STR 1 exhibited the highest vs. removal, followed by digesters 2, 5, 6, 3 and 4. This trend corresponds to the specific methane production of the reactors. Both digesters 1 and 2 exhibited better vs. reduction and higher methane production because they were impeller mixed. This improved homogeneity and reduced dead zones in the digesters. STR 1 exhibited higher vs. removal compared to STR 2, despite the fact that it was fed with doubled TS content and produced double volumetric gas production. Consequently, this affirms the earlier statement that higher volumetric gas production does not improve mixing at higher OLRs, and hence specific methane production. The lower vs. removal in the unmixed digester and CDDs could be attributed to limited mixing compared to the digesters 1 and 2, because interrupted mixing or intermittent mixing has been reported to create hydraulic dead zones, which can reduce hydraulic retention time and cause effects on reaction kinetics [22,67]. Indeed, in this study, unmixed digesters exhibited a lower vs. reduction compared to the hydraulic mixed digesters (CDD) for both TS concentrations because the absence of "sufficient intermittent" mixing could have reduced the effective volume of the unstirred digesters and lead to poor vs. degradation, see Zabranska et al. [68]. In addition, the impeller mixed digester has less vs. reduction variation compared to the unmixed and Chinese dome (hydraulic mixed) digesters. The vs. reduction for the steady period are also shown in Table 5 for all the reactors.

The results of the analysis of variance (ANOVA) implies the reactors fed with single TS performed better than double-fed digesters in terms of methane recovery, with the exception of unmixed and CDD digesters at an OLR of 1.36 g VS/L/day (day 149–169) and CDD at an OLR of 1.67 g VS/L/day (day 199–159). Furthermore, digesters were compared and summarized in Table 6. Comparison of these values show that the difference between impeller and unmixed, impeller and CDD, and unmixed and CDD at both TS concentrations were significant also ($p < 0.05$).

Table 6. Methane production at steady state for single and double influent total solids (TS) concentration. The data used are presented in Table 5. p = differences in specific methane production.

% TS	Impeller and Unmixed	STR and CDD	UMD and CDD	All Reactors
Single	$p < 0.01$	$p < 0.01$	$p < 0.01$	$p < 0.0001$
Doubled	$p < 0.01$	$p < 0.01$	$p < 0.01$	$p < 0.0001$

4.2. Digesters Performance and Energy Consumption

The energy requirement for STR 1 and 2 were 10.5 and 42 kJ/day, while the P.E created in the CDDs were 0.74 and 1.2 J/day for reactors 5 and 6, respectively. STR 1 and 2 had higher power requirement because of the mechanical mixing with an electric motor. The applied shear would be higher in STR

2 because of the higher substrate viscosity as a result of higher TS concentration. It has been shown by El-Mashad et al. [61] and Karim et al. [22] that shear rates increase with total solid concentration. In the CDD, the energy consumed is very low because the energy is naturally created initially, and it is in the form of gravitational potential achieved by the hydraulic variation as a result of pressure build up in the reactors. The pressure increases as a result of gas production at the headspace (gas phase) and pushes some volume of the slurry, which is non-Newtonian and non-compressible material from the reactor into the extension chamber creates potential energy. The volume of slurry displaced will depend on the amount of gas produced and height of the extension chamber.

The potential and kinetic energies would depend on the viscosity or the percentage of the TS concentration of the applied feed. As discussed earlier, higher volumetric gas production did not improve mixing at a higher OLR, because slurry at higher viscosity (higher %TS) has lower velocity compared to slurry with lower %TS concentration. The mixing intensity in the stirred reactors is higher than the CDDs because the spatial coverage of mixing and duration is more than the hydraulic mixing in the CDDs. The hydraulic variation in the CDDs occurred once a day in this study. In fully operational household digesters, the hydraulic mixing occurs between two to three times a day depending on the frequency of cooking. Biogas and methane production are higher in the stirred reactors. However, the differences are not significant compared to the large difference in power utilization. As a consequence, the CDD is more energy efficient than the stirred reactors according to the results presented in this study. Both types of reactors were intermittently mixed but applying different types of mixing. The duration between each cycle in the stirred reactor was 50 min while 24 h for the CDDs.

5. Conclusions

The effect of mixing in three reactors designs using cow manure as substrate was investigated at the laboratory scale. Significant differences were observed among the three types of digesters at the different influent TS concentrations applied in this study. The impeller mixed digesters or the STRs exhibited better biogas and methane production, and treatment efficiency, followed by the Chinese dome digesters (CDDs) and the unmixed digesters (UMDs). STR 1 produced 20% more methane than the CDDs and 37% more methane than the UMDs, respectively, at steady state conditions. However, the CDDs were more energy efficient than the STRs. By applying double influent TS concentrations, the reactors showed lower specific biogas production and higher VFAs concentrations with few exceptions. The VFA accumulation was more pronounced in the unstirred digesters and Chinese dome digesters mainly because of insufficient mixing.

The results of double-fed TS concentration experiments did not produce better reactor performance (based on specific methane production and VFAs concentrations) in the CDDs despite higher volumetric biogas production rate. This implies that hydraulic variation induced by the natural mixing by biogas production at higher volumetric rate did not yield sufficient mixing and further studies could focus on improving this. The hydraulic variation in Chinese dome digesters may not suffice for the treatment of cow manure at TS concentration of 10% and above.

Author Contributions: Methodology: All authors; data curation: A.O.J.; writing—Original draft preparation: A.O.J.; writing editing: G.Z. & H.B.; funding acquisition: A.O.J. & G.Z.

Funding: This work was funded by the Netherlands Fellowship Programme (NFP), the Netherlands. Grant number: NFP-PhD 12/426.

Acknowledgments: Special thanks to the analytic/lab support team at Environmental Technology department, Wageningen University, The Netherlands. Special appreciations to the management and support staff of Biochemical Engineering laboratory, Centre for Energy Research and Development, Obafemi Awolowo University, Ile-ife Nigeria for their support.

Conflicts of Interest: The authors declare no conflict of interest.

References

1. Yumkella, K.; Bazilian, M.; Gielen, D. Energy for all. *Mak. It: Ind. Dev.* **2010**, *2*, 22–29. Available online: http://89.206.150.89/documents/congresspapers/325.pdf. (accessed on 1 November 2016).
2. Gautam, R.; Baral, S.; Herat, S. Biogas as a sustainable energy source in Nepal: Present status and future challenges. *Renew. Sustain. Energy Rev.* **2009**, *13*, 248–252. [CrossRef]
3. Tumwesige, V.; Fulford, D.; Davidson, G.C. Biogas appliances in Sub-Sahara Africa. *Biomass Bioenergy* **2014**, *70*, 40–50. [CrossRef]
4. Modi, V.; McDade, S.; Lallement, D.; Saghir, J. *Energy Services for Millennium Development Goals, Achieving MGDs. Millennium Project, UNDP*; World Bank and ESMAP: Washington, DC, USA, 2005.
5. Bajgain, S.; Shakya, I.; Mendis, M.S. *The Nepal Biogas Support. Program: A Successful Model. of Public Private Partnership for Rural Household Energy Supply. Report Prepared by Biogas Support. Program. Nepal*; Vision Press P. Ltd.: Kathmandu, Nepal, 2005.
6. Sun, Q.; Li, H.; Yan Ji Liu, L.; Yu, Z.; Yu, X. Selection of appropriate biogas upgrading technology—A review of biogas cleaning upgrading and utilisation. *Renew. Sus Energy Rev.* **2015**, *51*, 521–532. [CrossRef]
7. Chanakya, H.N.; Bhogle, S.; Arun, R.S. Field experience with leaf litter-based biogas plants. *Energy Sustain. Dev.* **2005**, *9*, 49–62. [CrossRef]
8. Lansing, S.; Martin, J.F.; Botero, R.B.; da Silva, T.N.; da Silva, E.D. Methane production in low-cost, unheated, plug-flow digesters treating swine manure and used cooking grease. *Bioresour. Technol.* **2010**, *10*, 4362–4370. [CrossRef] [PubMed]
9. Chen, L.; Zhao, L.; Ren, C.; Wang, F. The progress and prospects of rural biogas production in China. *Energy Policy* **2012**, *51*, 58–66. [CrossRef]
10. China, National Development and Reform Commission (NDRC). China's Medium to Long-Term Renewable Energy Development Plan 2007. Available online: http://www.chinaenvironmentallaw.com/wp-content/uploads/2008/04/mediumand-long-term-development-plan-for-renewable-energy.pdf (accessed on 14 March 2016).
11. Fulford, D. *Running a Biogas Programme. A Handbook*; Intermediate Technology Publication: London, UK, 1988.
12. Chen, Y.; Yang, G.; Sweeney, S.; Feng, Y. Household biogas use in rural China: A study of opportunities and constraints. *Renew. Sustain. Energy* **2010**, *14*, 545–549. [CrossRef]
13. Ghimire, P.C. SNV Supported domestic biogas programmes in Asia and Africa. *Renew. Energy* **2013**, *49*, 90–94. [CrossRef]
14. Pérez, I.; Garfí, M.; Cadena, E.; Ferrer, I. Technical, economic and environmental assessment of household biogas digesters for rural communities. *Renew. Energy* **2014**, *62*, 313–318. [CrossRef]
15. Arthur, R.; Baidoo, M.F.; Antwi, E. Biogas as a potential renewable energy source. A Ghanaian case study. *Renew. Energ.* **2011**, *36*, 1510–1516.
16. Kanwar, S.S.; Guleri, R.L. Performance evaluation of a family-size, rubber-balloon biogas plant under hilly conditions. *Bioresour. Technol.* **1994**, *50*, 119–121. [CrossRef]
17. Singh, S.P.; Vatsa, D.K.; Verma, H.N. Problems with biogas plants in Himachal Pradesh. *Bioresour. Technol.* **1997**, *59*, 69–71. [CrossRef]
18. Deublein, D.; Steinhauser, A. *Biogas from Waste and Renewable Resources*; Wiley-VCH Verlag GmbH& Co. KGaA: Weinheim, Germany, 2008.
19. Jegede, A.O.; Bruning, H.; Zeeman, G. Location of inlets and outlets of Chinese dome digesters to mitigate biogas emission. *Biosyst. Eng.* **2018**, *174*, 153–158. [CrossRef]
20. Bridgeman, J. Computational fluid dynamics modeling of sewage sludge mixing in an anaerobic digester. *Adv. Eng. Softw.* **2012**, *44*, 54–62. [CrossRef]
21. Gerardi, M. *The Microbiology of Anaerobic Digesters*; John Wiley and Sons, Inc.: Hoboken, NJ, USA, 2003.
22. Karim, K.; Hoffmann, R.; Klasson, K.T.; Al-Dahhan, M.H. Anaerobic digestion of animal waste: Effect of mode of mixing. *Water Res.* **2005**, *39*, 3597–3606. [CrossRef]
23. Conklin, A.S.; Chapman, T.; Zahller, J.D.; Stensel, H.D.; Ferguson, J.F. Monitoring the role of aceticlasts in anaerobic digestion: Activity and capacity. *Water Res.* **2008**, *42*, 4895–4904. [CrossRef]
24. Halalsheh, M.; Kassab, G.; Yazajeen, H.; Qumsieh, S.; Field, J. Effect of increasing the surface area of primary sludge on anaerobic digestion at low temperature. *Bioresour. Technol.* **2011**, *102*, 748–752. [CrossRef]

25. Gomez, X.; Cuetos, M.J.; Cara, J.; Moran, A.; Garcia, A. Anaerobic co-digestion of primary sludge and the fruit and vegetable fraction of the municipal solid wastes-conditions for mixing and evaluation of the organic loading rate. *Renew. Energ.* **2006**, *31*, 2017–2024. [CrossRef]
26. Kim, M.; Ahn, Y.H.; Speece, R.E. Comparative process stability and efficiency of anaerobic digestion: Mesophilic vs. thermophilic. *Water Res.* **2002**, *36*, 4369–4385. [CrossRef]
27. Ward, A.J.; Hobbs, P.J.; Holliman, P.J.; Jones, D.L. Optimisation of the anaerobic digestion of agricultural resources. *Bioresour. Technol.* **2008**, *99*, 7928–7940. [CrossRef] [PubMed]
28. Kaparaju, P.; Buendia, I.; Ellegaard, L.; Angelidakia, I. Effects of mixing on methane production during thermophilic anaerobic digestion of manure: Lab-scale and pilot-scale studies. *Bioresour. Technol.* **2008**, *99*, 4919–4928. [CrossRef] [PubMed]
29. Ike, M.; Inoue, D.; Miyano, T.; Liu, T.T.; Sei, K.; Soda, S.; Kadoshin, S. Microbial population dynamics during startup of a full-scale anaerobic digester treating industrial food waste in Kyoto eco-energy project. *Bioresour. Technol.* **2010**, *101*, 3952–3957. [CrossRef] [PubMed]
30. Hamad, M.A.; Abdel Dayem AM El Halwagi, M.M. Evaluation of the performance of two rural biogas units of Indian and Chinese design. *Energy Agric.* **1983**, *1*, 235–250. [CrossRef]
31. Polprasert, C.; Edwards, P.; Rajput, V.S.; Pacharaprakiti, C. Integrated biogas technology in the tropics 1. Performance of small-scale digesters. *Waste Manag. Res.* **1986**, *4*, 197–213. [CrossRef]
32. Anjan, K.K. Development and evaluation of a fixed dome plug flow anaerobic digester. *Biomass* **1988**, *16*, 225–235.
33. Xavier, S.; Nand, K. A preliminary study on biogas production from cow dung using fixed-bed digesters. *Biol. Wastes.* **1990**, *34*, 161–165. [CrossRef]
34. Singh, R.; Anand, R.C. Comparative performances of Indian small solid-state and conventional anaerobic digesters. *Bioresour. Technol.* **1994**, *47*, 235–238. [CrossRef]
35. Singh, L.; Maurya, M.S.; Ramana, K.V.; Alam, S.I. Production of biogas from night soil at psychrophilic temperature. *Bioresour. Technol.* **1995**, *53*, 147–149. [CrossRef]
36. Kalia, A.K.; Singh, S.P. Horse dung as a partial substitute for cattle dung for operating family-size biogas plants in a hilly region. *Bioresour. Technol.* **1998**, *64*, 63–66. [CrossRef]
37. An, B.X.; Preston, T.R. Gas production from pig manure fed at different loading rates to polyethylene tubular biodigesters. *Livest. Res. Rural Dev.* **1999**, *11*, 11.
38. Ferrer, I.; Garfí, M.; Uggetti, E.; Ferrer-Martí, L.; Calderon, A.; Velo, E. Biogas production in low-cost household digesters at the Peruvian Andes. *Biomass Bioenergy* **2011**, *35*, 1668–1674. [CrossRef]
39. Tamkin, A.; Martin, J.; Castano, J.; Ciotola, R.; Rosenblum, J.; Bisesi, M. Impact of organic loading rates on the performance of variable temperature biodigesters. *Ecol. Eng.* **2014**, *78*, 87–94. [CrossRef]
40. Hashimoto, S.; Fujita, M.; Teral, K. Stabilization of wasteactivated sludge through the anoxic–aerobic digestion process. *Biotechnol. Bioeng.* **1982**, *24*, 1335–1344. [CrossRef] [PubMed]
41. Stroot, P.G.; McMahon, K.D.; Mackie, R.I.; Raskin, L. Anaerobic codigestion of municipal solid waste and biosolids under various mixing conditions I: Digester performance. *Water Res.* **2001**, *35*, 1804–1816. [CrossRef]
42. Bouallagui, H.; Ben Cheikh, R.; Marouani, L.; Hamdi, M. Mesophilic biogas production from fruit and vegetable waste in a tubular digester. *Bioresour. Technol.* **2003**, *86*, 85–89. [CrossRef]
43. Ong, H.K.; Greenfield, P.F.; Pullammanappallil, P.C. Effect of mixing on biometha-nation of cattle-manure slurry. *Env. Technol* **2002**, *23*, 1081–1090. [CrossRef]
44. Vesvikar, M. Understanding the Hydrodynamics and performance of Anaerobic Digesters. Ph.D Thesis, Washington University, St. Louis, MO, USA, 2006.
45. Sulaiman, A.; Hassan, M.A.; Shirai, Y.; Abd-Aziz, S.; Tabatabaei, M.; Busu, Z. The effect of mixing on methane production in a semi-commercial closed digester tank treating palm oil mill effluent. *Aust. J. Basic Appl.* **2009**, *3*, 1577–1583.
46. Rico, C.; Rico, J.L.; Muñoz, N.; Gómez, B.; Tejero, I. Effect of mixing on biogas production during mesophilic anaerobic digestion of screened dairy manure in a pilot plant. *Eng Life Sci.* **2011**, *11*, 476–481. [CrossRef]
47. Chen, J.; Li, X.; Liu, Y.; Zhu, B.; Yuan, H.; Pang, Y. Effect of mixing rates on anaerobic digestion performance of rice straw. *Trans. Chin. Soc. Agric. Eng.* **2011**, *27*, 144–148.

48. Hoffmann, R.A.; Garcia, M.L.; Veskivar, M.; Karim, K.; Al-Dahhan, M.; Angenent, L.T. Effect of shear on performance and microbial ecology of continuously stirred anaerobic digesters treating animal manure. *Biotechnol. Bioenergy* **2008**, *100*, 38–48. [CrossRef] [PubMed]
49. DTU. *Anaerobic Biodegradation, Activity and Inhibition (ABAI). Task Group Meeting 9–10 October 2006, Prague*; Institute of Environment & Resources, Technical University of Denmark: Lyngby, Denmark, 2007.
50. Wu, B. CFD simulation of mixing for high-solids anaerobic digestion. *Biotechnol. Bioenergy* **2012**, *109*, 2116–2126. [CrossRef] [PubMed]
51. Liu, B.; Zhag, Y.; Chen, M.; Li, P.; Jin, Z. Power consumption and flow field characteristics of a coaxial mixer with a double inner impeller. *Chin. J. Chem. Eng.* **2015**, *23*, 1–6. [CrossRef]
52. APHA. *Standard Methods for the Examination of Water and Waste Water*, 22nd ed.; American Public Health Association: Washington, DC, USA, 2006.
53. APHA. *Standard Methods for the Examination of Water and Waste Water*, 21st ed.; American Public Health Association: Washington, DC, USA, 2005.
54. Hill, D.T.; Cobb, S.A.; Bolte, J.P. Using volatile fatty acid relationships to predict anaerobic digester failure. *Trans. Asabe* **1987**, *30*, 496–501. [CrossRef]
55. Mills, P.J. Minimisation of energy input requirements of an anaerobic digester. *Agric. Wastes* **1979**, *1*, 57–79. [CrossRef]
56. Sung, S.; Dague, R.R. Laboratory studies on the anaerobic sequencing batch reactor. *Water Env. Res.* **1995**, *67*, 294–301. [CrossRef]
57. Lin, K.C.; Pearce, M.E.J. Effect of mixing on anaerobic treatment of potato-processing wastewater. *Can. J. Civ. Eng.* **1991**, *18*, 504–514. [CrossRef]
58. Stafford, D.A. The effects of mixing and volatile fatty acid concentrations on anaerobic digester performance. *Biomass* **1982**, *2*, 43–55. [CrossRef]
59. Vavilin, V.A.; Angelidaki, I. Anaerobic degradation of solid material: Importance of initiation centers for methanogenesis, mixing intensity, and 2D distributed model. *Biotechnol. Bioenergy* **2005**, *89*, 113–122. [CrossRef]
60. Linke, B. A model for anaerobic digestion of animal waste slurries. *Environ. Technol.* **1997**, *18*, 849–854. [CrossRef]
61. El-Mashad, H.M.; van Loon, W.K.-P.; Zeeman, G.; Bot, G.P.A.; Lettinga, G. Design of a solar thermophilic anaerobic reactor for small farms. *Biosyst. Eng.* **2004**, *87*, 345–353. [CrossRef]
62. Zeeman, G. Mesophilic and Psychrophilic Digestion of Liquid Cow Manure. Ph.D. Thesis, Sub-Department of Environmental Technology, Wageningen University, Wageningen, The Netherlands, 1991.
63. Lindmark, J.; Thorin, E.; Fdhila, R.B.; Dahlquist, E. Effect of mixing on the result of anaerobic digestion: Review. *Renew. Sust. Energy Rev.* **2014**, *40*, 1030–1047. [CrossRef]
64. Banks, C.J.; Chesshire, M.; Stringfellow, A.A. pilot-scale comparison of mesophilic and thermophilic digestion of source segregated domestic food waste. *Water Sci. Technol.* **2008**, *58*, 1475–1481. [CrossRef] [PubMed]
65. Angelidaki, I.; Chen, X.; Cui, J.; Kaparaju, P.; Ellegaard, L. Thermophilic anaerobic digestion of source-sorted organic fraction of household municipal solid waste: Start-up procedure for continuously stirred tank reactor. *Water Res.* **2006**, *40*, 2621–2628. [CrossRef] [PubMed]
66. Ghanimeh, S.; Fadel, M.E.; Saikaly, P. Mixing effect on thermophilic anaerobic digestion of source-sorted organic fraction of municipal solid waste. *Bioresour. Technol.* **2012**, *117*, 63–67. [CrossRef] [PubMed]
67. Elnekave, M.; Tufekcj, N.; Kimchie, S.; Shelef, G. Tracing the mixing efficiency of a primary meshophilic anaerobic digester in a municipal wastewater treatment plant. *Fresenius Environ. Bull.* **2006**, *15*, 1098–1105.
68. Zabranska, J.; Dohanyos, M.; Jenicek, P.; Zalpatilkova, P.; Kutil, J. The contribution of thermophilic anaerobic digestion to the stable operation of wastewater sludge treatment. *Water Sci. Technol.* **2002**, *46*, 447–453. [CrossRef]

Article

Development of an Optimised Chinese Dome Digester Enables Smaller Reactor Volumes; Pilot Scale Performance

Abiodun O. Jegede [1,*], Grietje Zeeman [2] and Harry Bruning [2]

[1] Centre for Energy Research and Development, Obafemi Awolowo University, Ile-ife, Nigeria
[2] Sub-Department Environmental Technology, Wageningen University and Research, 6708 WG Wageningen, The Netherlands; grietje.zeeman@wur.nl (G.Z.); harry.bruning@wur.nl (H.B.)
* Correspondence: jegedeabiodun@yahoo.com or ajegede@cerd.gov.ng

Received: 7 April 2019; Accepted: 3 June 2019; Published: 11 June 2019

Abstract: Chinese dome digesters are usually operated at long hydraulic retention times (HRT) and low influent total solids (TS) concentration because of limited mixing. In this study, a newly optimised Chinese dome digester with a self-agitating mechanism was investigated at a pilot scale (digester volume = 500 L) and compared with a conventional Chinese dome digester (as blank) at 15% influent TS concentration at two retention times (30 and 40 days). The reactors were operated at ambient temperature: 27–33 °C. The average specific methane production, volatile fatty acids and percentage of volatile solids (VS) reduction are 0.15 ± 0.13 and 0.25 ± 0.05L CH_4/g VS; 1 ± 0.5 and 0.7 ± 0.3 g/L; and 51 ± 14 and 57 ± 10% at 40 days HRT (day 52–136) for the blank and optimised digester, respectively. At 30 days HRT (day 137–309) the results are 0.19 ± 0.12 and 0.23 ± 0.04 L CH_4/g VS; 1.2 ± 0.6 and 0.7 ± 0.3 g/L; and 51 ± 9 and 58 ± 11.6%. Overall, the optimised digester produced 40% more methane than the blank, despite the high loading rates applied. The optimised digester showed superior digestion treatment efficiency and was more stable in terms of VFA concentration than the blank digester, can be therefore operated at high influent TS (15%) concentration.

Keywords: Mixing; optimised; household digester; Chinese dome digester (CDD); self-agitation; blank

1. Introduction

Energy is a vital component needed to improve quality of life, reduce poverty and for the promotion of socio-economic activities. However, there is still global uncertainty in the energy sector because of the declining quantity of fossil fuel reserves coupled with crude oil price instability. This global energy situation requires alternative or renewable sources of energy and a review of current technologies. It is vital to focus not only on the sustained economic usage of current finite resources, but also to identify and develop renewable technologies and resources that possess the potential to provide for the increasing energy demand. These resources and technologies should also be sustainable, clean, globally available and easy to exploit and operate, while contributing towards the materialisation of the United Nations millennium development goals (MDGs) [1].

The growing demand of energy due to population growth, the lack of clean energy, and the inadequate availability of natural resources have led to growing demand for anaerobic digestion technologies in rural areas of developing countries. Anaerobic digestion is a biochemical process that is applied for the efficient transformation of, for example, manure and other agricultural residues into biogas, a renewable energy source and biofertiliser. Biogas is rich in methane (50–70%), and CO_2 (30–50%) with traces of water vapour (1–6%) and H_2S. Biogas is a renewable, clean and efficient source of alternative energy which can be used as a substitute for fuels such as firewood, charcoal and cattle dung, usually used by rural dwellers [2–4].

The household or domestic digester is an example of an anaerobic digester usually applied in a single decentralised system mostly in rural areas of developing countries. It serves as an energy producer and sometimes, when coupled to the toilets, as a sanitation system. Household digesters are non-mechanically mixed and non-heated reactors [5]. Domestic biogas plants can serve as an energy solution to meet the cooking needs of rural households. Various governments and international funders support this system, and if managed well [6], it is a cost-effective way of mitigating greenhouse gas emissions from animal dung [7]. This is accomplished by replacing fossil fuels with biogas and reducing methane emissions during manure management, starting from its production to the final application on agricultural lands [8].

Among all household digesters, the Chinese dome digester (CDD) is the most popular and most widely implemented reactor because of its reliability, low maintenance requirement, and long lifespan [9,10]. The CDD is a basis for the development of recent household digester designs; for example, the Puxin digester is an example of a prefabricated digester [10].

In CDDs, gas pressure is created at the top of the reactor as a result of the biogas produced. The stored biogas pushes part of the slurry into an extension chamber, since the chamber is usually opened. During gas use, pressure is released, and the slurry flows back into the main reactor, creating a mixing regime or cycle (Figure 1a,b). Therefore, CDD depends on the hydraulic variation, i.e., the change of slurry level in the digester and extension chamber during gas use and could be regarded as intermittent natural mixing [6]. Mixing is an important process in anaerobic digestion for establishing contact between micro-organisms and feed for homogenisation of temperature throughout the digester and prevention of settling and floating layers [11].

The effect of mixing on the anaerobic digestion process has been studied by various authors. The requirement for digesters to be adequately and sufficiently mixed has been supported by many authors [12–15], and challenged by many other authors [16–19]. Lindmark et al. [20] concluded in a review that an intermittent mixing mode is better than continuous mixing, and shorter mixing durations are preferable for higher biogas production and from an energy point of view. Intermittent mixing can result in a similar quantity of gas to that obtained with continuous mixing. In addition, Chinese dome digesters are generally operated at long hydraulic retention times (>70 days) and low influent total solid (TS) concentrations (≤7%) when compared to mixed reactors (intermittently or continuously). Mixing is limited, leading to a large reactor volume and higher cost [6].

Operating anaerobic digesters at high TS content (>10%) could present a better alternative to anaerobic digestion (AD) systems operating at lower TS content (< 7%) because of the reduced reactor volume when applying the same HRT [21,22]. A major advantage of this approach is the smaller digester and the eventual reduction in reactor cost. Applying this approach to CDD could help in the reduction of both the water required for dilution and the reactor size; however, the mixing in CDDs that can be achieved during feeding, biogas production and use would be limited. In addition, Jegede [6] investigated the impact of different influent TS (3–15%) concentrations and the related volumetric biogas production on mixing in lab-scale CDDs. The results revealed that mixing in CDDs, due to biogas production and reactor feeding, is not sufficient at high (>10%) TS concentrations. Indeed, the rheological properties of manure are affected by water content or percentage of TS [23,24]. The lower the water content, the higher the yield stress, because manure becomes viscoelastic material at high TS. The yield stress is directly proportional to the required force to make manure flow. Because of this property, the increased volumetric biogas production at high (>10%) influent TS is not sufficient for mixing CDDs.

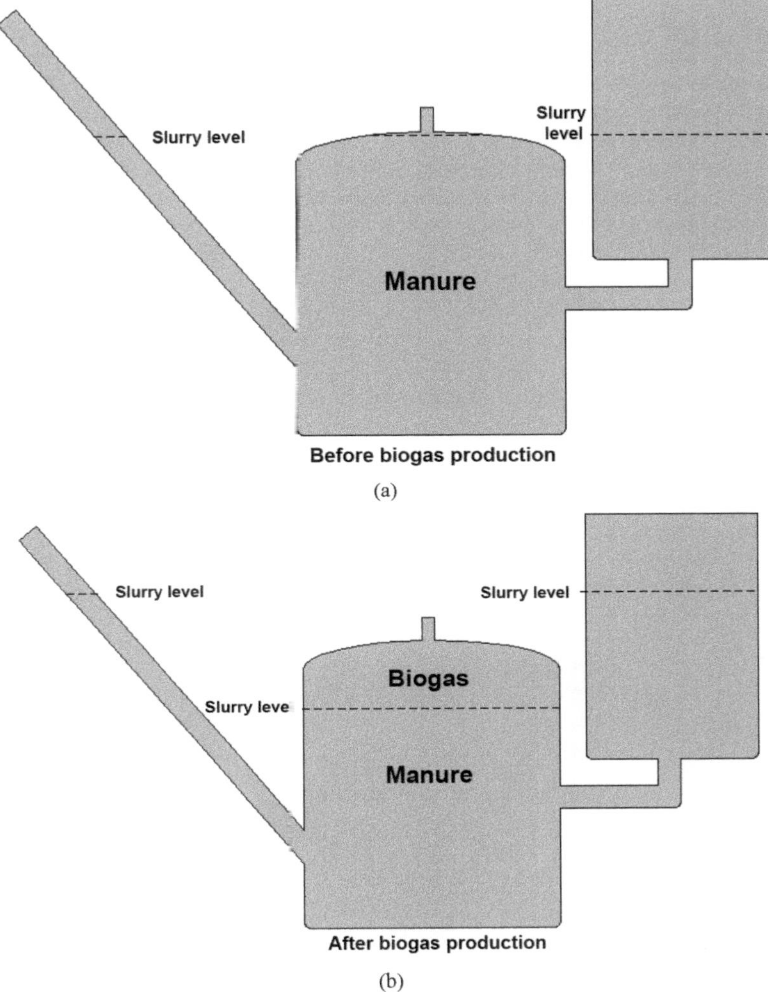

Figure 1. Schematics of the blank digester (**a**) before and (**b**) after gas production (Adapted from ref [25]).

Therefore, to reduce the digester volume viz. reduction of HRT and operation at higher influent concentration rate (ca. 15% TS), the Chinese dome digester has been optimised to improve mixing via a self-agitating mechanism, using the produced gas, while still being simple, cheap, easy to build and maintainable at low cost. The objective of this research was to evaluate the performance of this optimised self-agitating CDD at a pilot scale in continuous operation in comparison to a regular Chinese dome digester at higher organic loading rates (15% TS, corresponding to 2.6–4 g VS/m^3/d) at HRTs of 40 and 30 days. To visualise the mechanism of the self-agitating process, a demo was done using a 19 L transparent plastic bottle containing water, and biogas production was simulated with the injection of air.

2. Material and Methods

2.1. Reactor Design and Setup

In this study, two pilot-scale Chinese dome digesters, optimised and blank, were operated at a relatively high loading rate. The reactor volume, which is also the active volume, was 500 L, with expansion chambers of 250 L each for effluent variation and outlet. The expansion chamber was open, and was not closed throughout the study. A scheme of the two pilot-scale reactors is shown in Figures 1 and 2. The difference between the blank and the optimised digester was the inclusion of two baffles at the top and bottom of the optimised digester, as shown in Figure 2. The upper baffle, the main baffle, divides the headspace of the reactor into two compartments, A and B (Figure 2). The length of the baffle, which should not reach the same level as the effluent outlet pipe of the digester, was calculated based on the previous work of Jegede et al. [25]. The gas outlet is located in compartment B, while compartment A has no gas outlet. The biogas produced is stored in both compartments before gas collection, while the slurry level in the extension increases due to the pressure build up caused by the produced biogas. After gas collection or gas use, the pressure in compartment B will decrease to atmospheric pressure, while the slurry flow back into the reactor and the level of slurry increases in compartment B; however, the pressure in compartment A remains stable, with some slight variations when the level reaches the tip of the baffle. The lower baffle helps to prevent short circuiting of influent by creating mild hold-up and help to improve mixing of the influent with the reactor contents.

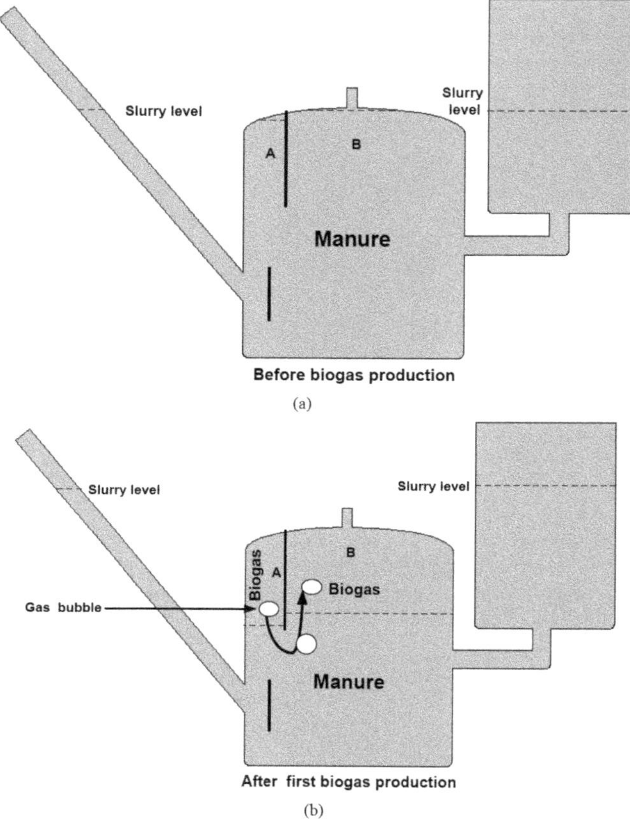

Figure 2. *Cont.*

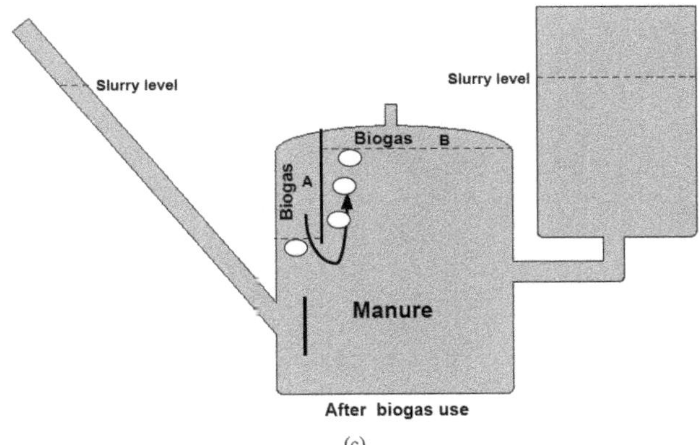

(c).

Figure 2. Scheme of the biogas production in the optimised digester (**a**) before biogas production. (**b**) Biogas production and flow of gas into compartment B. (**c**) Biogas production continues, gas flows to compartment B after gas is used.

Further production and release of biogas into compartment A will increase the pressure in the compartment and pushes the slurry below the baffle. This will cause some biogas to be transported below the baffle into compartment B; immediately afterwards, the slurry level in compartment A will return to the initial level. This process creates a self-agitation cycle. Septums were incorporated on top of the reactors to allow pressure measurements.

2.2. Manure Collection and Preparation

The inoculum used for digester seeding was collected from a 10 m^3 Chinese dome digester treating cow manure with a TS concentration of 8% operated at an average ambient temperature of 30 °C at 40 days HRT. The inoculum was collected on the same day the reactors were started, and 300 L was added into each reactor as seed. The cow manure used as feed in this study was collected freshly at the agricultural farm, Obafemi Awolowo University, Ile ife, Osun, Nigeria. Each batch collected was refrigerated at 3 °C prior to use and later diluted up to 15% TS prior to feeding at ambient temperature. The mean characteristics of the feed are given in Table 1.

Table 1. Feed characteristics and operating parameters; average values of input feed.

Parameter	Both Reactors
Total solids influent (TS) (% after dilution)	15
Volatile solids influent (VS) (% of TS after dilution)	73 ± 0.04
Organic loading rates (OLRs), g VS/m^3/d (40 days)	2.6–3
Organic loading rates (OLRs), g VS/m^3/d (30 days)	3.5–4
NH_4^+-N (g/L)	2.1 ± 0.6
Hydraulic retention time (HRT)	40, 30

2.3. Operation

The reactors were operated throughout the study period at the same conditions. After seeding with 300 L, the reactors were fed at a TS of ca. 15% corresponding to approximately an OLR of 2.6–3 g VS/L/d, without effluent withdrawal, until the digesters were filled up. Effluent withdrawal from the expansion chamber started on day 32, and HRT was 40 days from day 32 until 136 and 30 days (HRT) from day 137 until 319. The digesters were operated at ambient temperature 26–33° C.

2.4. Monitoring and Analytical Methods

The ambient temperature of the shed where the reactors were located was monitored using an EL-USB digital temperature logger. pH of effluents was measured using a table top pH meter with a probe, Ohaus Starter 2100. The total solids (TS) and volatile solids (VS), ammonium nitrogen (NH_4^+-N) of influents and effluents were determined according to standard methods as described by APHA [26]. Biogas volume was collected in a gas bag three times a day and methane percentage was measured using an Ultrasonic biogas meter BF-2000, PUXIN, Longgang, Shenzhen, China. The gas concentrations of volatile fatty acids (VFAs), mainly acetate, propionate, iso-butyrate, butyrate, iso-valerate valerate in effluent samples, were determined in triplicate using a 7890 B gas chromatograph (Agilent Technologies) equipped with an HP-5 column (30 m × 0.32 mm × 0.25 μm, Agilent Technologies) and a flame ionisation detector (FID). The carrier gas was nitrogen with a flow rate of 6.5 mL/min. The operating conditions were as follows: injector temperature, 120 °C (splitless); detector temperature, 250 °C; an oven temperature program initiating at 40 °C, followed by three sequenced temperature increases (i) at a rate of 60 K/min up to 100 °C, (ii) at a rate of 50 K/min up to 150 °C and, finally, (iii) at a rate of 90 K/min until 240 °C was reached. Calibration stock solution and sample preparation where done according to standard methods for the examination of Water and Wastewater [27]. The pressure measurement was done with a Greisinger GMH 3151 digital pressure meter with logger and was done after feeding representing steady state period for both reactors. The specific biogas and methane yields, were calculated as daily biogas methane produced, divided by the amount of VS fed to the reactors, were used to monitor the digestion efficiency of the digesters.

Biogas production rates were calculated as volume of gas produced per litre of digester volume (500 L) per day. Methane yields were calculated as the volume of methane produced per unit mass of VS added. All measured gas volumes were converted to standard conditions (273 K, 1 bar). A steady state condition is assumed when methane production is within 15% variation [28].

3. Results and Discussion

3.1. Reactor Performance

The specific gas production, VFA concentrations and VS reduction in time for both the optimised and blank reactors are presented in Figure 3, showing a start-up period with relatively low gas production for both reactors up until ca. 80 days of operation. VFA concentrations were relatively low for the whole period in the optimised reactor, while the blank reactor showed much higher VFA concentrations up to 1 g/L.

Steady state at an HRT of 40 days was achieved for the optimised reactor in the period between 82 and 137 days, characterised by an average specific methane production of 0.32 ± 0.05 L CH_4/g VS, a VFA concentration of 0.7 ± 0.2 g/L and a VS reduction of 63 ± 4%. Even at a relatively long HRT of 40 days, gas production and VFA concentration in the blank reactor fluctuated considerably, characterised by an average specific methane production of 0.19 ± 0.14 L CH_4/g VS, VFA concentration of 1 ± 0.3 g/L and a VS reduction of 54 ± 17% during the period in which the optimised reactor operated rather stably.

The decrease in HRT from 40 to 30 days on day 137 to 309 (end of experiment), resulted in the decline of gas production in both reactors.

In the blank, methane dropped rapidly to approx. 0.01 L CH_4/g VS, while the average VFA concentration increased to about 2.3 g/L, but VS reduction remained high, at an average of 56 ± 3%. This might be a result of inhibition of methanogenesis arising from system's instability and reactor overload because of an absence of sufficient mixing. Gas production increased once again and peaked at 0.51 L CH_4/g VS on day 182, with VFAs dropping to 0.5 g/L before becoming a bit more stable again from day 183 to day 270. Apart from the instability caused by the change in HRT from 40 to 30 days, the blank digester experienced instability for most of the time except for the period from day 183 to 270. The accumulation of VFAs, mainly acetate, was due to the change in HRT indicating system overloading. The high gas production recorded during short periods, viz., days 80–85, 176–195, and 253–258, might be attributed to the degradation of accumulated VFAs. The VFA concentration observed in the effluents is shown in Figure 3.

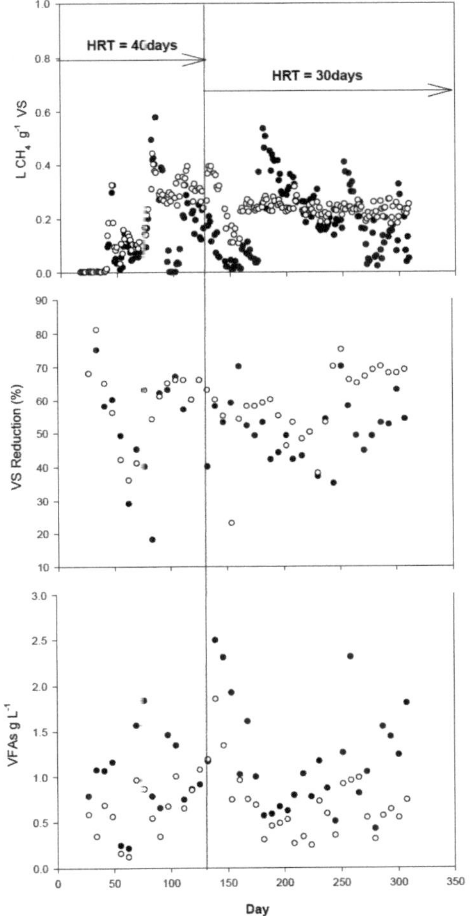

Figure 3. Methane production, volatile solid reduction, and total volatile fatty acids. Blank digester (●), optimised digester (○).

Similarly, a rapid decrease in methane production from 0.35 to 0.10 L CH_4/g VS and increase in VFA concentration from 0.7 to approx. 2 g/L was noticed in the optimised reactor when decreasing the HRT to 30 days (days 138–158). Afterwards, the reactor recovered and remained stable until the end of the study (day 309).

3.2. Steady State Period

In Table 2, the average results of the two reactors are presented over the periods days 82 to 137 (HRT 40 days) and days 160 to 309 (HRT 30 days). Based on the set criteria, 'steady state' could be assumed for the optimised reactor for both periods, but not for the blank reactor. The average methane production for the optimised reactor at an HRT of 40 days was 0.32 L CH_4/g VS, with a small variability (std = 0.05), compared to the blank, which had an average methane production of 0.19 L CH_4/g VS, but with a large variability (std = 0.14). Similarly, at an HRT of 30 days (days 160–309), average methane production was 0.23 L CH_4/g VS, with small variability (std = 0.02), compared to 0.21 L CH_4/g VS, large variability (std = 0.12) in the blank digester. This trend also holds for VFAs, as shown in Table 2. In addition to stability, the optimised digester produced 50% more methane than the blank at an HRT of 40 days.

Table 2. Average operating conditions, gas production and effluent values of the optimised and blank CDD reactor under 'steady state' conditions in the optimised reactor.

Reactor	HRT	Period	OLR	Biogas Prod. Rate	Sp. Methane Prod.	VFAs	VS Red.	Effluent NH_4 +N	Effluent pH
	day	Day	g VS/L d	L/L/d	L CH_4/g VS	(g/L)	(%)	g/L	
Blank	40	82–137	2.6–3	0.91 (0.6)	0.19 (0.14)	1 (0.3)	54 (17)	2.5 (0.10)	7.5 (0.2)
Blank	30	160–309	3.5–4	1.16 (0.6)	0.22 (0.12)	1 (0.47)	50 (9)	2.1 (0.4)	7.8 (0.1)
Optimised	40	82–137	2.6–3	1.40 (0.2)	0.32 (0.05)	0.7 (0.2)	63 (4)	2.05 (0.16)	7.6 (0.12)
Optimised	30	160–309	3.5–4	1.30 (0.12)	0.23 (0.02)	0.58 (0.23)	60 (10)	2.1 (0.5)	7.8 (0.07)

NB: Standard deviation values are in parenthesis.

As expected, some lower gas production was recorded at and HRT of 30 days as compared to 40 days. This is also true for the VS reduction in both reactors. In the optimised digester, the average specific methane production and average percentage of VS reduction during the second steady state period is much lower compared to the first steady state. However, the average VFA concentrations (0.7 ± 0.3 g/L) during the first steady state period is slightly higher than VFA concentrations (0.6 ± 0.2 g/L) during the second. This implies that the change in HRT from 40 to 30 days affected the digester performance. A similar trend was observed by Zeeman [29] in a continuously stirred tank reactor (CSTR). When HRT was changed from 25 to 10 days during anaerobic digestion of cow manure at 30 °C, the methane production observed was 0.13 and 0.10 L CH_4/g VS for 25 and 10 days, respectively. Also, Ghanimeh et al. [30] observed a reduction in methane production from 0.32 to 0.21 L CH_4/g VS when HRT was changed from 67 to 40 days during the thermophilic anaerobic digestion of source-sorted organic fraction of municipal solid waste. Indeed, digesters treating slurries and solid wastes might have their HRT similar but not equal to the solid retention time (SRT), depending on how well the reactor is mixed. During the anaerobic digestion process of solid wastes, the hydrolysis stage of particulate matter is usually the rate-limiting step [31], and therefore long SRT is often required. Long solid retention times are of interest and advantageous because they increase the conversion capacity, giving buffering capacity against shock loadings and toxic compounds [32,33]

The lower and more stable VFA concentration in the optimised reactor may be a result of relatively longer SRT in comparison to the blank digester, assuming that Monod kinetics prevailed. The higher gas production in the optimised as compared to the blank reactor is mainly due to improved hydrolysis, as the VFA concentration is relatively low and acidogenesis is generally not limiting in manure digestion [29]. However, when first-order kinetics are valid, smaller differences in hydrolysis are expected at longer HRTs. The large difference in methane production and therefor hydrolysis at a theoretical HRT of 40 days between the optimised and blank CDD digester suggests a very large difference in real SRT. The blank reactor might suffer from extensive dead zones, as shown in laboratory experiments by Jegede [6]. It is not clear why these differences are less distinct when operating the CDDs at a theoretical HRT of 30 days. Nonetheless at both theoretical HRTs, the optimised CDD, with inclusion of two baffles, performs better as compared to the blank. Details of the impact of the baffles are discussed in the next section.

3.3. *Effect of Baffles on Mixing*

3.3.1. Self-Mixing Cycle

The results of in situ pressure measurement to confirm and evaluate the self-agitating mixing cycles are presented in Figure 4 for the baffled and unbaffled sides of the optimised and blank digesters. The pressure measurements were performed after the optimised reactor achieved a steady biogas production; however, the gas production in the blank digester did not stabilise throughout the study period. The aim of the pressure measurement was to confirm if the reactor under goes a self-mixing cycle and to determine the frequency of the pressure variation in the reactors.

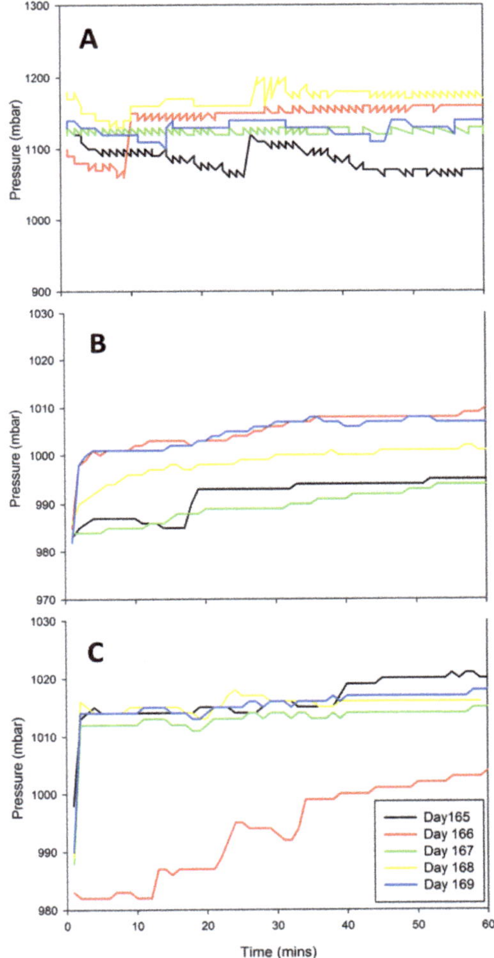

Figure 4. Pressure reading on different days, black—day 165, red—day 166, green—day 167, yellow—day 168, blue—day 169. (**a**) Measurements taken at the top of baffled side—compartment A of the optimised digester; (**b**) measurement taken at the top of the unbaffled side—compartment B of the optimised digester; (**c**) measurement taken at the top of the blank digester. The blank digester has only one compartment.

Figure 4a presents pressure measurements of the baffle side of the optimised reactor on five different days. The pattern for each measurement is almost the same. The patterns show pressure fluctuations at an average of two minutes interval for all measurements indicating a self-agitating mixing cycle when gas flows from compartment A to B. Meanwhile, in Figure 4b,c, which shows the measurements for the unbaffled side of the optimised and blank digesters, respectively, a gradual step-wise increase of the pressure is noticed. This means a gradual pressure build-up due to biogas production.

3.3.2. Effect of Baffles on Mixing and Reactor Performance

Since the upper baffle of the optimised digester results in self-mixing of the top layer of the reactor, the latter might have improved the overall performance of the digester despite high applied loading rates. In addition, improved mixing was also achieved by prevention of short circuiting of the incoming substrate to the outlet/expansion chamber by the lower baffle. Short circuiting of the incoming substrate will lead to the reduction of the real HRT and eventually low digestion performance. The lower baffle helps the incoming substrate to mix with the reactor content and therefore prevent stratification and eventually mitigating the dead zones at the bottom of the reactor. Dead zones impact negatively on overall digestion performance because the active volume in anaerobic digesters is reduced. For example, in previous studies, Jegede [6] showed that laboratory-scale conventional CDD systems have dead zones resulting in shorter real HRT as compared to the theoretical HRT at 15% influent TS. 23% dead zone was estimated based on the residence time distribution (RTD) technique, which impacted negatively on the amount of methane generated and digestion efficiency. Furthermore, the improved mixing in the optimised digester prevented the accumulation of scum or floating layers on top of the reactor compared to the blank digester. This might be another reason for the large fluctuations of the biogas production in the blank digester, because the produced gas might be trapped in these floating layers. Chinese dome digesters have a cylinder shape with a large surface area to allow large storage volume of biogas. The large surface area aids the build-up of suspended solids or scum if mixing is limited [34]. Moreover, mixing is important to (i) provide even temperature distribution and proper diffusion or spread of metabolic intermediates [13], (ii) enable sufficient contacts between micro-organism and nutrients [18], and (iii) improve hydrolysis and increase substrate surface area [15].

There are no significant differences in the average NH_4^+ -N concentrations in the influent and effluent from both digesters; however, concentrations fluctuated over time (shown in Figure 5). In the blank, the average concentration of NH_4 $^+$-N concentration fluctuated between 1.1 and 2.6 g/L, with an average of 2.3 ± 0.5 g/L at 40 days HRT. The concentration varied between 1.2 and 2.6 g/L and averaged 2 ± 0.3 g/L in the optimised digester. After the change of HRT to 30 days and an increase of OLR, NH_4^+-N concentration peaked at 3.6 g/L in the blank and 3.1 g/L on day 140, but later dropped and averaged 2.2 ± 0.5 and 2.1 ± 0.5g/L until the end of the experiments in the blank and optimised digester, respectively. Generally, these moderate NH_4 +N concentrations did not cause adverse inhibition in the digesters because of the acclimatisation of feed and digesters operating temperature (mesophilic). The VFA accumulation (2.5 g/L, day 140) in the blank digester during the experiment might have been induced by moderate inhibition of NH_4^+-N concentrations at 3.4 g/L–3.6 g/L. However, overall there is no clear pattern between NH_4^+-N and VFA in either digester.

After the change in HRT and the increase in OLR, higher concentrations of NH_4^+-N were recorded in both reactors, with 3.6 and 3.1 g/L in the blank and optimised digesters, respectively, but no significant change in the pH. The higher values could be a result of accumulation of NH_4^+-N during the hydrolysis of protein after the increase in OLR. This trend is similar to the results of a previous study by the authors [30], in which an increase in ammonia concentration was recorded with increasing OLRs. Sufficiently high ammonia concentration to result in inhibition was not recorded in this study, because cow manure has an optimum C:N ratio (25-30:1)

The pilot-scale reactors unfortunately could not be tested for mixing behaviour using tracers, but the reactor performance results clearly show the better stability, higher gas yield and lower effluent VFA concentration for the optimised reactor in comparison to the conventional CDD reactor. The letters indicate that the real HRT in the optimised CDD approaches the theoretical HRT predicted in the model study. The latter should be tested in practice in future research.

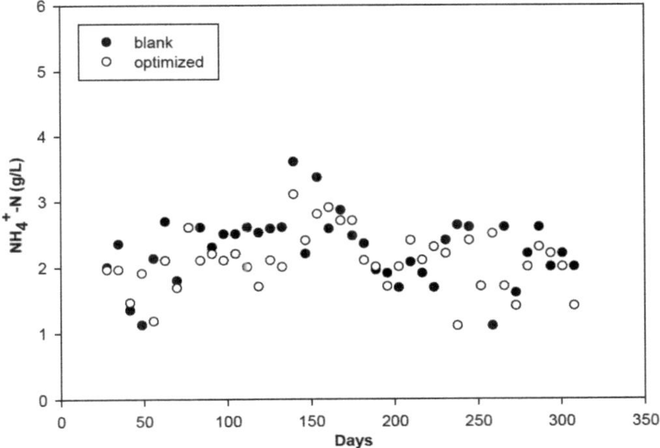

Figure 5. Concentration of ammonium nitrogen in effluent of both digesters over time.

3.4. Implication of Optimisation on Reactor Size and Cost

The impact of the improved mixing in the optimised Chinese dome digester on size, cost and water reduction is evaluated in this section. The improved Chinese dome digester design parameters and output were compared with the optimal design of a conventional Chinese dome digester for rural Kenyan and Cameroonian households [35]. The main parameters are presented in Table 3. Both digesters presented, for Kenya and Cameroon [35], have similar characteristics, except that the Kenyan digester was operated at a lower OLR. The estimated size for the improved system in this study is based on the reduced HRT (30 days) and increased influent TS. The cost of the improved CDD, using 50% of the Cameroonian reactor volume, was estimated, after consulting W. van Nes of SNV, to be $450 plus 7.5% for the added baffles.

The two digesters (Kenya and Cameroon) were made from cheap stabilised interlocking soil blocks constructed by the organisation SNV. From the table, it is obvious that the improved mixing created by the addition of baffles made a positive impact on the problems highlighted in the review at the beginning of the article—long hydraulic retention time, high water dilution, reactor size and capital cost of the system.

The size of the improved baffled CDD is half of that of the Kenyan and Cameroonian digesters, while the volumetric biogas production is higher. The major differences between the improved and conventional digester [35] are the higher OLR, higher biogas production (more than two-fold) and the improved mixing conditions. In addition, the requirement for water is reduced by half. The applied influent VS concentration for the Cameroonian was calculated to be 88.2 g VS/L, while it was 108 g VS/L for the improved CDD, respectively. Assuming the manure was 25% TS and 18% VS (73% of the TS in this study) before water dilution, 1.1 and 0.7 litre dilution water per litre manure is needed for the Cameroonian and improved CDD, respectively. The lower water requirement will positively impact the application of the improved system because of limited water availability in arid regions.

The similar biogas production at a quite different organic loading rates applied in the Kenyan and Cameroonian systems indicates that mixing is limited in the conventional CDD. The results show that the established intermittent mixing cycle in the improved CDD considerably improves the conversion efficiency as compared to that in the conventional CDD.

Table 3. Optimal digester design for rural Kenyan and Cameroonian households, and optimised digester in this thesis.

Digester Design Details	Kenya [35]	Cameroon [35]	Improved Design (This Study)
Digester size (m^3)	6	6	3
Material	Stabilised soil block digester	Stabilised soil block digester	Stabilised soil block digester
Hydraulic retention time (d)	62	49	30
Organic loading rate (OLR) (g VS/L/d)	0.81	1.8	3.6
Estimated daily biogas production (m^3)	1.48	1.43	3.9 [a]
Estimated capital cost (USD)	684	628	484 [b]
Estimated annual running cost (USD)	2.85	2.75	2.85 [c]
Supplier contact details	Uganda Domestic Biogas Programme/ SNV (The Netherlands Development Organisation	Uganda Domestic Biogas Programme/ SNV (The Netherlands Development Organisation	

[a] Based on volumetric biogas gas production. [b] Based on exchange rate of 1 USD = 361 Naira (May 2018); capital cost estimated 450$ + 7.5% for baffles (pers. communication W. van Nes).
[c] Assumed same as Kenya.

4. Conclusions

The improved Chinese dome digester with two baffles was evaluated in a pilot study and compared with the conventional (blank) digester in continuous operation at higher organic loading rates (ca. 15% TS, corresponding to 2.6–4 kg VS/m3/d) at HRTs of 40 and 30 days. The optimised Chinese dome digester has a two-minute self-mixing or agitation cycle of the top layer of the digester content, using the produced gas without a moving device. Since the optimised digester is self-mixed with the aid of baffles, the overall performance of the digester did improve and produced 40% more methane than the blank, despite the applied high loading rates. The optimised digester showed superior digestion treatment efficiency and was more stable in terms of VFA concentration than the blank digester, and therefore could be operated at high influent TS (15%) concentration.

The results revealed that mixing becomes important at high influent TS (15% TS influent), because influent rheological property requires more mixing to achieve similar results. The experimental results showed that the self-agitation cycle contributes to the improved performance of the optimised digester. The specific biogas production from the optimised reactor is comparable to results from continuously stirred reactor tanks (CSTR). As a consequence, a smaller reactor volume could be achieved at high loading rate at reduced HRT ($\leq$ 40 days) and eventually reduction in reactor cost as compared with the two digesters (Kenya and Cameroon). High-input TS concentrations mean lower water dilution; therefore, the lower water requirement will positively impact the application of the improved digester because of limited water availability in arid regions where Chinese dome digesters are used for cooking.

Author Contributions: Methodology: All authors.; data curation: A.J.; writing—Original draft preparation: A.J.; writing editing: G.Z. & H.B.; funding acquisition: A.J. & G.Z.

Funding: This work was funded by the Netherlands Fellowship Programme (NFP), the Netherlands. Grant number: NFP-PhD 12/426.

Appreciation: Special thanks to the analytic lab support team at Environmental Technology department, Wageningen University, The Netherlands. Special appreciations to the management and support staff of Biochemical Engineering laboratory, Centre for Energy Research and Development, Obafemi Awolowo University, Ile-ife Nigeria for their support.

Conflicts of Interest: The authors declare no conflict of interest.

References

1. Nzila, C.; Dewulf, J.; Spanjers, H.; Kiriamiti, H.; Langenhove, H. Biowaste energy potential in Kenya. *Renew. Energy* **2010**, *35*, 2698–2704. [CrossRef]
2. Sasse, L. Biogas Plan, GTZ Publication. 1988. Available online: http://www.susana.org/en/knowledge-hub/resources-and-publications/library/details/1799 (accessed on 20 February 2016).
3. Yu, L.; Yaoqiu, K.; Ningsheng, H.; Zhifeng, W.; Lianzhong, X. Popularizing householdscale biogas digesters for rural sustainable energy development and greenhouse gas mitigation. *Renew. Energy* **2008**, *33*, 2027–2035. [CrossRef]
4. Bond, T.; Templeton, M.R. History and future of domestic biogas plants in the developing world. *Energy Sustain. Dev.* **2011**, *15*, 347–354. [CrossRef]
5. Qi, W.-K.; Toshimasa, H.; Li, Y.-Y. Hydraulic Characteristics of an innovative self-agitation anaerobic baffled reactor (SA-ABR). *HoshiTech* **2013**, *136*, 94–101. [CrossRef] [PubMed]
6. Jegede, A.O. Optimization of Mixing in a Chinese Dome Digester for Tropical Regions. Ph.D. Thesis, Sub-Department of Environmental Technology, Wageningen University, Wageningen, The Netherlands, October 2018.
7. Sommer, S.G.; Petersen, S.O.; Møller, H.B. Algorithms for calculating methane and nitrous oxide emissions from manure management. *Nutr. Cycl. Agroecosyst.* **2004**, *69*, 143–154. [CrossRef]
8. Bruun, S.; Jensen, L.S.; Vu KV, T.; Sommer, S. Small Scale household biogas digesters: An option for global warming mitigation or a potential climate bomb. *Renew. Sustain. Energy Rev.* **2014**, *33*, 736–741. [CrossRef]
9. Ghimire, P.C. SNV Supported domestic biogas programmes in Asia and Africa. *Renew. Energy* **2013**, *49*, 90–94. [CrossRef]

10. Cheng, S.; Li, Z.; Mang, H.P.; Huba, E.M. A review of prefabricated biogas digesters in China. *Renew. Sustain. Energy Rev.* **2013**, *28*, 738–748. [CrossRef]
11. Deublein, D.; Steinhauser, A. *Biogas from Waste and Renewable Resources*; Wiley-VCH Verlag GmbH & Co. KGaA: Weinheim, Germany, 2008.
12. Bridgeman, J. Computational fluid dynamics modeling of sewage sludge mixing in an anaerobic digester. *Adv. Eng. Softw.* **2012**, *44*, 54–62. [CrossRef]
13. Gerardi, M. *The Microbiology of Anaerobic Digesters*; John Wiley and Sons, Inc.: Pennsylvania, PA, USA, 2003.
14. Conklin, A.S.; Chapman, T.; Zahller, J.D.; Stensel, H.D.; Ferguson, J.F. Monitoring the role of aceticlasts in anaerobic digestion: Activity and capacity. *Water Res.* **2008**, *42*, 4895–4904. [CrossRef]
15. Halalsheh, M.; Kassab, G.; Yazajeen, H.; Qumsieh, S.; Field, J. Effect of increasing the surface area of primary sludge on anaerobic digestion at low temperature. *Bioresour. Technol.* **2011**, *102*, 748–752. [CrossRef] [PubMed]
16. Gomez, X.; Cuetos, M.J.; Cara, J.; Moran, A.; Garcia, A. Anaerobic co-digestion of primary sludge and the fruit and vegetable fraction of the municipal solid wastes-conditions for mixing and evaluation of the organic loading rate. *Renew. Energy* **2006**, *31*, 2017–2024. [CrossRef]
17. Kim, M.; Ahn, Y.H.; Speece, R.E. Comparative process stability and efficiency of anaerobic digestion: Mesophilic vs. thermophilic. *Water Res.* **2002**, *36*, 4369–4385. [CrossRef]
18. Ward, A.J.; Hobbs, P.J.; Holliman, P.J.; Jones, D.L. Optimization of the anaerobic digestion of agricultural resources. *Bioresour. Technol.* **2008**, *99*, 7928–7940. [CrossRef] [PubMed]
19. Ike, M.; Inoue, D.; Miyano, T.; Liu, T.T.; Sei, K.; Soda, S.; Kadoshin, S. Microbial population dynamics during startup of a full-scale anaerobic digester treating industrial food waste in Kyoto eco-energy project. *Bioresour. Technol.* **2010**, *101*, 3952–3957. [CrossRef]
20. Lindmark, J.; Thorin, E.; Fdhila, R.B.; Dahlquist, E. Effect of mixing on the result of anaerobic digestion: Review. *Renew. Sustain. Energy Rew.* **2014**, *40*, 1030–1047. [CrossRef]
21. Jha, A.K.; Li, J.Z.; Nies, L.; Zhang, L.G. Research advances in dry anaerobic digestion process of solid organic wastes. *Afr. J. Biotechnol.* **2011**, *10*, 14242–14253.
22. Karthikeyan, O.P.; Visvanathan, C. Bio-energy recovery from high-solid organic substrates by dry anaerobic bio-conversion processes: A review. *Rev. Environ. Sci. Biotechnol.* **2012**, *12*, 257–284. [CrossRef]
23. Battistoni, P. Pre-treatment, measurement execution procedure and waste characteristics in the rheology of sewage sludges and the digested organic fraction of municipal solid wastes. *Water Sci. Technol.* **1997**, *36*, 33–41. [CrossRef]
24. Benbelkacem, H.; Garcia-Bernet, D.; Bollon, J.; Loisel, D.; Bayard, R.; Steyer, J.-P.; Gourdon, R.; Buffière, P.; Escudié, R. Liquid mixing and solid segregation in high-solid anaerobic digesters. *Bioresour. Technol.* **2013**, *147*, 387–394. [CrossRef]
25. Jegede, A.; Bruning, H.; Zeeman, G. Design considerations for inlet and outlet of Chinese dome digesters to mitigate methane loss. *Biosyst. Eng.* **2018**, *174*, 153–158. [CrossRef]
26. APHA. *Standard Methods for the Examination of Water and Waste Water*, 22nd ed.; American Public Health Association: Washington, DC, USA, 2006.
27. APHA. *Standard Methods for the Examination of Water and Waste Water*, 21st ed.; American Public Health Association: Washington, DC, USA, 2005.
28. Karim, K.; Hoffmann, R.; Klasson, K.T.; Al-Dahhan, M.H. Anaerobic digestion of animal waste: Effect of mode of mixing. *Water Res.* **2005**, *39*, 3597–3606. [CrossRef]
29. Zeeman, G. Mesophilic and Psychrophilic Digestion of Liquid Cow Manure. Ph.D. Thesis, Sub-Department of Environmental Technology, Wageningen University, Wageningen, The Netherlands, 1991.
30. Ghanimeh, S.; Fadel, M.E.; Saikaly, P. Mixing effect on thermophilic anaerobic digestion of source-sorted organic fraction of municipal solid waste. *Bioresour. Technol.* **2012**, *117*, 63–71. [CrossRef] [PubMed]
31. Sanders, W.T.M. Anaerobic Hydrolysis during Digestion of Complex Substrates. Ph.D. Thesis, Sub-Department of Environmental Technology, Wageningen University, Wageningen, The Netherlands, 2001.
32. Climenhaga, M.A.; Banks, C.J. Anaerobic digestion of catering wastes: Effect of micronutrients and retention time. *Waste Sci. Technol.* **2008**, *57*, 687–692. [CrossRef]
33. Wang, Z.W.; Ma, J.; Chen, S. Bipolar effects of settling time on active biomass retention in anaerobic sequencing batch reactors digesting flushed dairy manure. *Bioresour. Technol.* **2010**, *102*, 697–702. [CrossRef]

34. Tchobanoglous, G.; Burton, F.L.; Stensel, H.D. *Wastewater Engineering: Treatment and Reuse*, 4th ed.; Metcalf and Eddy Inc., McGraw-Hill: New York, NY, USA, 2003.
35. Rupf, G.V.; Arabzadeh, B.P.; De Boer, K.; McHenry, M.P. Development of an optimal biogas system design model for Sub-Saharan Africa with case studies from Kenya and Cameroon. *Renew. Energy* **2017**, *109*, 586–601. [CrossRef]

MDPI
St. Alban-Anlage 66
4052 Basel
Switzerland
Tel. +41 61 683 77 34
Fax +41 61 302 89 18
www.mdpi.com

Energies Editorial Office
E-mail: energies@mdpi.com
www.mdpi.com/journal/energies

www.ingramcontent.com/pod-product-compliance
Lightning Source LLC
LaVergne TN
LVHW071005170726
843515LV00006B/1074

* 9 7 8 3 0 3 6 5 3 2 3 7 0 *